THE MASTER MARINER

BOOK I

RUNNING PROUD

NICHOLAS MONSARRAT

THE
MASTER
MARINER

BOOK I

RUNNING
PROUD

'Nothing of him that doth fade,
But doth suffer a sea-change,
Into something rich and strange.'

William Shakespeare, *The Tempest*

CASSELL
LONDON

CASSELL LTD.
35 Red Lion Square, London WC1R 4SG
and at Sydney, Auckland, Toronto, Johannesburg,
an affiliate of
Macmillan Publishing Co., Inc.,
New York.

First edition September 1978
First edition, second impression October 1978
Maps and ship drawings by Peter McClure

ISBN 0 304 29609 0

Printed in Great Britain by
Richard Clay (The Chaucer Press) Ltd
Bungay, Suffolk

THE MASTER MARINER

One of several versions of the 'Wandering Jew' legend tells how a Jewish merchant, on whose doorstep Jesus paused on his way to Calvary, pushed the Master away roughly and told him to be gone. Jesus replied: 'I am going, and quickly—*but tarry thou until I come again.*'

THE MASTER MARINER
BOOK I
RUNNING PROUD

ONE
THE COWARD

1588

'Who would, therefore, repose trust in such a nation of ravenous strangers, and especially in those Spaniards which more greedily thirst after English blood, than after the lives of any other people in Europe, for the many overthrows and dishonours they have received at our hands, whose weakness we have discovered to the world, and whose forces at home, abroad, in Europe, in India, by sea and land, we have even with handfulls of men and ships overthrown and dishonoured.'

Sir Walter Raleigh: *A report of the Truth of the Fight about the Isles of Azores*

FRANCE

Dunkerque
Calais

The Naze
North Foreland
Dover
Rye
Portsmouth
Isle of Wight

Spurn Head
Hull

LONDON

Bristol

Southampton

ENGLAND

ENGLISH CHANNEL

Sleeve

Exeter

Plymouth

Lizard Point
Rame Head

IRISH SEA

DUBLIN

IRELAND

Bantry Bay

The Islands of Scilly

Wrecks

Wrecks!

The Passage of The ARMADA
Actions between the English and Spanish Fleets

THE MAN AT THE EDGE of the tide-mark stood a little apart from his fellows, not choosing to join the talk of the ship-boat's crew, wanting only to enjoy the hot July sun and the brief peace of the Plymouth shore. His ship-boat, bearing the Vice-Admiral and a half-score of his volunteer gentlemen, had grounded on the beach some way from the landing-steps of Sutton Pool, to avoid the throng which pressed round their commander wherever he went at this time, in Plymouth or throughout the West Country; and when he and the other great lords had gone about their business in Plymouth, the crew had beached the boat and laid aside their oars.

Now they stood idling, glad of the sun, the firmness of the sand under their bare feet, and their shelter from the forcible wind which even now was whipping Plymouth Sound into a grey-white fury.

The man standing apart was of the middle height, broad, slow-moving, with the brown-grained skin and the spread stance of a West Country seaman. He was bravely dressed, like all his companions, as fitted the Vice-Admiral's own ship-boat's crew; his quilted coat was decked with bold Tudor ribbons of green and white, and the boat itself wore the Vice-Admiral's standard, presented not long before by the Lord High Admiral of England.

The man at the tide-mark, by name Matthew Lawe, was entitled to stand apart, as chosen coxswain to the Vice-Admiral, and a young man in earned authority. His crew were all trained men of his own making, and though he stood aloof he kept them within the corner of his eye, ready to check their talk if it grew rough or boisterous.

At his back lay Sutton Pool, the inner harbour of Plymouth; this Pool, and the nearby Cattewater, and the whole Sound itself, was one forest of ships, a woven pattern of drying sails, steep oaken hulls, rocking masts, and flags and banners streaming in the brisk southerly wind.

Before him lay the town, clamorous and astir with men, and the green slopes of Plymouth Hoe. But Matthew Lawe was preoccupied with the smaller things, and his main gaze rested on the tide-mark: the rime of flotsam on the sand, the seaweed, the chippings of wood, the pale husks of shrimp and crab, the rope-strands, the polished shells of periwinkle-blue, blood-red, dawn-yellow.

To Matthew Lawe, this tide-mark spelled peace—almost the sole peace in England on that day. For all elsewhere, in Plymouth, and the West Country, and indeed the whole realm, there was clamour, martial show, boasting, striving, and fear.

By now they had lain nearly eight weeks in harbour, and it had proved little enough time to deal with all that had to be done, to make the *Revenge*, and all the other ships of the English fleet, ready to meet the might of Spain and the Dons' vainglorious vaunting. They had to draw on the blood and the mettle of Plymouth; and Plymouth, finest of English harbours, could scarcely contain the men and ships that daily and nightly thronged about.

Every young man within the space of twenty miles, and every grandfather but lately dreaming by his chimney corner, was now labouring on board the ships, or ashore on business of the same sort.

By day the ship-boats, barges, cockboats, and pinnaces sped to and fro, bearing men and weapons; the fly-boats loaded with provisions went about their dutiful business, the powder-hoys with their red flags of danger wove carefully through the fleet; on shore, the shipwrights and the sailmakers and the armourers toiled against the run of the clock, to fashion whatever the captains and the sailing-masters and the gunners demanded—whether it was new spars, flaxen ropes, brass culverin and basilisk, fresh-water casks, or cross-bar shot for the cannon-royal.

By night these same men wandered ashore, crowding past the corbelled houses, thronging the Barbican, drinking in the taverns at the sign of the Turk's Head in St Andrew's Street, and the Mitre, and the Pope's Head in Looe Street, where the Vice-Admiral himself had his household. They drank deep, in courage or in fear, as sailors always do on the eve of battle; or they drew apart, and wrote crabbed letters to their wives; sometimes they would go wenching with the guinea-hens, if these had not been whipped from the town by that upright citizen, His Honour the Lord Mayor, Mr William Hawkins.

But always, on the morrow, they returned to work, in their ships or at their forges; making the whole containment of Plymouth, from St Nicholas Island to the Hoe, a cauldron of warlike preparation.

So it had been for the past two months or more; and now, when all was done that could be done, and this spear-point of England on the threshold of war was thus sharpened and poised, the ship-boat's crew was waiting, as they had waited on a hundred strands, from Plymouth to Brazil, from Nombre Dios to the Magellan Strait, for the greatest sailor in all the world.

They knew him as a man, yet saw him also as the gleaming pennant of achievement: Sir Francis Drake, sometime Lord Mayor of Plymouth, Member of the Parliament, Vice-Admiral of England, and the first Captain-General that ever sailed about the globe, turning up a furrow around the whole world.

To suit this corsair laden with honours, the most famous culprit alive, knighted by the Queen's own hand on his return with staggering treasure, his immortal crest was thus described:

'Sable, a fess wavy between two pole stars argent, the helm adorned with a globe terrestrial, upon the height whereof is a ship under sail trained about the same with golden hawsers by the direction of a hand appearing out of the clouds, all in proper colour with these words, *Auxilio Divino*—By God's Help' . . .

It was a name to enthral the simple and the common sort, and to inspire even the jealous noble; among the unmatched and glittering array of the great lords who were his peers, they being seamen, men-at-arms, and venturers all—Raleigh, Frobisher, Humphrey Gilbert, their lordships of Essex and Leicester and Howard of Effingham, Richard Grenville, John Hawkins—even among these, Sir Francis Drake was the *nonpareil*.

It was small wonder that, when he needed volunteers to man the ships against the threatening Armada of Spain, men and boys had flocked to his standard from every part of the West of England. There was one such among this very ship-boat's crew: a farm-lad who had walked a full hundred miles from Shepton Mallet in Somerset, and (though bred to the farm like his great-grandfather before him) had taken to the oar like a Turk at the galleys.

Indeed, Matthew Lawe himself, ten years earlier, as a runagate orphan boy from the Poor School at Barnstaple, had been cozened by the same magnet and had made the same translation, from the plough to the sail. Drake, the great captain for whom the narrow seas had proved too small a prison, drew men after him to the very end of the world, like a woman her lover or a sheep-herd his flock.

There were some indeed who had striven to set up a quarrel when, by the Queen's command, almost on the eve of battle, Sir Francis had been supplanted in the rank of Admiral by this same Lord Howard of Effingham. There were many who swore wounds and blood that this was a matter not of merit but of rank; and that when the Duke of Medina Sidonia was appointed to lead the Spanish fleet, it was for this paltry reason thought necessary to oppose him with a man of noble family—as if one coat-of-arms could fight against another.

Why, they asked in hot dispute, should a captain of Sir Francis' quality, a man of good fortune reputed to be fatal to the Spaniard, thus lower his standard, when his gifts and his exploits had entitled him to command the whole enterprise? But Drake would have none of these rebellious mutterings: he led his squadron out to welcome the new Lord Admiral with an excellent grace, firing the salute to the *Ark Royal* and the other great Crown ships, sounding the trumpets and drums, lowering his own flag and putting in its place the Vice-Admiral's standard. Not for him the disobedience of lesser men. . . .

Later, he had let it be known that no man of his own fleet, from great officer to ship's boy, was to make this an occasion for complaint; and he himself and Lord Howard had gone together in state to receive the Blessed Sacrament on Whit Sunday, and in all other things he had shown himself to be of one mind with his appointed Admiral.

This had always been his quality, a faithful mirror of the obedience and self-command of the ancient Romans; and those who called him pirate and greedy freebooter were left to scratch their heads.

II

There was a stir among the ship-boat's crew, a general pointing to seawards; and Matthew Lawe, preoccupied with a strangely-wrought shell in the shape of a pearly breast-plate, was recalled to the world by one of his crew, shouting:

'Matthew! The scouting pinnace!'

The crew-man who called out was John Waggoner, a man valued for his keen eye; it was he who, on that voyage round the globe, had ever been the first to spy their looked-for landfall, whether on the coast of Brazil, or in the bleak latitudes of the Magellan Strait, or among the treacherous Molucca Seas.

Now John Waggoner, shading his eyes which had the penetration of an eagle, was watching and pointing at the same time; and presently they saw what it was that had caught his attention, to the southward at the very margin of Plymouth Sound—a small two-masted ship, under full sail, that had the look of a pinnace in hot haste.

Matthew Lawe walked back a pace to the higher ground, to give him a longer range.

'Is it truly she, John?'

'Aye . . .' John Waggoner's voice was contented in his certainty. 'I can see His Honour's banner at the fore-mast . . . 'Tis the *Golden Hind*, come back again.'

The *Golden Hind*—it was a name of omen for them all; not the famous ship of Drake which had made the circumnavigation, but one of the same baptism, under Captain Thomas Fleming, despatched these many days towards the Sleeve between Ushant and the Scillies, to scout for the coming of the Dons.

They watched her bearing towards them up the Sound, scudding before the frank wind: threading her way past the great ships of the fleet, lowering her flag to the Admiral in the *Ark Royal*, making a dancing courtesy to the bristling guns of St Nicholas Island (which Drake himself had fortified against the Spaniard), dousing her taut sails one by one as she neared the confines of Sutton Pool.

She had a breathless, hasty air; as she drew near they saw that her mainsail was wet as high as the cross-yard, and when she turned to come up to her anchorage they could catch a flash of gleaming, watery deck. Clearly she had news that would allow no delay.

There was a creamy splash at her forefoot as the anchor plunged down; the last of her sails was loosely furled. Then, after some confusion and shouting, borne on the wind, two small cockboats were put over her side, and made ready to leave.

'All at six and seven!' the youngest of the ship-boat's crew, the farm-lad from Shepton Mallet, said scornfully.

'The *Revenge* could give them a beating,' John Waggoner confirmed, 'at putting a cockboat away, or any other matter.'

'Belike they are dog-weary,' Matthew Lawe said. He felt some foreboding. 'Or short of men because of wounds.'

The two cockboats drew away from the *Golden Hind* one after the other. The first of them buffeted its way out of the inner harbour again, making for the flag-ship lying in the Sound; the other came towards Sutton Pool quay, driven down-wind with ragged, lusty strokes.

It had a crew of four oars, and an officer of quality sitting in the stern, looking this way and that. It was steering for the steps of Sutton Pool, but at the last moment altered its course towards their own boat that bore the Vice-Admiral's standard, and presently grounded on the sand a few paces from them.

The officer, borne ashore on the bowman's back, jumped down in haste and strode towards them. He was a tall young man, with a weary face and yet a proud bearing.

Matthew Lawe came forward, lifting his cap to make his duty.

'Sir!' he said, stiffly, as he had been schooled. 'Coxswain to the Vice-Admiral!'

The officer looked from him to the Vice-Admiral's standard, and back again.

'Is His Honour still ashore?'

'Aye, sir,' Matthew answered. 'Up to the Hoe, he said.'

'How long since?'

'This two hours.'

'I'll go seek him . . .' The officer turned towards his own boat, and called out commandingly: 'Await me here!' Then he made as if to go on his errand, moving slowly as a man plainly short of strength.

Matthew Lawe summoned his courage. 'Is there news, sir?'

The officer turned back, losing some of his haughtiness with his will to be the first to impart.

'Yes, lad. The Spaniard is come at last.'

'Where away, then?'

'Off the Lizard, when we left them. Nearer now, by God's blood!'

The Lizard. . . . Fewer than twenty leagues distance from Plymouth. . . . As the young officer turned again towards the Hoe, Matthew felt, not for the first time in that year nor on that day, the mortal chill of fear.

III

Fear. . . . It was something that a man must come to feel, who
kept the company of Sir Francis Drake; after a time, no matter
how bold one's spirit, nor how much a man confided in his Captain's skill and fortune, fear crept in apace, to confound judgement
and turn the knees to water. For that famous skill must some time
fail, that luck must run out, like sand within the glass.

Fear of this sort was what Master Thomas Doughty had felt, at
Port St Julian near the Magellan Strait, when in rebellion he had
tried to leave their squadron and put back to England. It had cost
him his neck, as he knew it must; but he had dreaded the other
perils more.

For Matthew Lawe, fear had begun slowly, a single strand at a
time. The sickly orphan boy of eleven years, who had run from the
bleak and cruel Poor School in Barnstaple, and taken to the sea
because it was his dream, had become the youth who had followed his Captain round the world; and now, when he was twice
eleven years, and grown to broad manhood in the service of Sir
Francis, he could not truly say at what moment of time he had felt
himself more afraid than valiant.

Certainly it was not at the beginning, when the *Pelican*, with
five other barks and pinnaces, and near one hundred and sixty
men, had left the Cattewater nearby, to try a course around the
globe under a Captain already renowned for his daring and good
fortune. Then, it was all magic and young ecstasy, a voyage like
the golden dream itself.

The *Pelican* and her train of pretty chickens—the barks *Marigold*
and *Elizabeth*, the pinnaces *Christopher* and *Benedict*, the fly-boat
Swan—had set sail from Plymouth in stormy seas which presently,
as they neared the Coast of Barbary, grew kind. From thence
forward, for the space of three years, not one day of that voyage
was without its beauty, its strangeness, its excitement and its
terror.

In Barbary they tasted the flesh of coconuts; in Mogador, they
landed and broke down certain crosses with evil-faced pictures of
Christ. On the line of the Equator, they ate dolphins, bonitoes, and
flying fish.

Then they voyaged sixty-three days without sight of land, and
when land came up, it was Brazil. Here the rude inhabitants,

sighting their ships, lit fires and danced to conjure up shoals and storms; and presently they were struck by a gale so fierce that the English vessel was in truth a *Pelican* in the wilderness.

Here they saw grossly-formed women who carried their children on their backs and, needing to suckle them, rolled back their breasts over their shoulders. Here they killed three hundred sea-wolves with cudgels, and saw huge footprints ashore. Here they navigated, for a space, the great River del Plata.

Down south now for many more days, and presently to Port St Julian, where the treacherous Patagonians slew some of their company, and they found a gibbet made of a spruce mast, used by Magellan near sixty years before to execute some of his rebellious crew. At its foot were their dry bones. Here also, Thomas Doughty, shipmaster in their own squadron, met a like fate.

He had expressed the wish to go no further, but to return home, and had sought secretly to suborn others to this mutiny. Drake, learning of this, flew into a cold rage, damned them all for a clutch of caterpillars and idle-heads, and had a public inquiry made, and asked for their verdict.

Without delay, they pronounced Thomas Doughty the child of death; and when he and Sir Francis had taken the Sacrament together, and drunk heartily to each other at dinner, Doughty's head was struck from his body.

He had faced his death in quiet mind. When it was over, Drake took the head by the hair, shook it in the face of the company, and cried: 'Long live the Queen of England!' Then he led them all in the holy Psalms. He afterwards called the place the Island of Justice and Judgement, though his enemies called it the Island of Blood.

Now (some said as a punishment for this murderous spite) came the dark part of their journey, the fearful passage through the Magellan Strait. Often it was as if a blind man were threading an endless forest of sail-maker's needles: every passage was stopped, like a fox his earth, and every lead that promised them progress grew narrow and came at last to a wall of rock.

It seemed that they were doomed never to burst out into the Pacific Ocean, but would turn and twist till their ships fell to pieces and they themselves, grown gaunt, were food at last for the fish.

There were spouting volcanoes in the Straits as they struggled onwards, and the inhabitants lit flickering fires on either side, like evil eyes jealous of their course. There were indeed few such living creatures: this was a desolate, deserted land, where (it seemed) birds might have perched and preened the world's first feathers. But they put ashore many times; to climb peaks and survey their onward passage, to gather fuel and forage for wretched sustenance, to take possession of the land with turf and twig on behalf of their Queen.

Once they killed three thousand birds that could not fly; and once again, when they landed for water, they found a skeleton, of a man so long dead that his bones would not hold together. Of what man, what miserable outcast? The doubt pursued them, like a black bird of ill-omen, as they strove to press westwards.

It was in this same Strait of Magellan that they changed the name of *Pelican* to that of *Golden Hind*, to honour the Lord Chancellor of England, Sir Christopher Hatton, whose crest was (in the old language) a *hind trippant or*.

There were some who muttered about the ill-luck attending, and others who hoped for better things from this change; but Sir Francis listened to no one, and perhaps he had mingled a good and a bad potion—for shortly they found a way through the Strait, and came out at last into the liberal sea of the Pacific, only to meet there a furious storm, unsurpassed for violence, which endured not five but fifty days.

In this storm, when the moon itself was for a time awefully eclipsed, they lost the last of their consorts. The *Marigold* perished with all her souls aboard; the *Elizabeth* disgracefully turned her tail and fled back to England. (Of the others, the *Benedict* had parted company near Plymouth, the *Swan* and the *Christopher* had grown leaky and been broken up for firewood.) Now the *Golden Hind* was alone, with half the world behind her, and half before.

It might have been thought that Sir Francis Drake, infected by this lonely peril, would now creep quietly homewards, having voyaged enough for one man and one ship, and given ample proof of his valour. But such was not his mettle: he drew only strength and further ambition from adversity. He now embarked upon a fresh and glittering adventure, with all the stormy brilliance of a corsair who chances on a new prey. Perhaps it was now that the young Matthew Lawe, a lad rising thirteen years, felt his first foreboding.

Sir Francis, their Captain-General, hated Spain, with all his heart and mind; this hatred had pricked a spirit already rash enough. It grew partly from his devotion to his Sovereign, whom he wished to see Queen without peer: partly from his loathing of the Popish contrivances of the Spaniard; and partly from misliking their cruelty and treachery, proved before all the world.

In particular, it enraged him that they showed neither honour nor mercy to their prisoners; his friend, John Oxenham, was after capture publicly hanged at Lima, and Robert Barrett, of the company of John Hawkins, had been burnt alive in the market-place of Seville.

Now was his time of revenge. He was loose in an ocean where no Englishman had sailed before; no man knew him to be there, nor ever suspected it; the rich plunder of Peru and New Spain was his, if not for the taking, then for the fighting. He fell upon the Spaniard like a hawk unleashed.

They took ship after rich ship, they surprised garrison after garrison, they stormed ashore and wrought havoc, from Valparaiso to the coast of Panama. Especially did they prey upon the ships of lading that carried the wealth of Spain, easing them of their burdens with jovial readiness, so that by the end of it, no ship of Spain would put to sea, but had to be boarded and captured within its nest.

From these argosies they took every kind of spoil—emeralds, gold, linen, rope, silks, china dishes, chests of *reals* of eight. They sacked the town of Saint Iago, taking from the church a silver chalice, two cruets, and an altar-cloth, all decently given over to their own chaplain, Mr Fletcher.

They landed at Tarapaxa, and found by the seashore a Spaniard fast asleep, who had lying by him thirteen bars of silver, which weighed 4000 ducats Spanish; they took the silver, *and left the man*. To the north, they captured eight llamas, each carrying two leather bags of fifty pounds of silver. They seized 1700 jars of sack and Canary wines in one place; and in another, a falcon of gold, handsomely wrought, with a great emerald set in its breast.

Lastly they took a huge carrack of Spain, the *Cacafuego*, by a subterfuge: hanging oil-pots over their stern to deceive her as to the speed of the *Golden Hind*, then boarding by stealth and killing their prey with great shouts of 'Go down, dog!' She was the richest prize of all: her prime lading was twenty-six tons of pure silver, worth half a million English pounds.

There were some who later called her, in jest, not *Cacafuego* (meaning, Spit-Fire), but *Cacaplata*—that is, Spit-Silver; for she set no great price upon her capture. But jesting or not, they made her their own, and everything of value in her holds. Then they quit the havoc they had wrought, along three thousand miles of pliant coast, and sailed northwards again on their lawful way.

Now, after these hot encounters, it grew cold; and as they voyaged forever northwards, seeking a passage which might trend east towards England and home, it grew colder still—a nipping frost, a pinching biting air which infected every part of the *Golden Hind*, and daily blighted their hopes. Presently, having reached the forty-eighth parallel north without respite either from the cold or the forbidding barrier of land, they put back again; and venturing inshore on a favourable slant they came to an entrance so broad and fair that Drake named it the Golden Gate.

They took possession of this land also, calling it Nova Albion; they found it full of strange conies with handsome skins and broad tails, and great herds of fat deer roaming like sheep upon the English Downs. There were natives also, decked in skins and feathers, who cut and sliced their flesh to do homage to their god-like visitors.

But the *Golden Hind* still stood at the greatest cross-roads of the world. Drake their captain, barred from a north-east passage home, had choice of two pathways: to retrace the track by which he came, or to voyage onwards round the globe. Once again, he chose the unknown and the valiant path; and from then onwards their journey grew fantastic, never to be forgotten by any who made it.

Seeking China, or a way past it, they sailed seven thousand miles across the Pacific Ocean, where no English ship (and only one ship of any nation) had ever ventured; they were sixty-eight days without sight of land, divining their course by astrolabe, cross-staff, and compass, until they came to the Pelew Islands.

There they were received by the King with presents, and loving homage. For many months they wandered in that Archipelago, seeking a way westwards, passing by the island named Mindanao, and the Moluccas, and Celebes, and Java. They found a strange land that had canoes full of savages with long nails, and blackened teeth, and holes in the lobes of their ears whence they hung their ornaments.

In the Moluccas (where they ran on a rock, but floated free) the King did them homage in rich attire, decked with gold chains, diamonds, rubies, emeralds: he sat in state under a costly canopy, guarded by twelve lances. In the Celebes, they observed fiery worms flying in the air, and bats as big as hens. In Java were five Rajahs, and a diet of boiled rice to live on, and idle natives who sunned themselves to cure the French pox.

Free at last of the confusion of these islands, they crossed another great ocean, and presently came up with the Cape of Good Hope (that which Diaz had first rounded), the fairest cape they saw in the whole circumference of the earth. But they passed by without a respite, turning their prow homewards. In Sierra Leone there was (for Matthew) the strangest sight of all that voyage—great elephants at work in the forest. Then they came home to Plymouth Sound again, having traversed the whole world and taken three years in the task.

It was a wild and wondrous voyage for a boy, and for Sir Francis a resplendent feat of daring which resounded throughout the realm. The Sun himself (they said in homage) cannot forget his fellow-traveller.

IV

Above them, on the well-trodden pathway leading to the Hoe, there was a stir, and as they turned, the Lieutenant from the cockboat came into sight again, making his return journey with resolute strides. He had the look of a young man out of humour: his brow was creased, and the silver-chased sword at his belt slapped his thigh angrily as he descended.

'Look closely,' John Waggoner, the sharp-eyed look-out and also a licensed droll, murmured, 'and you will perceive the flea in his ear.'

'How so?' Matthew Lawe said, taken aback. 'With such news of the coming of the Spaniard, he should have been received like a king!'

'Sir Francis is used to kings,' Waggoner answered. 'And this one looks to be the smallest one on earth.'

The Lieutenant drew near them, his expression sour. His cockboat's crew stood to their oars, with the usual ceremony, waiting for his embarkation. Matthew Lawe doffed his cap afresh, and the Lieutenant acknowledged with a brief salute.

'Did you find His Honour, sir?' Matthew asked.

'Aye,' the Lieutenant answered curtly.

'Does he come, then?'

'No.'

The Lieutenant stepped towards his boat. It seemed for a moment that he was so angry that he would keep his counsel, but (being a young man) he could not forbear to tell his story.

He turned. 'I found your *Vice-Admiral*,' he said, with cutting emphasis. 'Playing at bowls, up on the Hoe. . . . I told him that the Spaniard was near—as near as the Lizard itself. . . . He seemed to find it no greater matter.'

'How so, sir?'

'He said there was time to win the game, and thrash the Spaniards afterwards.'

At that, there was an intake of breath, and a great murmuring, and laughter here and there among the two boats' crews. But now it was easy to see the difference between them, a difference made by the quality of one man. The cockboat's men, who had driven fast and far to bring their news, only to have it thus despised, were astonished, then full of derision; preparing to return to the *Golden Hind*, they jested that if some people were so slow into battle, they themselves must needs do the work of two. Perhaps, they said further, it were best done that way.

Drake's own men were equally astonished, but staunch in their loyalty: if the Vice-Admiral had said this thing, then the Vice-Admiral was right, the thing was the truth, and that was the end of the matter. A Captain-General did not sail round the world, and then turn fool or boaster when his foot touched land. . . . They answered the men of the *Golden Hind* by bidding them go on ahead; the *Revenge* would follow shortly, to bind up their wounds and turn defeat into victory.

The Lieutenant, waiting while the cockboat was pushed to the water-line, checked his men's talk, though not firmly; from his face it was clear that he could have added a jest or two on his own account, if he had had a mind to. Then presently they were gone, with a flurry of oars and a final parting, and Drake's men were left to wait at the tide-mark, under the hot sun.

'Fools!' John Waggoner said, looking after them with contempt. 'They do not know His Honour. He will come when the time is ripe.'

'If the Spaniards are up to the Lizard, it is ripe enough now,' muttered James Weaver, another of the crew, an older man. 'What was it he said—"win the game of bowls"? What sort of talk is that, when we fight and sweat for our very lives?

'Sir Francis fights alongside us,' John Waggoner said sternly. 'Sweats with us also. . . . And we all know it.'

'But they say he trusts his luck too far,' said a third man, newly joined to them.

'Luck!' Waggoner countered again. 'Who talks of luck? Did luck bring us round the world? Did luck take the *Cacafuego*, and all the rest? Skill, and preparation, and valour—those are his qualities. Luck is for children, and women!'

'Who talks of women?' another voice said. 'Belike His Honour plays his bowls between the sheets. . . . Shall he take time to finish *that* game? The Spaniards could be in London town, and the game still open.'

There was a roar of rough laughter, and Matthew Lawe rounded on them.

'Curb your tongue!' he said harshly, to the man who had last spoken. 'What are you looking for—a lashing at the capstan? Say the word, and you can have it, till the skin is off your back.'

'Easy, Matthew,' John Waggoner interposed mildly. 'We did but jest.'

'Then keep your jests away from His Honour. He does not roll in the same filth as you.'

There was a wondering silence. Matthew Lawe had their respect, and they knew it was his duty to keep order, being the Vice-Admiral's coxswain—to command his crew (as his commission stated), to see his boat trimmed with carpet and cushions, to cheer his gang with his silver whistle, and with his hand to steer his boat. . . . Matthew had their respect, and His Honour's confidence; but he had never yet used his authority so roughly, nor (as it seemed) for so slight a reason.

'Easy, Matthew,' John Waggoner said again. 'There's no man here would slander Sir Francis.'

'The first Captain-General to sail round the world,' James Weaver said.

'The second,' John Waggoner corrected him. 'Though that is no dishonour. . . . Magellan's ship was the first.'

'Magellan died on his journey.' James Weaver, a man of exactitude, had his discourse ready. 'Only his Lieutenant made the whole voyage. Sir Francis was the first commander to sail around the globe. In one navigation, he did all that Diaz, and Vasco da Gama, and Magellan, did severally. He——'

It was a familiar argument, and Matthew Lawe turned away from it. He was uneasy, and secretly afraid, which was the reason why he had rounded on one of his men for a careless joke. The reported talk of 'winning the game of bowls' was what had moved him; it was just such a thing as Sir Francis would say, and it seemed to be forcing his famous luck too deep into peril.

In truth, as John Waggoner had so stoutly said, no one could pretend that luck was all that their Captain trusted; it was not even the main meat in this mighty dish. Voyaging, Drake left nothing to chance. He carried carpenters and artisans to see his ships seaworthy, no matter for how long. In unknown waters, he seized fishermen as pilots, and bade them use their best skill to guide him—with a pike at their backs.

He even carried 'painters'—not to make pretty his ships, but to paint pictures of whatever land was seen, so to serve for the next time. And going into battle, he made such nice preparation that his men came to groan each time they heard the boatswain's whistle.

'What now?' Matthew had once heard a gunner's mate grumble. 'Do we tie ribbons on the cannon-shot?'

If ribbons would have served his purpose, Drake would have had them tied. . . . Yet when all this was done, and every chance foreseen, it could not be denied that their Captain swallowed huge hazards in his appetite for conquest.

One such hazard, which might have been the ruin of them all, was the late attack on Cadiz. 'Shall we squat like chickens in a run, waiting for the fox?' Drake had demanded, when news of the Great Armada's preparation reached England. 'Do we drop an egg in terror, or thrust up a torch to smoke out the robber? By God's leave, I will take my torch and singe that Spanish King's beard!'

For this 'singeing', he had chosen Cadiz, the heart of the enemy's might, where armed ships and warlike stores were thick as fleas on a dog. He had sailed into their lair like a lion, sunk or burned thirty-three of the Spanish fleet, sailed forth again without a scratch, and so delayed their stroke that the Armada only set out now, a year late by the proud calendar of Spain. But the chances they had faced in this enterprise. . . . Matthew Lawe, even on a warm day in Plymouth Sound, could run with sweat at the mere remembrance.

A further taunting of the enemy, fresh from this shame of Cadiz, seemed to give yet more hostages to fortune. This time it concerned a single ship, boarded and taken in the Azores; but she was a royal East Indiaman, and a prize of such magnificence as to wound her nation, the Portuguese, past endurance. The huge carrack *San Felipe,* thus seized in a hot encounter, was found to be laden with wealth beyond any dream of capture: her dazzling cargo, when the pirate's reckoning came, was thus described:

'Bullion and jewels, gold chains, crystals, uncut stones, tons of spice and ambergris, huge chests of china and raw silk and velvet, bales of lawn starched and unstarched, sarsenet, calico, carpets and pieces of taffeta and coloured buckram: all to the value of one hundred thousand pounds English.'

The *San Felipe* was not the greatest prize of Drake's life, nor was her commander his most noble prisoner. But at such times her capture was an affront to the whole manhood of Portugal. Why provoke, by such wanton stealing? Why make, of a backward enemy, a foe pricked to the full raging fury of revenge?

The offspring of this brazen insolence now towered above them all: the Armada stood at the great door of England—while Sir Francis Drake, who had so boldly tempted them, delayed like a truant boy on Plymouth Hoe.

At his back Matthew Lawe heard young Jem the farm-lad, who had learned his boasting in this same stable, say as he took a swig from the water-keg: 'Tomorrow we shall be quaffing Spanish blood!'; and Lawe, even as he smiled—for the boy was a favourite of them all—thought, with a sick heart: 'This is the great Armada, not a harvest-home. You will not be so bold, when you spill out your guts on the deck.'

Matthew's private fear was the fear of all England: a country long convinced by rumour that she was to be invaded by a mighty foe, and long suspecting that cowards in high places, and a niggardly spirit in the highest place of all, had left them defenceless. It had already reached the broadsheets that even so loyal a courtier as Lord Howard of Effingham, infuriated by royal parsimony, had dared to write to his Queen: 'For the love of

Jesus Christ, Madam, awake and see the villainous treasons round about you!'

The reply (it was said) had been female, dissembling, and for-ever miserly. Alas, the royal coffers were shrunk to nothing, and the fleets must make do with what they had. Surely brave men would not be the less brave, for a mere lack of powder and shot. . . . So the rumours could only feed upon themselves, and grow like giants of despair. Men now knew for certain that thousands of armed invaders were hell-bent on biting into English soil.

The Armada was to embark thirty thousand of these soldiers, under the Duke of Parma, at Dunkerque across the Channel. They would come in boats, in barges, in wicker coracles: they would be launched through the air like birds: they would swim under the water; and when they touched the coast of Kent they would blacken the sky with smoke, and sear it with flame and red-hot cannon balls, and thus set whole acres of downland afire, crops and villages and citizens alike.

Then they were to march inland, and join hands with the Catholic north, and bring down the whole commonwealth . . . England, like a starveling orphan without a rag to cover her nakedness, lay at their mercy already.

But Drake, who had the ear of great connections, and must have known his country's peril, would have none of it. His motto matched the scornful challenge of his kinsman John Hawkins: 'Came we to the south seas to put out flags of truce?' The simple spur of anger was enough to spark him, beyond prudence and into a ferment of endeavour: to hear his Queen called 'a poisonous Strumpet' by the Dons of Spain was enough to wash out all charges against her, and to double his loyalty—and his rashness.

Such hatred, such contempt of man for man . . . Sir Francis could carry it into every corner of thought, whether it concerned the best way of boarding a fat vessel-of-lading or the best way of approaching the sacred Body of Christ. His loathing of the Catholic religion had become a by-word, even among a nation split like a log by the axe of this division.

He lost no chance of preaching his crusade; he could write bitter words of accusation, and when he read them aloud, as he sometimes did, he could make them sting like scorpions, even when sailing with a happy wind.

Matthew Lawe could recollect one such reading, on their

voyage round the globe, when the embrace of the warm South Pacific Ocean might otherwise have persuaded his crew that they sailed onwards in gentle bliss and love. At the end of prayers, Drake had suddenly commanded them to remember their faith— and to remember, especially, those whom he called its enemies.

His words had a cruel, relentless spite: though he was reading from the tattered day-book in which he confided his thoughts, he seemed to speak from a spleen which envenomed the night:

'For as it is true that in all parts of America, where the Spaniards have any government, the poisonous infection of Popery hath spread itself; so, on the other side it is as true, that there is no city—yea, no house almost in all these Provinces—wherein not only whoredom, but the filthiness of Sodom, not to be named among Christians, is not common without reproof; the Pope's pardons being more rife in these parts than they be in any part of Europe for these filthinesses, whereout he sucketh no small advantage.'

He had said these last words—'*Whereout he sucketh no small advantage*'—with a hissing snarl, a seeming relish, enough to chill the blood.

Perhaps he meant no more than to hearten them against their coming trials. Perhaps he wished to hearten himself. . . . But could not such hatred lead a man too far? Could not courage itself?

Now warned by sudden noise and movement, Matthew Lawe looked up to the heights of Plymouth Hoe, and saw that he might soon discover his answer.

A great press of people was advancing towards them, down the fair green slope which led to the beach. In the van, to judge from colour alone, was all the quality of Plymouth: clothes gaily adorned, ruffs and ribbons fluttering in the breeze, the sun gleaming on polished blades, silver buttons and buckles, and the gold chain of office of the Lord Mayor of the city. At their back was a throng of townsfolk, less glittering than these birds of paradise, but not less jaunty in their attendance.

High spirits were in the air, as if they were all on holiday; for the man they escorted, the man leading them all, was one who met only smiles and shining faces.

Matthew Lawe needed no second glance at this figure in the centre of the crowd. He rounded swiftly on his crew, who were staring, open-mouthed, as if at a peep-show.

'Launch your boat!' he said, in crisp command. 'Then stand to your oars! Jem—at my side, and ready to make a back for His Honour!'

Their great Captain drew near.

V

He was forty-six, at the peak of manhood, with a countryman's stalwart body and a sailor's ruddy colour in his face. The red beard, which seemed to gape like a cavern when he was merry, and to jut like a rock when the mood—or the wind—turned contrary, was trimmed with as much care as one of his own sails. He was a man who could look the whole world in the face, and had indeed done as much, and either find it to his satisfaction, or make it so.

A fine friend, a foul enemy, he delighted to call his officers into conference when any doubt arose. He would listen. He would nod agreement, or tap his clay pipe-stem against his teeth—perhaps to show a wakeful presence. Then he would act just as he pleased, and neither man nor mouse would squeak thereafter.

If they ever dared, the arrow-scar on his right cheek, the remembrance of an Indian skirmish ashore during that voyage around the world, sounded the alarm. It flamed scarlet, like a pennant unfurling, and after that the tempest raged. Matthew Lawe had only once or twice seen it so. He recalled it most awefully when certain rash mariners, at a time of great privation, had complained of the food they were given to eat.

'Eat!' their Captain-General had roared. 'You will eat what falls from my tail onto the anchor fluke, if I say so!'

It was enough that they believed him.

Sir Francis Drake was all energy—and he was all energy now, as he strode resolutely down the hill at the van of his escort. His appearance was not brilliant, when set against the curtain shining at his back—the doublets slashed and laced, the velvet cloaks, the satin hose and feathered hats which were the modish colour of the age. But his presence, like his humour, good or bad, was enough.

On the firm sand of the beach, with his boat rock-steady and waiting, he said his farewells. There were many flourishes of courtesy, many compliments, and much huzza-ing from the crowd: the last link with the land was to be severed in all formality.

His Honour the Lord Mayor of Plymouth, who wore the chain which Drake himself had adorned, began a speech but was cut short—with every politeness, but cut short none the less.

The freshening breeze, or the Dons, or Drake's own will, could not wait for windy civic pomp.

With no more delay the Vice-Admiral, and his small company, were borne on board their ship-boat. The throng on the beach gave a hearty cheer; the quality saluted with their feathered hats, like an aviary disturbed. Then the boat's crew bent to their dripping blades, and Matthew Lawe to his stern-oar, and the sailors were alone on the business of the sea.

In the cushioned stern-sheets, Drake sat frowning; but Matthew, who knew a frown from a scowl, and could gauge both as closely as any wench with her heart on her sleeve, was not put out. His Honour was thinking deep, as well he might; and that was all.

When at last the Vice-Admiral spoke, it was in one of those catch-thoughts which made men love him.

'Did you eat while you waited, Matthew?'

'Aye, your Honour, and thank you. They sent down bread and beef from the inn.'

'With a pretty wench to serve it?'

Matthew caught his humour. 'Aye. Save she was a pot-boy with a squint.'

'Then praise God for the beef.'

The scudding wind tugged at their clothing as the boat rose to the crest of a wave. They were clearing the inner harbour, and setting course for the Sound and the great fleet of Crown ships at anchor there. Matthew sighted his eye on the *Revenge*, their home and refuge, and bore down on his stern-sweep to bring the ship-boat on a strict course. He had no need to urge his crew, nor Drake to correct him.

Presently, when they were set, Matthew Lawe was made bold by his Captain's humour to ask:

'Did your Honour win the game?'

Sir Francis Drake gave him an instant hard look. 'What game is this?'

'Why, bowls, sir. Or so we heard.'

'What else did you hear?'

On a sudden, it was no moment to be bold. 'Naught else, your Honour.'

'You heard that I delayed.' And as Matthew sat dumb and confused, fearful that he had ventured too far, Drake continued with unaccustomed freedom: 'Why do you think I delayed?'

Matthew made himself busy with his stern-oar. 'Sir, I do not know.'

'The tide, lad, the tide!' Drake spoke loudly, as if to benefit some of his gentlemen attending, who sat in the stern-sheets with him. 'The Spaniards are up to the Lizard. Can I prevent them? I could not, if they were past Dover at this very hour! Against this wind, we cannot leave harbour till the tide ebbs, and lets us go. It will be God's own task, even so. We cannot sail, with the wind blowing up our nose; we can only pull ourselves out with hawsers and capstans, or tow with ship-boats. Or *swim*, with a rope in our teeth!' He pronounced 'Swim' as once he had hissed 'Sucketh no small advantage', with a coarse snarl of contempt. 'Our fleet will not move an inch towards the open Channel, for four hours. So I stayed to finish the game.' He expelled a breath of air. 'God save me from blockish fools who cannot tell a jest from the truth!'

'Sir, we did not know.'

Instantly Drake smiled, like a shaft of sun through a lowering cloud. 'No matter, boy! Not a Christian soul knows, seemingly. *But I know* . . . Now tell your lads to pull, if it breaks their backs. We have work to do, and at last the time is right.'

They rowed through the fleet right manfully, as much to take heart from this huge display of sea lions as to give courage to those who spied the Vice-Admiral's standard, and knew that his squadron was now helmeted again. The mustered fleet was the bravest sight in England that day, and each single ship enough to give backbone to a butterfly.

The tall hulls, of blacked oiled oak, rose to glittering paint-work above: topping this were the shields with their gaudy heraldic badges—the Tudor rose, the cross of St George, the fleur-de-lys; and at the very peak, the noble flags and silken banners, that could be a hundred feet long, embroidered in silver and gold, streaming landwards as they were buffeted by the wind.

Their names, which were like another banner of glory, were strangely assorted: some were pretty, some seemed to threaten, some had a warmer clasp, as if to remind a sailor of his home and hearth.

Six of them were Lord Howard of Effingham's own, others were furnished by the London merchants; and the roll-call rang like a trumpet-blast—*Ark Royal* the flag-ship, *Lion*, *Bear*, *Elizabeth Jonas*, *Victory* which was John Hawkins' own command, *Triumph*, *Merchant Royal*, *Centurion*, *Margaret and John*, *Mary Rose*, *Golden Lion*, *Dreadnought*, *Swallow*, *Galleon of Leicester*, *Mayflower*, *Rainbow*, *Vanguard*, *Bonaventure*, *Frances of Fowey*, and *Delight*.

Now Matthew Lawe was steering his boat towards Drake's private squadron, set apart since it had been ready and victualled for some days; and soon they were among the family which the Vice-Admiral called his own. Men crowded the bulwarks and set up a cheer as the ship-boat with its golden standard passed them by: stalwart men of the *Hope*, the *Nonpareil*, the *Swiftsure*, the *Advice*, and the *Aid*.

Drake their commander sat in stern and solemn state as he received this tribute; though Matthew, in a sidelong glance at his Captain's face, thought that he glimpsed a glistening eye. Then their brisk voyage was over, and no ship in the fleet could match the welcome which waited for them as they drew alongside the *Revenge*.

It was as if their vessel were dead when their Captain was absent, and came to thrusting life when he set his foot upon the deck. But this Queen's galleon *Revenge* was a platform of courage in her own right. She sat squat in the water, because of the cauldron of weapons which she bore; yet she could glide like a seagull in the lightest air, and ride an opposing sea like a war-horse mounting a parapet.

From her sharp projecting beak, which carried the figure-head, to the tall sharp-cut stern, was one hundred and fifty feet. The high after-castle was not so towering as the Spaniards used; they chose to build their ships like houses piled one on another, as if to flaunt the prodigious wealth of Spain. The English mode was more humble, though it carried the gold-embossed stern-gallery where the Vice-Admiral could walk at ease, across the whole breadth of his ship.

From this vantage he could, with a little craning of the neck, look boldly across quarter-deck, main-deck, and forecastle, and the two tiers of guns which were the spit-fire teeth of the *Revenge*.

While the rapture of welcome to the Vice-Admiral reached its height as he mounted to the quarter-deck, Matthew Lawe saw his ship-boat secure and his men released from duty, and then climbed the long swaying ladder to the oaken lair which was his home. Vaulting the bulwarks, he was met, as usual, by a fierce figure, Tuke the boatswain.

Tuke, whose body was monstrous, wore a leather jerkin, soaked in brine to harden it against the enemy, and a red woollen cap copied from an Algerine corsair. He also carried his tool of office, the silver whistle which was one of the three allowed to the crew: to the master, to the boatswain, and to Matthew himself, as the Captain's coxswain. But Tuke needed no whistle to confirm his rule over six hundred men, from swabbers to quartermasters.

It was in his bulk, larded with fat, muscled like a plough-horse: it was in his face, which bulged like a keg of ale with a mouth for a bung-hole, and in his manner, which proclaimed aloud: 'I am below every officer, and above every man. *Do not let it slip your memory!*'

Of all people, it was Jem the farm-lad who had once passed verdict on Boatswain Tuke—a judgement so apt that it had gone round the fleet.

'We had a breeding bull like him once, at home,' he had said, in his simple West Country sing-song. 'But we had to make an ox of it.'

'Why, lad?'

'For our own lives!'

Now Tuke confronted Matthew Lawe, with his usual show of authority. 'Is the ship-boat secure?'

'Aye.'

'See it hoisted inboard, when His Honour has no further use.'

'Aye.'

'You were long enough returning.'

'Aye.'

It was best to deal with him so. The soft answer turned away wrath; no answer gave it naught to feed on, and thus it starved. . . . Matthew could see why Tuke had grown such a merciless rogue; he had his duty, though he need not make such a hearty meal of it. He had his duty, which was to stand between the men above— captain, lieutenant, master, sea-corporal, master gunner—and the men below: if he did not skin their backs, his own might be judged the culprit.

Indeed, it might be flayed. For through all this shouting and cursing and loud bombast there ran a gory thread of punishment: the harsh command which governed a Crown ship at sea. Tuke had his own Holy Bible, and Matthew had once seen it; it was the Scale of Punishment for sailors who ran foul of the law, and against this law there was no appeal—a man might be dead before he could open his mouth to cry his innocence.

The least burden was a whipping at the capstan, for insolence, or disobedience, or petty pilfering. From thence it rose through a fearsome catalogue of woe.

For stealing on board: PENALTY, ducking at the yardarm, and putting ashore, dead or alive. For sleeping on watch: PENALTY, once, a dousing: twice, a double dousing: thrice, bound to the mainmast for a day and a night: four times, hung at the bowsprit-end till he cut himself down, or starved to death. For drawing a weapon to strike the captain: PENALTY, loss of the right hand. For desertion: PENALTY, hanging or keel-raking, which was drawing underneath the ship. For murder: PENALTY, bound to the corpse, and thrown overboard. For mutiny: PENALTY, hanging overboard by the heels, till his brains be beaten out.

Penalties, penalties, penalties. . . . Were there no rewards, beyond beef and beer and the commander's glory? For if a man escaped the wrath of the great, there was still Tuke's fist, as hot as the pox, and just as ready to strike. There was still death or maiming, on any tomorrow. Was it for this that a man went to sea?

It was. When Boatswain Tuke, baulked of argument, told him with an oath to quit his airs as His Honour's coxswain, and join the men hauling shot from the orlop deck, and to *jump*!, Matthew knew that after a half-day's blessed peace on the tide-mark, he was home.

Down in the nether dungeon of the orlop deck, which was little more than a store-pit set above the bottom frames of the ship, it took time to grow accustomed to the gloom, barely lit by a single lantern lashed to a deck-beam overhead. But after a moment or two, Matthew Lawe found his sight, and with it his friends.

Under the eye of the master gunner, they were lifting round-shot from the racks, and passing it from hand to hand up the ladder to the gun-tiers. With room only to crouch, it was back-breaking work, but it went with a will. There was no need, on this evening and at this task, for Tuke's cursing and buffeting; the men of the *Revenge* need not be driven to this, their last preparation.

On it their lives might depend—and after that, their honour as men of Drake's squadron which would, so rumour ran, be given the vanguard of attack against the Great Armada. So close to battle, there could be no skulkers; if there had been one man such, his own mess-mates would have cuffed him into willingness.

But they found time to talk. As well as the shot, there was a call for tallow, sides of raw beef too mouldy to eat, canvas strips, and the wood plugs called stopples, all of which would prevent their leaking if they were holed. It was old James Weaver, pausing to stretch his back, who said:

'I would rather be sending up shot than stopples. Smite them first, I say! Then they can staunch their own wounds!'

'When we were holed at Hispaniola,' Matthew told him, 'you were glad enough of stopples. The sea was running in like a sluice-gate!'

'Aye,' John Waggoner said, out of the gloom of the orlop, 'he would have stuffed his bum in the hole that day, if all else failed.'

''Tis big enough,' a nameless voice proclaimed, and laughter rang round the shadows of their cave.

'Now hark 'ee——' James Weaver began, offended.

'Or even what lies to the front of his breeches.'

'Too small,' the unknown taunter declared.

Old Weaver was growing angry. 'Who said that?' he demanded.

A falsetto voice, which they all knew to be Jem's, gave the answer: 'Big Annie at the Turk's Head.'

At that, the deck echoed with laughter again. Matthew joined in gladly. From long experience, he knew this mood, which was poised on the knife-edge of fear. They were cheering themselves with foolish, feeble jests, to avoid the grimmest jest of all—that tomorrow they *would* be ripped and shredded by a hundred shots, that the sea would flood in like a cataract, that this orlop deck, and the tiers above, and their own shattered bodies, would join the fish, and presently feed them.

Many of his mess-mates must have been as fearful as Matthew was himself. Stouter men, great men who could make great phrases, might pronounce that England's hour was about to dawn, as the bell tolled for Spain. It was well enough for them, snug in their palaces as the sailors stripped for death. . . . Above his head, a horrid voice snarled a message down the hatch-way:

'Below there! What are you laughing at? Have you no work?'

Silence fell, like a doused sail. Their jesting was not two-sided; it had three elements now—friends, enemies, and Tuke the boatswain. As frightful threats of whippings and duckings rang through the gloomy air, they bent their backs once more.

Towards ten o'clock of the evening, the stubborn tide turned, and it was time to make their move.

While the Vice-Admiral, as always, supped to the music of viols, the *Revenge* began her laborious journey out of Plymouth Sound, followed by her consorts. With the ebb tide now behind them, they raised anchor and drew themselves inshore; then, with ropes and hawsers, capstan and brawny arm, they crept from quay to quay, and then from rock-face to rock-face, sending out new warps by carrying them ahead, making them secure to any ring-bolt, any projection of stone, and then hauling ever southwards towards the open Channel.

There was no space to sail, while the wind continued as stubborn against them as the tide had been. They must lift themselves out, as if by their own boot-straps.

It took all the evening, and all the warm July night, before they were clear, and into deep water; at anchor off Rame Head, with room to tack and wear, waiting to see what the dawn would bring. But what a sight met their eyes as they looked landwards!

The whole coast of southern England, headland by headland, shire by shire, was aflame with giant beacons; their sparks whirled towards the black heavens, making the expected signal, spreading the alarm along the coast, and then marching inland as they were sighted from distant hamlets and villages, and the torch was taken up.

England, with thanks to God, was at last awake.

Dawn came at four o'clock, as gentle as the bird-song which blessed it; and then and there, south-west beyond the Eddy Stones, they saw their enemy.

VI

The alarm raced round the ship, spreading as swiftly as the red glare of the beacons. Men, exhausted by their brutish work at the capstans, knuckled their eyes wearily, then sprang up as other men called: 'Come look—the Great Armada!' On deck, fine gentlemen in ruffs and doublets, and common sailors in stained worsted, ran to any vantage point in the rigging, to see what could be seen.

No man could have guessed what this Armada would look like; nor had any man in England ever set eyes on its like before.

Matthew Lawe, astride the bowsprit which was his duty-place when handling sail, peered into the hazy morning; and as the dawn-light gained, and a brisk wind dispersed the mist, the amazing sight came up, in all its glory and terror. Away to the south-west, not yet levelled with Rame Head, the Spanish fleet was running free up the Channel, like a Roman triumph on the march.

But this was more than a fleet; it was a city of huge ships, with here and there a true cathedral afloat—one of the great galleons of Biscay or Seville or Tarragon—towering monstrously above its fellows. They were set in the form of a crescent, a demi-lune seven miles across, with the horns curving astern; and they sailed in order and discipline such as no free-moving English fleet could have matched.

It was like a single fist, clenched tight, aimed at the mouth or the heart or the guts, and nothing seemed more sure than that it must find its target, and pierce it, and bring it to bloody ruin.

Matthew, appalled and sick at heart, began to number them and then lost his count—the distant mass of ships, and the bustle close round him, were too monstrous for a whirling brain. All he could think of now, as he gripped the sprit with trembling knees, was the rumour of what lay within that mortal crescent of the Armada.

It was said that all these ships were larger, and stronger, than anything the English could put against them—and so it looked, at this distance, on this morning. It was said that they carried two thousand and five hundred pieces of ordnance, eight thousand sailors, and twenty thousand soldiers.

It was said that they had within them people and things more wicked, and cruel, and treacherous than any sailors or men-at-arms. There were ships with engines for storming castles: ships with false banners, displaying the cross of England to confuse the people they sought to kill: secret ships carrying women, who would poison men or infect their manhood: ships full of priests ready with branding irons, and scourges for heretic backs—and other priests, more serpent-like, with jewelled swords for the Catholic peers of England, when the Spaniard joined hands with these treasonable rogues.

It was said—he came out of his dream of horror to feel a power-ful nudge in the small of his back. He turned, to find another of the sprit-sail men, Thomas Berry, grinning at him.

'Wake up, Matthew! Do you want to fall in the sea, before they send us a shot?'

Matthew smiled back, though wanly. 'I do not want to fall in any fashion. . . .' He pointed towards the south-west, with its threatening cloud blown not by God but by man. 'Can you count them, Thomas? We need John Waggoner here, to number these vultures.'

Thomas Berry, late arriving, had important news. 'John is sent back to the quarter-deck, helping His Honour. He says, more than a hundred and twenty sail.'

'And we have but eighty—and no more than thirty are great-ships.' Matthew shook his head, not hiding his discomfort. 'What does His Honour plan?—that we should make a charge at them, like a pack of little hounds?'

Thomas Berry looked at him curiously. 'What's your matter, Matthew? Cheerly, man! All you lack is your break-fast! They are but Spaniards—women in armour. We have given them a thrash-ing round the world!'

'Not in these numbers.'

'Then His Honour will find a plan, and choose his moment. Do you think he is suddenly asleep?' Berry shaded his eyes, looking first at the crescent of the Armada, then landwards along the coast. 'They have the wind of us, if we go out now. And there's little room to work between them and the land. If they choose to attack us here——'

He was interrupted, as if Drake their Captain had heard his warning, and honoured it. The boatswain's whistle made its shrill call; then there was a shout, which rang the length of the *Revenge*. Tuke had lungs of brass, and did not spare them.

'Heave up short to the anchor!' And then, with scarcely a pause: 'Make sail!'

Lord Howard in the *Ark Royal* led the way; Drake's squadron followed, obedient, and at one mind with the Lord Admiral. Then they divided: the *Ark Royal* to the south, round the furthest horn of the Spanish crescent, the *Revenge* to the west and north, to slip— if she could—between the land and the Armada, and so gain the weather gauge.

It was all done in a brilliant show of the sailor's art, though the work, and the daring needed, were beyond estimation.

All that day, and for most of a pitch-dark night, the *Revenge* led her consorts as close inshore as ships, and pilots, and skilled men taking soundings every minute of every watch, could manage. These were Drake's own waters, from boyhood on; but where he used to play with row-boats and hand-lines for minnows, he now shepherded and put at risk Queen's galleons and men's lives.

For hour after hour they tacked, and pinched the wind, and cheated rocks and shoals, and drove to exhaustion the men who must change sail a score of times in a single hour; and laboriously they made their gain. It was said later that if a few ships of the Armada had struck them then and there, on a lee shore with as much sea-room as in a closet for clothes, Drake would have been destroyed, the city of Plymouth taken, and Lord Howard's squadron overwhelmed at will.

But the Duke of Medina Sidonia had his own orders, which were to take the Great Armada up the Channel, come what may, and join with the Duke of Parma and his 30,000 invaders gathered ready at Dunkerque; no other bait was to tempt him, nor any other plan put in its place.

Thus he did this, and Drake did that; and on the morrow the whole battle-ground was reversed. While the Armada bore on- wards like a great travelling circus, Lord Howard and Sir Francis joined hands behind it, with the wind at their backs and the enemy before. A pair of crouching sheep-dogs could now turn themselves into wolves.

'Now we have them!' were Drake's constant words during the next few days, the confident boast of a sailor who had won the weather gauge on an enemy, and was never to lose it till the last shot was fired. But there was no haste to engage closely; it was a time for darting in and out, for many a gun-duel to probe the strength of the Armada, for pouncing upon stragglers and teaching them the first lesson of warfare—to stay with their fleet.

He learned two things during this time: that his ships, smaller and lighter than the imposing foe, were also more nimble, and better to manoeuvre; and that the Spaniards had the utmost discipline in their ship-handling, and could swiftly reform their ranks and concentrate their might, at any moment required.

He also observed that a hurt ship was received into the centre of the crescent, like a sheep into the fold, while the huge protecting horns closed in on either side. But this could not always be done; and there came a day—the third day, off the Race of Portland— when the English drew their first blood; when the Vice-Admiral, looking at a certain Spanish ship, could exclaim:

'And now, by God's teeth, we have *her*!'

The unfortunate was a strong galleon, the *Nuestra Señora del Rosario*, flag-ship of the Andalusian squadron. She had been in collision with one of her consorts; she lost her bowsprit, and presently her foremast; and though she was shepherded into the centre of the fleet, she soon became no more than a nuisance to them. For she could not sail, and the Armada could not wait: better, they must have thought, to lose a ship—even such a great one as this—than to delay their plans.

So the laggard *Rosario* was left to fall behind; and when the *Revenge* came upon her, she was wallowing, dismasted, beyond control, an orphan lacking either friends, or strength of her own, or mercy from the world.

Perhaps, thought Matthew Lawe, who was kept busy in his ship-boat during the next few hours, it was best to start this mortal battle with a jest. For the plight of the *Rosario* was a jest, and a rich one at that—one of the richest, for Drake the conqueror, of all the fight. Seeing that nothing stirred on board the wreck, and that no guns menaced the *Revenge*, Drake sent an emissary by boat, to inquire if the *Rosario* surrendered.

When Matthew returned, with this English gentleman, he bore a new passenger, a Spanish grandee who, with great show of politeness, said that he represented the Admiral of the *Rosario*, Don Pedro de Valdés.

'Well,' Drake asked him curtly, 'does he surrender?'

The noble Spaniard, who was exquisitely dressed in silk and gold-laced ruffles, with a silver-fluted breast-plate to give him a martial air, had not, it seemed, come to answer so rough a question.

Instead, he asked a polite one of his own. Might he know, he inquired, with many a flourish, whom he had the honour of addressing?

Sir Francis Drake told him.

'Ah,' the Spaniard answered. 'That may well make this a different matter.'

'How different?'

'I believe that Don Pedro de Valdés would think it an honour to surrender to such a valiant man.'

'That is good news,' Drake told him.

'On the other hand——'

'There is no *other hand*!' the Vice-Admiral roared out suddenly. 'I cannot waste my time. I have other ships to take! It is either surrender, or fight. Which?'

'I will inform Don Pedro de Valdés,' the Spaniard said, mortified at the cold breath of what seemed to him a monstrous breach of courtesy.

'In my exact words . . .' Drake turned, to where Matthew was waiting at the head of the ladder. 'Matthew! Carry this gentleman back. Then wait awhile. You will have a passenger on the return journey.'

'What passenger?' asked the Spaniard, for whom events were moving far too swiftly.

'Don Pedro de Valdés.'

And so it proved. The Spanish admiral, with a glance at the vanishing top-sails of his Armada, struck his flag, and returned with Matthew in the ship-boat. ('Thanks be to God!' John Waggoner muttered under his breath, after enduring the long swell and the curling wave-tops of their second journey. 'My back is breaking, with all this to and fro!') Once on board the *Revenge*, Don Pedro repeated what his noble emissary had said—that he thought it no shame, but rather a pleasure, to surrender to a man of such quality as Sir Francis.

Drake replied that it was a pleasure shared, and he meant it so. The *Rosario*, with a prize-crew aboard, sailed for Tor Bay. She carried chests of gold escudos, up to fifty thousand pieces, and other riches unusual in a ship of war, and Don Pedro de Valdés himself would fetch a fat ransom. . . . The *Revenge* bore away, to take her station in the van, while the Spanish admiral was received below with the most liberal courtesy. Finding him agreeable for many reasons, Drake bade him eat at his own table, and sleep in his great cabin.

He knew that he would have little use for either, in the next days. 'We will take twenty ships like her!' he said loudly, in the hearing of all on the quarter-deck, and they believed him. The Vice-Admiral's famous luck. . . . Privately, they said that his pirate's nose had led him to this prize. Now it could only lead him to glory.

VII

The promise of glory was surely there, and it lay on this side of the horizon rather than the other. But who was to earn it, and who to pay for it, was something which vexed more people than Matthew Lawe, who feared that the answer, on both accounts, might be himself. Yet it was a great man, rather than a small, who must plan the outcome.

While such humble fry as Matthew brooded on their chance of seeing the next dawn, Lord Howard of Effingham, Admiral of this whole sea-province, strove to make sure that every next dawn would glow more brightly than the last. There was no other way to prosper, either in battle or in love. . . . First he had to smooth certain differences which already poisoned the thought of his commanders; then he had to direct their ardour towards the real enemy—not their fellow-Englishmen, but the hated Spaniard.

Lord Howard, a wily counsellor as well as a sailor of renown, knew that he did not command such a happy band of brothers as he would have wished. Already there was talk that Sir Francis Drake, leading the fleet at night, had doused his stern-light (on which they all depended for their station), and gone off into the dark without a word, in order to take the *Rosario* in the morning— for his own profit.

Lord Howard knew also that another of his Captains, Martin Frobisher, had for this reason named Drake a treacherous robber; and that John Hawkins, who had laboured to design the English fleet, thought that Drake was carving his reputation out of other men's brains. Even Lord Howard could see that his Vice-Admiral, in such seasoned company, was not the darling of all the world.

Yet he had to resolve this knot, in the midst of battle, or the battle itself would be lost. Warring admirals, like warring lords, or jealous kings and queens, made up no more than a nursery jousting-field, which must be quelled before it infected the realm outside.

It must be quelled by men who knew the use of power, and did not fear it. He needed these children of discord. Each had talent and courage. Together they vied with each other: separately they might prosper, and England with them.

Lord Howard had this power, and he wielded it. With no more delay than was needed to silence argument by direct command, he divided his fleet into four squadrons, under himself, Drake, Frobisher, and Hawkins. He bade them act, rather than talk or plot, and then he let them loose. He put his Vice-Admiral in the vanguard—and only God could have dared to speak the reason: that Sir Francis was the right man to triumph or, if not, the right man to be killed.

Thus did great officers dispose of those less great; while the little shrimps among these whales solaced themselves with their marvellous prospect—a blind chance of life or death. For after this prudent pause the true battle began, and though it proved Lord Howard's wisdom, it was the hottest work Drake's men had ever known.

The Armada was not directly engaged; this cliff of huge galleons could not be scaled with the little ladders which were the English ships. But it could be worn down, and undermined, and sucked to disaster, piece by piece: what the greedy sea did in its yearly toll of the Dover heights, the *Revenge* and her squadron copied to advantage.

It was hit-and-run-away, for the next four days and nights. Their ships proved quicker, and their gunfire faster, than anything the Spaniard could match. Drake, the point of this burning spear, drove himself, and all his men, with a fury nearer to lust. The *Revenge*, darting in and out, was holed many times; the Vice-Admiral's own cabin was breached and wrecked.

But though men were killed, no English vessels were forced from the fight, while ship after ship of the Armada was pounced upon, with all the malice of a pack of hounds snapping and snarling at the heels of a deer, until that deer was glad to limp into refuge.

Sir Francis Drake could report to his Queen: 'The feathers of the Spaniard are plucked one by one.' It was a true account of damage done, though some fifty of his men were plucked at the same time. A Spanish galleon's broadside was like an evil mouth opening, then venting the breath of hell. Men were slain at their ruined guns: men below were fired like torches: men on deck were swept to destruction in this iron hailstorm.

Boatswain Tuke, that man of harsh authority, was one of those killed. Between one oath and the next, a cannon-ball took off his head, and laid it on the main-deck as neat as a duck on a platter. There was a bright side, it was thought, to every threatening cloud.

The *Revenge*, and all the English squadrons, pursued their enemy with relentless, pounding blows, until the proud crescent of the Armada lost all shape, and became a formless rout. On the fifth day of this hot pursuit, they could only draw together and run for Calais Roads. There they anchored: to escape the foul weather, to lick their wounds and count the toll of their advance: and to make what repairs they could.

They still maintained their last forlorn ambition, which was to embark the Duke of Parma's host of invaders, now camped round Dunkerque, for their vaunted 'Enterprise of England'.

Soon it was dusk on this last day. While the Spanish fleet lay crouched at anchor, the beacons still flamed along the English coast, casting devilish shadows for fifty miles or more. It might have been this which gave Sir Francis, and then Lord Howard, the same concept of their next, perhaps their last stroke. When they came together in council, they both shared one war-like thought:

'There is only one thing now! Smoke them out!'

Even a full day later, at the same hour of dusk, Matthew Lawe still wanted sleep more than life itself. He lay face downwards in his hammock, slung in the stinking shelf which was the lowest tier of the forecastle: beset all about by men talking, men wolfing food, men groaning of their wounds; yet trying to shut out the *Revenge*, and all else which stood between himself and his dream of hopeful life.

The past week's fighting, the long running battle up the Channel, the brazen uproar of cannon and counter cannon, the hours of shifting sail and patching leaks, seemed to have sapped his strength beyond endurance. It had been crowned by the labour of today, when he and his crew had worked like dogs to prepare their fire-ship, one of the eight vessels which were to be loosed into Calais Roads, on this evening's tide.

She was Sir Francis Drake's own property: the *Thomas*, of Plymouth, 200 tons, a profitable ship which was now to be sacrificed to the fiery attack against the anchored Armada. They had sweated twelve hours to turn her into a torch: stuffing her with straw and brushwood and tarry rope-ends, dousing the spars and rigging with pitch, double-shotting the guns which would explode when the heat of the fire reached them: setting up barrels of powder, and preparing the slow-match which would spark the whole ship into a volcano.

And this was the Sabbath day . . . He turned on his back, and stretched his weary limbs, and closed his eyes for the hundredth time: still sleepless, and tormented by it: yearning to be dead to this world, but not yet ready for the next.

His thoughts warred with his hunger for oblivion. The gory death of Boatswain Tuke had been no jest for Matthew Lawe. Tuke, a tyrant to the last, had been breathing fire down Matthew's neckerchief when a more frightful fire was loosed upon himself. The cannon-ball had missed Matthew by the span of a hand, and when he swung about, it was to see a great gout of blood erupting from a headless trunk, and the head itself spinning away like a scarlet-dripping top before it settled on the deck.

If the *Revenge* had rolled to starboard a moment before, it would have been his own head. . . . This was the hazard of following the Vice-Admiral—the hazard which *must* claim him, perhaps within the next hour. . . . When presently he drifted into sweating sleep, it was only to dream that it *was* his own head, and as he advanced to claim his rightful property, the blood he slipped in was his own.

In his dream he crashed to the deck, bruising his shoulder—and he woke in abject terror to find his shoulder seized and shaken by a hand which would no more be denied than the hand of Death himself.

Would he never find sleep—even such evil sleep as this? . . . He turned over with a groan of despair, to meet the cheerful face of young Jem the farm-lad.

'What is it, Jem?'

'You are called.'

'But I want to sleep!'

'Am I to tell that to the Vice-Admiral?'

Jem's face had in it such pride and confidence that Matthew felt ashamed. He sat up: the hammock tipped, as every hammock had done since sailors were condemned to them, and he nearly fell. Jem, his strong arm ready, steadied him.

'What is this, for Christ's love?'

Jem was still full of the importance of his mission. 'A message from the master. The Vice-Admiral wishes to see his coxswain on the quarter-deck.'

Matthew collected his wits. 'Has something gone amiss? What have you heard?'

'I heard he wished to make you a knight.'

'*What?*'

But it was only a shipboard jest. During the past few days, both Martin Frobisher and John Hawkins, together with certain favoured kinsmen, had been knighted for their service by Lord Howard of Effingham on board the *Ark Royal*; so that the common sailors, muttering of this and that, had come to say: 'If we live long enough, we will all be lords.' It was not true. . . . But then, all the best jests were lies. . . . Matthew Lawe straightened himself, and his clothes, which were rumpled. He managed a grin for Jem the farm-boy.

'Thank you, Sir Jem.'

'Think nothing, Sir Matt.'

Then it was time to obey the summons.

When Matthew topped the ladder and reached the quarter-deck, he found it thronged with all the quality on board, from Drake himself to the pages of the volunteer gentlemen, all dressed in their best finery, as if this were a Palace audience rather than a pause in battle. Such an array of peacocks, waiting to sup in splendour, with sweet music to speed the wine. . . . He hesitated to advance, conscious of his rough clothes; but Sir Francis, who had eyes everywhere, sighted him, and beckoned.

He approached, and made his salute with his cap. Should it have been a curtsy, in this perfumed air? . . . Yet the company round the Vice-Admiral made place for him, and Drake's first words, as so often, were considerate and kind.

'Have you slept, Matthew?'

'Aye, your Honour. Enough.'

Drake smiled. 'When a sailor admits enough sleep, it is the end of the world.' He waited for the laughter of those round him; then he grew serious, between one sentence and another, as was his habit.

First he pointed landwards, through the last dusk of the evening, to the Spanish fleet at rest in Calais Roads. A gleam from the beacon lights flamed red among their topsails. 'Mark the Spaniard,' he said harshly, 'whom we mean to burn!' Then he swung about, and pointed again, towards the cluster of fire-ships prudently set apart from his own squadron, and growled a question. 'What do you see there?'

'The fire-ships, your Honour.'

'And the nearest?'

'The *Thomas*, sir.' Matthew was puzzled, and afraid, and even a thought rebellious. Had he broken his sleep to answer children's riddles? 'Is aught wrong with her?'

'No—and nothing is going to be wrong! If I give my little ship to the Queen, I will take care she is not wasted. *You* will take care.'

'Your Honour?'

'When you see the *Thomas*, you see your new command.'

There were smiles all round him at the words, which Drake's companions must have been prepared for. They could not warm a sinking heart and spirit. Matthew wanted no new command such as this floating firebrand. He wanted to walk ashore, and find peace. Any roadside ditch in Devon would serve.

The Vice-Admiral took up his instruction. The jutting red beard was flying all the signals of determination.

'I want to make sure that the *Thomas* keeps her true course, and strikes home. Like we did at Cadiz. So she will be manned, and directed, till the very last moment. Take the best of your ship-boat's crew back on board. Two more will suffice, if they are lively and determined. At my signal, cut the cable, and trim your sails. See that she is well aimed, for the greatest Spaniard at anchor. Then light the brushwood in the lower hold. *Then* light the slow-match for the powder-kegs. But wait till the *Thomas* is full ablaze. Then slip back into your boat, and leave her. . . . Is that clear?'

'Aye, your Honour.' Many things about it were clear. 'Save when are we let go? How near to the Armada?'

'When you can smell their fear! When you are roasting your-selves!' But this was a joke, a merry jest before supper. 'How near, is something in your own hands. That is why I send you, and not another man. And, Matthew——'

'Aye, sir?'

Drake was smiling now, a smile for a favourite, for the man he could trust. 'I look to see you again.'

It was a moment when Matthew knew, within his cold bones, that he would never see Drake again, in the living world.

As dusk deepened into night, and the south-west wind blew fresh, and the tide began to flood into Calais Roads, Matthew Lawe obeyed his orders, with a sullen will. Work he must: work was all—if he paused to think, he began to tremble. . . . With the two men he had chosen to help him, John Waggoner and Jem of the farm, he made ready the *Thomas* for her last voyage.

The light working-sails were set to draw, as soon as their ship turned her bows off the wind. There was an axe handy to cut the cable. The ship-boat was moored snugly under the stern, ready for their escape. Each man had a lantern for the firing—first the hold full of brushwood, then the trusses of straw on deck and the standing rigging, then the powder-barrels which would take the *Thomas* to her grave—and God-knew-what with her.

All was done that could be done. Now they sat on the after-deck, in the cold black of night; watchful for the signal from the *Revenge*, and falling silent more often than they spoke. It had been a killing day, for work and sweat; now its fearful peak drew near.

John Waggoner looked up at the moon, and then at a star he could trust. 'See how she swings,' he said presently. ''Tis a full flood tide already.'

Matthew, whose teeth were desperately clenched, grunted his answer. But Jem's young spirits could never be quenched. He beat the palms of his hands together, and called out:

'Give us the signal, your Honour! We aim to fire the whole Armada for you!' Then, to Matthew at his side: 'Shall they try a broadside at us, do you think?'

Matthew opened his mouth to answer, but it was his teeth which chattered instead. 'Curse this cold!' he mumbled. 'A man cannot speak.'

'We will be hot enough by and by!'

John Waggoner came suddenly to attention. 'There it is!' he said. 'They hoist the signal!'

They looked towards the *Revenge*. Slowly, and swinging as the flagship rolled, a red light rose above the poop lantern, and there remained fixed in its position. It was time to go.

They had their plan, and their orders, and swiftly they moved.

As Jem cut the anchor cable with lusty strokes of the axe, it splashed in the water and was lost. The bows fell off the wind, and one after another their topsails filled. Matthew tugged at the steering whip-staff and then steadied it, bringing the *Thomas* onto her course. A glance astern showed that they had other black ghosts in company. But the *Thomas* led the way.

Steering down the wind, Matthew gazed fearfully at his target, the careless cluster of lights which the Spanish fleet was showing. They were still far off, but already there was something in his taut breast which could hardly wait a moment longer. He had only one clutching thought: never to approach too near the Armada, but to burn and begone! . . . To John Waggoner, standing at his elbow, he said:

'Set the fire below.'

'So soon?' Waggoner objected, astonished. 'We have a mile or more to go yet. If we——'

Matthew made a supreme effort, and broke in roughly: 'Light it, I say!'

Waggoner picked up his lantern, and went forward without a word.

Presently there was a burst of fire, and then flickering flames within the forecastle. John Waggoner leapt up from below, and, following his orders, began to touch his lantern to the standing rigging in the bows, and then to some bales of straw high on deck. The fire crackled and spread, while Matthew watched it, appalled. It was so huge and fierce. . . . It was a signal to the whole world. . . . His guilt began to match his fear. They had fired too early, and every man in the fleet would know it.

Standing at the whip-staff which steered them, he could almost feel the Vice-Admiral's bleak eyes on the back of his skull. Drake might even see him, black-outlined against the flames, and be raging and cursing him already. For the *Thomas*, in full blaze, must be alerting the Armada, and betraying all the other fire-ships, which had hoped to be dark as the night. She might even sink disgracefully, far from the enemy.

But 'Burn and begone!' remained his craven watchword. Somewhere in that raging fight up-Channel, he had had enough of war; and all that followed now was the child of this sickness.

He must continue to the end, wherever it lay. . . . He called to Jem in the shadows:

''Tis your turn, lad. Take your lantern down to the waist, ready for the slow-match. I will tell you when.'

'But Matthew——'

'What is it?'

Jem, simple yet lively, had become confused. 'We are so far off! We can hardly see the Armada, but by God they can see us! What use to touch off the powder now?'

'I did not say *now*!' Matthew snarled at him. 'Christ's blood, do you think I want to start the slow-match *now*? The powder-kegs are ten feet from us! I mean, go down and stand ready. Cannot you feel the wind? We are running in like a storm.'

'Like a storm. . . .' Jem turned to look at the fore-part of the ship. It was now beset by mounting flames: already the rigging dripped great gouts of blazing pitch onto the deck below. 'It is a storm for us, not for the Dons. I say we are too early.'

'And I say, do what I order!' Between shame and fear, Matthew's voice was almost a scream. 'Do you want me to tell the Vice-Admiral, one of my crew refused to obey?'

'Tell him what you choose.' From Jem, these were the boldest words of his life. Then he shrugged his big shoulders, picked up his lantern, and was away. Another friend was lost.

The *Thomas* plunged onwards, a fiery beacon alone upon a tossing sea. The glare of flame on the water shone red as blood. It was not possible to see where they were headed. All was wasted, all was lost. Yet somehow a little mouse of profit—perhaps a single man's life—might be saved.

The ship heeled over suddenly to a lurching wave. A spar, loosed from the blazing foremast, came crashing down into the waist. A scream in the darkness told him that it had struck Jem. Perhaps it had pinned him down; but the lantern in his hand made its own wild arc in the air, and plunged into a stack of straw. On the instant, the dry kindling took fire.

Matthew, distraught, left the whip-staff, which was balanced and easy, and advanced a step at a time towards this new sea of leaping flame. 'Jem?' he called out fearfully.

'I'm trapped!' Jem's voice was full of his agony. 'Christ's sake, my leg is gone! Help me, Matt!'

Already, in the fiery glow, he could be seen, beating at the flames round him, like a man caught in a monstrous tide-rip. But Matthew knew only one thing: that the fire would be among the powder-barrels in a moment. The cordage of the mainmast, above his head, was now pouring down the same cascade of molten pitch.

He began to back away again.

Within the circle of fire, beyond his flailing arms and smouldering jacket, Jem saw him retreat, and could still be astonished at his friend. 'Matthew! You run away! Don't you hear me?'

There was another, far-off cry from the bows. John Waggoner was also trapped.

Matthew Lawe looked this way and that. The powder-kegs would burst at any moment. Or the double-shotted cannon would spray the world with grief. There was no time, not even to loose the ship-boat and climb into it. He ran for the ship's side.

Jem's wailing voice followed him. 'Matthew! The flames! My leg! Curse you for a coward! *Do you want to live forever?*'

But Matthew Lawe was deaf to all save his thudding heart, ready to crack from fear. Now an outcast beyond honour, beyond the sailor's creed, he took leave of his shipmates, vaulted the bulwarks, and threw himself into the sea.

VIII

The proud galleon *San Vigilio*, finest daughter of the Biscayan squadron, and pay-ship of the same sisterhood, was now no more than a fugitive drab. Her commander, Don Iago de Olivarez, knew it, and feared it, and was shamed by it. But day by day he had grown reconciled to the knowledge that he could do nothing to mend it.

He sat alone in his day-cabin, forty feet above the waves, hungry and cold like all his crew—did the English really prosper in these harsh northern airs, and if so, what was their cursed secret?—and drew his cloak about his thin shoulders, and wondered, as all Spain would soon wonder, together with her monarch and perhaps God himself, what had gone amiss with their enterprise.

They had started out with such happy sails, such high hopes. The Grande Armada Felicissima would carry all before it: they would thrash El Draque, the pirate who presumed to call himself an Admiral of England, in his own waters; they would invade his pest-house of a country, depose *that woman* who went by the name of Queen Elizabeth, join hands with their multitudinous friends, and bring an erring nation back into the true fold of Christ, and of Spain.

High hopes indeed. . . . What had turned them to such starve-ling despair? It had been blamed on their failure to take Plymouth, when that city was bare of its defences. But how could Medina Sidonia, Captain-General of the Ocean Sea, turn aside and dis-obey his royal orders? Others had questioned their fighting skill in the long battle up the Channel. But what could the ponderous Armada do against little agile ships which darted in, loosed off a broadside, and scuttled away like dogs which snarled once and then turned tail?

As a last excuse, it had been assigned to the weather, which had been so fair for the English, so foul for the ships of Spain.

Why so? Was it thus the finger of God? But the *San Vigilio* alone carried ten priests, and prayed in company five times a day! His chaplain and confessor (who did not have to govern a fleet, nor so much as a supply-hulk a-stink with rancid olive oil) had told him to search his heart for any stain of sin. He had searched it until it was sore, and empty as their water-barrels, and had found nothing.

He could hazard no answer except the sin of pride—the pride with which they had set out from Corunna, the pride which had thought to make a mock of the little ships of England, and had now reduced a royal Armada to a scattered handful of shrimp-boats, running for shelter like the shrimps themselves.

Don Iago de Olivarez huddled deeper into his boat-cloak. He was cold and hungry. They were all cold and hungry. They were all so thin. . . . Certainly matters had gone amiss, in terrible measure, as soon as they had anchored in Calais Roads.

They had thought themselves safe, at least for the night; who would dare to attack the massed power of the Armada, securely at anchor, in the dark? There could be no more hit-and-run-away, on this blessed, easeful Sabbath. . . . Fire-ships were the only danger, if the English could assemble them in time; there had been rumours of devilish machines, barges stuffed with diabolical bombs which could strew ruin and destruction for a mile around.

So the Armada made ready for fire-ships, with grappling pinnaces to drag them aside; and fire-ships it was, and the plan with the pinnaces failed because of the foul dark weather, and the whole fleet had to cut their cables and scatter before the threatened burning, like chips of wood in a mill-stream.

It was true that no single one of these flaming attackers found its mark, thanks to one foolish fire-ship which gave the alarm while they still had time to escape; but the escape was as costly as any other assault. Their swift departure, after midnight, with a veiled moon and a scudding wind, was chaos unfolded.

Ships missed their course and ran aground; ships bore down on other ships, and tangled in confusion, and were lost; cut cables became wrapped round rudders, so that a galleon tethered like a cow at pasture could make no escape, even when the next day dawned.

As these fiery cats were put among the pigeons, the unseemly flutter did more harm than all the English fleet combined.

Olivarez, with his *San Vigilio* which he handled like a dashing light horseman, was one of the fortunate. Having seen the errant fire-ship in flames while it was still distant, he cut and buoyed his cable without waiting for his Admiral's orders. Then he cleared the anchorage ahead of the fleet, and laid a course north-eastwards up the Flemish coast.

There he anchored, and at dawn found his flagship *San Martin*, and joined her company with five others, and sent out scouting pinnaces to look for stragglers.

By noon they had knitted these fragments together, and were an Armada again, though shrunken and battered. Then the wind turned sour, blowing fiercely from the same southerly point. They could only forget the Duke of Parma's soldiery, and all the dreams of invasion from Dunkerque, and continue on northwards, wherever this might lead.

Before dusk the pursuing English, led by El Draque, were on their heels once more.

Still ready to make a fight of it, the Armada had formed again into that demi-lune which had shown such strength before. There was a fine running battle from the coast of Europe, all the way up the North Sea—past the level of the North Foreland, and the Naze, and Spurn Head, and the mouth of the Tyne. But it was finer for El Draque and the other English commanders than for anyone else in the world.

This time the pursuing gunners were at their best. Ship after Spanish ship was holed, or its rigging was shot away, while its luckless hulk sagged down the wind, to be lost to the battle and to life. Try as they would, they could not get to grips with these nimble adversaries, who held the weather gauge from the Point of Margate to Newcastle and beyond.

Soon the English could close the range, and fire at their will. For besides being short of food and water, the Armada now encountered the worst famine of all. They were the first to run out of shot; not enough gold had been expended on this, before they were told by their King to be on their way. At the last, all they could do was to flee, before a wind as merciless as their enemy.

They might have been ruined altogether, if their attackers had not suffered the same ill-luck. But the pitiless meanness of the English Sovereign, who counted each cannon-ball as if it were a ruby from the royal vaults, robbed them of a full victory. While they were still in hot pursuit, their guns also cooled for lack of powder.

They could only follow, dogged and silent, as far as Berwick Head; and there, having sturdily shown their foe the back door of England, they bore away for the Firth of Forth. The Armada, scattered but still numerous, vanished below a troubled horizon.

Though they would never join hands with the Duke of Parma, nor anyone else, until they all met on Judgement Sunday, Spanish sea-might had *not*, as England had hoped, been ripped from the face of the ocean for an age and a day.

As one of this brotherhood of sailors, with shared knowledge of royal neglect and avarice, Don Iago de Olivarez would not have been astonished at another wicked turn of fate.

While Spanish prisoners-of-war, haltered like cattle, were being herded down the length of England, English ships were already being paid off, to save the Queen's purse; and such was this cruel haste that sick and starving English mariners, put ashore with a 'Thank you, and farewell', now lay dying on the streets of Dover and Margate and Harwich, and of any other sea-port which had sent its sons to fight.

Presently the *San Vigilio* found herself alone, an orphan and martyr of this storm.

First they had clung together, sixty ships and more; and though they never formed their demi-lune again, nor any other manageable shape, they were at least in loose company, and could on each cold dawn spy a companion topsail and imagine a friendly face.

But in the weeks of wicked weather which beset them as they neared the top of Scotland, they were forced into scattering. Now it was each one for himself; and in this lonely wilderness, ships began to die.

Olivarez, fighting his groaning ship through the worst seas in the world, past the foul headlands of Caithness and then between the Orcades and the Shetland Isles, had glimpses of disaster which he could only pray to his Saviour would never afflict himself.

He was forced to watch loyal men whom he loved, and stout ships which had shared all their trials, cast ashore as if they were sodden driftwood. Stricken vessels, holed by cannon-shot, and so broken by the weather that not a stick showed above their decks, drove ashore, to become splinters without a mortal soul among them. Others drifted down the bitter wind, laden with doomed and starving men, to find their graves in Norway, or Russia, or the limitless ice of the Arctic.

To save precious water in ships still capable, horses and mules were cast overboard. At one awful moment, the *San Vigilio* found herself sailing through a sea of these tormented creatures, whose screams were more piercing than the wind.

Their last consort, the galleon *San Josef*, they watched tossed ashore onto the rocks of the Orcades. Even as she struck, bands of murderous men dashed out of caves and began to kill her people as they crawled ashore. Sailors in the last extremity of terror fought for their lives in the surf; but the isle-men were too skilful, and no living man ever stood upright more than a few feet from the tide-mark.

She was the last friend they ever saw. After her death, the *San Vigilio* was alone in a wilderness of howling seas, without a rag of other sail in sight.

For Olivarez, it was the loneliest hour of his life; as well as the shame of defeat which would ring round the world, there was the desperate condition of his ship, which was in his sole charge. The *San Vigilio* lacked any food save rotting meat and salt fish, and mouldy bread as hard as stone; the water in the casks was foul beyond human bearing. Men were dying because they could only retch on this evil swill.

Her rigging was tangled in ruins; and exhausted sailors who worked every watch of the day and night at the pumps, to keep her afloat, had no spirit or strength to repair it. Below decks there was a multitude of sick and wounded, beyond any help but the shriving of their priests.

From his station on the poop deck which he scarcely left, Olivarez saw nothing save a labouring ship near to her doom, and felt nothing except the storm, blowing ceaselessly against them with a fury not to be described, and the stinging salt spray which could whip a face till it bled raw. It was a month before the *San Vigilio* turned Cape Wrath, as horrible a headland as had ever met Spanish eyes; and there she faced the full force of the gale, and mountainous seas which had crossed the whole Atlantic to vent their spite upon this last survivor.

He was now cleft by doubt. Though he *must* make the coast of Spain, the wind howled its refusal. Though his starving men *must* be fed, where could a scrap of bread or meat be found, in this waste of waters? Clinging to the rail of the poop deck, side by side with his wrestling helmsman, Olivarez made his decision. A Captain must choose, and bear the outcome, whether on his stained honour or his tormented soul.

Whatever happened when he put ashore, and no matter how murderous their welcome, he could not venture out into the wild Atlantic before his crew had rested, and made their repairs, and loaded enough stores—however mean, however rank upon the tongue—to turn the last page of their fearful journey.

So, at dusk on a bitter September evening, his ship crept into a slit of harbour which the faded chart told him was part of the Isle of Mull, and there dropped anchor. But there was no trace of comfort or hospitality to meet them; only a coarse shingle beach, and a countryside of stark cliffs and wind-swept trees. The few small cottages, black-browed, were fashioned of stone, and thatched with torn heather. Though there might have been many eyes watching them, not a face showed and not a dog barked.

What haven for them was this? What had happened to their dreams? Was God asleep? Don Iago de Olivarez stared at the land, and the land stared back, seeming to hate the *San Vigilio* and all her men. His gaunt face was bitter with disappointment, and his heart bare of hope.

IX

His servant Carlos, a pinched scarecrow like himself, woke him from a fitful doze, as the pale sun topped the horizon of their retreat, and pierced the windows of his cabin. Olivarez found himself stiff and cold; he had slept in his clothes, as he had done for many nights, the more ready to greet a miserable dawn.

For a moment he was bemused; there was something lacking in the *San Vigilio*. Then he discovered it. The ship no longer rolled and plunged and lurched from wave to wave, like a drunkard making his evil way home. She was still.

So did a prisoner miss his fetters. . . . At least they were in shelter.

There were things he could do before a trusted body-servant which no grandee of Spain would ever show the world. Olivarez yawned cavernously, and knuckled his rheumy eyes, and stretched limbs which had been aching for as far back as his memory could reach.

'What is the time, Carlos?'

'Near nine, my lord.'

'And what of the day?'

Carlos, long reconciled to an unfriendly world, could still salute the birth of a new day with the melancholy it deserved. His waxen wizened face expressed a fresh tide of ill-humour.

'Cold as charity, my lord. There was even *snow* in the night.' He said 'snow' as if he were pronouncing a word which should never pass Christian lips. 'How people can live in such purgatory . . . Men need not seek the end of the world . . . It is here, before their eyes! . . . Will you eat now?'

Don Iago, after many days, was conscious of appetite. 'Yes. What have you for me?'

Carlos pursed his thin lips. 'Mulled wine, my lord. Bread toasted with oil. The fish, you should not eat any more. Even the fish would not eat it.'

It was the same wretched fare as always, and after a night in shelter Olivarez could not understand it. 'Is not the forage-party come back yet?'

'No, my lord. And what they will find here, God alone knows!'

'But what time did they put ashore?'

'Two hours past, or more.'

There was something here which needed a closer look. 'Bring the wine. Then call Don Alonzo to my cabin.'

'He waits already, sir.'

'Summon him.'

Don Alonzo, his nephew and secretary, chosen rather to please a favourite brother than by any strict measure of merit, had lost much of his fire and zeal since he first joined the company of the *San Vigilio*, on the warm shores of Corunna. Now he could best be described as a young gallant in decay.

Nothing of his splendid garde-robe had weathered their voyage without ruin, and the soft leather thigh-boots which had once set off his manly legs now hung forlornly, like Easter decoration a month after the triumph of Our Lord's rising.

He mourned constantly for the sunshine of his beloved Córdoba. He mourned the food and the wine which had once made life so tolerable. He lacked the companionship of his fellow gallants, and the dark eyes of the girls; and he who had sworn to spit ten English before break-fast would now have embraced these same foul enemies as blood-brothers, if they would give him a little peace, and some fresh linen, and above all a swift passage home through calm seas.

There, he could discourse on how near to glittering victory his Armada had come.

As soon as he entered the great cabin, and had made his salute, he began to chafe his hands together. Don Alonzo had suffered the miserable cold for upwards of two months, and he wished the world to take note of it.

Olivarez was in no mood for such antics. 'What news of the forage-party?'

'Nothing, Uncle, to my knowledge.' Don Alonzo, not expecting happiness until his private world was set to rights, made the report as if it were of no consequence.

'But they have been ashore since seven o'clock!'

'So they tell me.'

'Are they not in sight anywhere?'

'I believe not.' The words needed a trifle of explanation. 'It is so *cold* on deck, I was waiting for news in my cabin.'

If this were not my brother's son, Olivarez thought, I would *hang* him for an idle, useless rogue! But he had borne this itching so long that it could not be mortal. The mosquitoes on the coast of Panama had bitten deeper . . . He asked:

'Why then are you here?'

Don Alonzo widened his brown gazelle's eyes. 'I thought you might wish to write something.'

'Mother of God!' His uncle was truly stung at last. 'What should I write today?'

His nephew was offended. 'Indeed, Uncle, I do not know.'

'Then we are both in the same case!'

But Don Alonzo, a hind at bay, manfully conjured up one useful office to perform. 'Now I recall—Fray Bernardo wishes to see you.'

'What is it now? More blasphemy?'

There was a strict rule on board all their ships that no blaspheming would be tolerated, in case it harmed their holy cause; and the *San Vigilio* had seen more sailors whipped for calling on the name of God than they carried rats in their holds. His chaplain, Fray Bernardo, was the zealous intendant in all this.

Don Alonzo tossed his head, still mortified. 'It is no matter for me, I am sure. But he thinks there is a curse on us.'

One such curse on us, his uncle thought, stands before me now. . . . Do I drive body and mind to distraction, to carry such butterflies home to their silken sheets? . . . A Captain's answer could only be Yes. . . .

'Go tell him to wait on me.'

'Yes, Uncle.'

'And then make inquiry about the foraging.'

'Yes, Uncle.'

'In your best fur cloak, on the poop deck.'

There was no answer save retreat.

Fray Bernardo, a black friar of the Dominicans, had always been thin; he had been a thin man at home, in the sun-warmed castle on the slopes of the Sierra de Guadarrama, and he was the same thin man on board the *San Vigilio*, marooned in this icy hole of Scotland. But wherever he found himself, he never failed to wield the same rod, with the same fierce piety.

Don Iago de Olivarez sometimes wished that he might love his chaplain more, as a good son of the Church should; but all he could feel was respect for his office and the way he discharged it, and perhaps a touch of fear—the fear of a man who had the power of prince-hood for one who had the power of God.

There were times when this thin man of God seemed thin of soul as well. Without taint of heresy, one could feel that such holy power need not strike so cruelly those who fell by the way, or those who had never found it.

On this cold grey morning, Fray Bernardo lacked nothing of his zeal.

'Good morning, my son.'

'Good morning, Fray Bernardo. I hope you are glad to be at rest, for a change.'

'I am not at rest. Rest is sloth. Sloth is sinful.'

Well, well, Olivarez thought: God pardon us sinners. . . . To bring relief to a severe subject, he said:

'I am sorry we have not found fresh provisions yet. But perhaps our fortune will turn soon.'

'Perhaps it will never turn.'

Olivarez surveyed the bony face before him, the face carved from a dry olive tree, the face which never relented. 'Why do you say that?'

'Because there is still a poison in your ship.'

Olivarez sighed. So this was it. Even in the midst of all their other troubles, the chaplain still thirsted after his prey. . . . 'Is he well?'

Fray Bernardo answered sardonically: 'Perhaps as well as he will ever be.'

'What is it you want, then?'

The priest's affected surprise was well done. 'The just punishment of a heretic! What else?'

'Then why did you not let him die?'

'That is not punishment. It is escape.'

Don Iago sighed again. These Dominicans were so strict! Not for nothing did they wear their punning nickname—*Domini Canes*— the Hounds of God. But why need they fatten up the sacrifice before the slaughter? Why must they yearn to tear to pieces a poor rabbit of an Englishman, who had fallen into their hands by such a chance, and might have fallen out again, to the mercy of release?

'Has he spoken?'

'Many times. It is always the same story. But now it should have an end. I must do my duty, and you must do yours.'

'He is a prisoner of war.'

'He is a heretic, an infection, an offence to the Faith! Can you wonder we have been so punished? I would examine him formally in your presence. I will establish his guilt before God, and you will punish him before the world. Only so can God be honoured—and this ship cleansed!'

There would be no end to this until Fray Bernardo won his way. Thus it must be resolved, and the sooner the better. . . . It was something they had carried with them for many weeks: the question of the Man—the man in the boat.

'Very well.' Don Iago was resigned—and with a little artifice he could pretend it was a small resignation, when set against their great hazards. 'Have him brought up.'

Matthew Lawe awoke from a miserable sleep to the usual rough assault of one of his guards. These men, in wretched state themselves, could show their meagre authority by kicks and blows, cuffings and cursings; only thus might the gaoler be distinguished from the prisoner, the master from the slave, within their floating lazaretto.

But after this awakening, the man who had put his booted heel on Matthew's neck and wrenched it round to face a flickering candle, made signs and gestures, as well as the noises of an animal snarling.

These must all be interpreted as 'Get up! Move smartly! Follow me!' and when he did raise himself unsteadily to his feet, a final push completed the translation. For the first time in uncounted weeks, he was to travel elsewhere.

On trembling legs he followed his keeper up one ladder, and then another, and then a third. There he was forced to pause, and crouch on the top rung; and for once he was not cursed or kicked, but left for a moment of peace till he regained his strength in the first feeble daylight he had seen since Calais Roads.

All that time he had been lying half senseless, sick and starved, in the black dungeon below: suffering all things, and the fearful shocks of a sea-battle as well. A sailor's instinct told him that the ship in which he lay was enormous; yet she was tossed about all the time by mountainous waves racing up from astern, and bruised and holed and torn by cannon-shot, and turned into a hellish cauldron of noise and flame by her reply.

Then gradually the shocks receded, and a sort of peace had returned, after a number of days or weeks which he could not reckon in the darkness. The cannonade died to nothing, and the only enemy was the sea. But this continued to beat furiously at their shattered hull; and by the suck of the pumps, for watch after watch, day after day, the sea might prove the worst enemy of all.

His prison became the refuge, the last frail ladder between life and death, for the wounded and the sick. Countless men were laid down to die in his company, and when they ceased to groan, their miserable journey was over. In such a cage of torment, buffeted by vile weather, sustained by food which a rat would have walked over in search of something better, and a cup of slimy green water once a day—in this plague-pit of the damned, Matthew Lawe had time to consider his plight, and the guilt which had brought him to it.

During these days of torture, there had been certain Spanish sailors who, learning his condition, had strength to curse him before they died. One such had even tried to strangle him in his sleep, before the last sleep of all robbed him of his fury. But none of these enemies could have matched Matthew's hatred of himself, nor salved his own shame.

After his coward's flight he had swum for his life towards the shore of Flanders fields, and swum into a kind of madness which had a fellowship with his own. In the stark darkness, a panic fear was taking hold, as soon as the first fire-ship was sighted, and the threat to their anchorage struck the Great Armada. Borne on the wind were shouts of warning, and signal guns firing, and lanterns waving, and the thud of axe-blows, and all the sights and sounds of alarm.

Black shapes began to move between the sea and the land, while at his back the fire-ships—the true fire-ships, not the disgraced *Thomas* which had blown up in fiery ruin as soon as he quit her decks—bore down in flames upon the Spanish fleet. Matthew Lawe trod the water, half-drowned already, cold to the marrow of his bones, as the secure haven became the mortal trap, and sailors like himself fought and struggled for their lives.

A vast bulk drew near him, and passed by, hissing its own desperation. The last effort he could recall was to grasp at a smaller shadow, a ship-boat forgotten in the swiftness of flight and still towing astern. He had hauled himself on board in frantic terror, and there collapsed. He had only wakened, at an hour which he guessed to be the next midday, in the stinking hole which was now below him.

In this black prison there had followed weeks, not to be counted, of the harshest suffering. Of all living souls, only one man had been kind. This was a priest, a stern priest who asked questions, and scanned the answers as if life might depend on it. But he seemed to wish Matthew to live, which was more than could be said for his gaolers. Matthew did not ask 'Why?' He could only love his only friend.

Yet he did not tell all of the truth in this noisome confessional. Need one be shamed before God as well as man? . . . He had said that he had fallen overboard from one of the fire-ships. He had said that he was not a spy, nor a man bent on some secret destruction, but only a sailor in misfortune. He acknowledged his faith, the Protestant faith of England, but added with toadlike humility that he had had no chance to learn any other.

Would he recant his sins? Matthew could answer, in the very garments of truth, that he wished that he could.

Now his prison guard, after staring down at him, gestured again. It was time to go. By another ladder, and yet another, he climbed painfully upwards. Was there no end to this wondrous ship, which he had guessed to be the biggest he had ever trodden, and was now displayed in all its pride?

Its decks were huge, and there was tier after tier of them, scorning the sea, mounting to the sky. As it grew lighter, and the pale gleam from the gun-ports became sunshine streaming through bayed windows, as if into some palace, Matthew could only be astonished that the *Revenge*, and all the others together, had dared to close with this monster and to challenge it.

But the men seen in passing did not match their vessel. Though many of them wore splendid clothes, and armour made by artists of their trade, all was rags and tatters, stains and filth, rust and decay. And they were so thin and forlorn. . . . He passed many of these walking scarecrows, and there was hardly a man among them who was not as miserable, gaunt, and lack-lustre as himself.

While they looked at him with a residue of hatred and contempt, their spirit seemed almost burnt out by their own desolation. It was as if they knew that, friend and foe alike, all were doomed. Could a Spaniard think himself above an Englishman, when they shared the same coffin and might soon be wrapt in the same great shroud of the sea?

Presently he emerged into cold sunlight, and then into a cabin more splendid than two of Lord Howard's own, in the *Ark Royal*. Here he found the priest whom he knew and trusted, and a great lord with a proud face, whom he did not.

The questioning began.

Fray Bernardo's English was excellent, and Don Iago's serviceable enough; together they made brisk work of Matthew Lawe's condition, and what had brought him to captivity. By mischance, he was their prisoner of war: an English sailor who had become a hostage from the sea. He owed his life to the *San Vigilio*. What that life was worth now could hardly be measured in the cheapest coin of the realm.

Don Iago de Olivarez, whom great rank had not made arrogant or cruel, surveyed this bruised and wretched man, decked in the foul clothes of captivity. All he could feel, in his own ill-fortune, was compassion.

He called to his servant: 'Carlos!' and when his man appeared, with the promptness of servants whose ears are never far from an open door, he said: 'Give him a cup of wine.'

Carlos, with a face sour as a quince, did as he was bid. His expression might have declared, 'Wine for this dog?' but his movements were deft as ever; it was his duty to obey, even to the benefit of such heretic scum as this. . . . Matthew drank thirstily, while Olivarez continued to survey his captive. Then his eyes, his sailor's eyes, narrowed, and he asked sharply:

'What is that circling your neck?'

Matthew, slow to understand, came to it at last. 'My whistle, your Honour.'

'So I thought. . . . I also carry a silver whistle, as commander of this ship, though naturally I do not use it. . . . Have you been lying to us? What are you? An officer?'

'No, sir. A coxswain.'

Olivarez turned inquiringly to Fray Bernardo, who made the best translation he could. 'A helmsman, I suppose.'

Matthew wanted to help them. He had little choice. He made the motions of a man steering a boat with an oar. 'No, sir,' he said again. 'I commanded a ship-boat. The whistle is the badge of office.'

'Such magnificence. . . . What ship-boat is this?'

'The Vice-Admiral's. Sir Francis Drake.'

A deadly silence fell. No Spaniard in the world—not even a priest with his soul above all earthly things—could hear the name of Drake without a shock of anger or fear. So here was a man with this taint of infamy, of piracy and treacherous evil, a man touched by the very monster of the ocean seas, infecting the *San Vigilio*. . . .

Olivarez, conscious of cold chill, came slowly back to his questioning:

'So . . . We call him El Draque. . . . We call him many things. . . . So you were his—how did you name it?—his coxswain. . . . But were you in command of a fire-ship as well? How so?'

Matthew was not strong enough, nor quick enough of wit, to dissemble. 'It was his own fire-ship, your Honour. The leader. Called the *Thomas*.'

'The leader? Was this the one which gave the alarm?'

'Yes, sir.'

Olivarez surveyed him grimly. 'Does El Draque fall overboard? Do his chosen men fall overboard? Why was your ship fired so early?'

'There was a mistake, your Honour.'

'Your own mistake?'

'Yes.'

'But what sort of mistake? You set the fire by accident?'

'No, your Honour. I thought—I thought we were close enough.'

'And did not wish to come closer?'

Matthew looked down at the rich carpet on the cabin deck, unable to meet this man's eyes. 'Yes, sir.'

'And after that, you—fell?'

Wretchedly, Matthew knew that this lord, who had a sailor's mind, had divined the truth. And now he was urgent to confess it—to purge all his guilt, before such a witness. He shook his head. 'No, your Honour. The *Thomas* was on fire. She was ready to burst! So—I jumped.'

When Don Iago said: 'We are grateful to you,' Matthew's shame was complete.

But now the silent man in all this, Fray Bernardo, stirred impatiently. He had listened to all he could endure, to the dull questions of how and why. But what did it matter? So an English sailor was a coward or a fool? There were more important things in the world—in this world, and above all in the next. A man's soul was at stake, not his repute.

With determination in his dry olive face, Fray Bernardo took command. The voice of a higher authority addressed Don Iago: 'Now, my son, let us go to the examination.'

He could not be denied, and only Matthew was left to stare as he heard the tone of an order he did not understand.

Fray Bernardo was a subtle and ardent questioner, as skilled in matters of the faith as was Don Iago in sea concerns. He touched no subject which he had not broached before; but his catalogue was now assembled in a swift indictment, with artful flourishes which at the end had stripped Matthew Lawe of all save the need for abject surrender.

'You have named El Draque,' he began, bearing down on Matthew. 'It is not a name to recommend you. . . . Tell me, what did El Draque think of the faith of Rome?'

Matthew still believed the priest to be a man who would honour the truth, even though it sounded harsh to his ear. 'He hated it.'

'Did he teach you to hate it?'

'He talked of such things.'

'And you listened and believed?'

'He was our Captain-General.'

'That was not my question. Did you believe him?'

'Yes.'

'So you hate Rome? You would destroy it?'

Already, aware of danger in the face of this fierce tone, Matthew was growing doubtful of his friend. 'Sir, I told you earlier, I know nothing save what I have been taught.'

'I know what you have been *taught*. I wish to hear what you believe, now.' And, as Matthew still hesitated, Fray Bernardo brought to his voice a silky menace: 'Have you been *taught* that we have whips and pinching-irons, to make men tell the truth?'

Matthew mumbled: 'Yes, sir.'

'Well, you may believe it. . . . Let us begin again, without deceit. Are you—*you*—an enemy of the Roman faith?'

After that, it was no match. While Don Iago looked on impassively, with no more cups of wine for a tormented man, Fray Bernardo thrust in with question after question, threat after threat. At the last Matthew could only answer what he believed the priest wished to hear.

Yes, he had hated Rome, as he had been taught. Yes, he must have been in error, because he did not know enough, and was deceived by lies. Yes, he believed what Fray Bernardo was telling him now. He freely acknowledged his sins, and wished to repent them. He was ready to recant. He would accept this new faith, because it was the only true faith. It was the gift of God to a heretic.

It was then that the fearful shock fell on his ears. Fray Bernardo sat back in his chair. He nodded almost gently at the abject man in front of him.

'That is well,' he said. 'You have now confessed all, recanted all, and found the truth at last. . . . By the infinite mercy of God, *you may die at peace.*'

A betrayed man could only stare back, confounded beyond belief as he took in what was now said to him. So it had all been a jest, a monstrous cruel trick, a game with cat and mouse which he could never have won, whatever he answered, whatever grovelling surrender he had made. So Sir Francis had taught them a true lesson: the Spaniards *did* torture and burn their prisoners, to the glory of their merciless god!

For all his humble duty, for all his profit as a turn-coat and traitor, he might as well have consigned the Pope and his tricks to purgatory!

In spite of his peril, Matthew was about to burst out with protest when he was forestalled. There were sudden shouts from outside, and a trampling over their heads, and then a brisk rally of musket shots. Was the *San Vigilio* under attack? He could wish nothing better. . . . Might she fry in hell, stoked by a whole boatload of treacherous priests! . . .

But the silence which now fell brought no such good news. When the sound of muskets ceased there were swift footsteps outside, coming nearer. Don Iago drew his sword, and stood ready. But the man who burst into the cabin, in gasping alarm, was his nephew, Don Alonzo.

'Uncle, Uncle!' he cried, clasping his hands as if they might otherwise fall to the ground. 'The forage-party!'

Without a word Olivarez sheathed his sword, wishing oddly enough that their English prisoner might have something better to see and hear than this pallid water-fly who was his own flesh and blood. . . . Then he asked curtly:

'What of the forage-party? What news have you?'

'The worst in the world!' Don Alonzo was suffering as no Spaniard had suffered before. 'They must all be dead—*butchered* by these heathen rogues!'

'Tell me. In swift words.'

'Oh Uncle, it was pitiful! My heart *bled* for those brave fellows! We saw but two of them left, running down to the beach as if the fiends were after them. And it was true! They were chased by men in skirts, who cut them down with great swords before they could reach the boat.' He turned aside, quite distraught. 'Carlos! A cup of wine, I beseech you.'

'What of the musket fire?'

'It was ourselves! I fired on those demons myself, and would have killed them all, save they were too far away. But the rest of our men—they must have been set upon and murdered!'

The news at last was clear. 'Put down that wine,' Olivarez commanded. 'Go tell Captain Barrameda to wait on me instantly.'

'But what can we do? There may be a thousand such devils lying hid, ready to cut our throats!'

'A thousand men in skirts?—get you gone!' Then Olivarez swung about, towards Matthew and the priest. 'We must close our business,' he said. 'I cannot deal with such matters now.'

Though a Captain's will should top all others at such a moment, Fray Bernardo was still intent on his particular pursuit. 'But your verdict, my son? The case is proved. This man is——'

'This man is nothing,' Olivarez snapped. 'We have our own fellows to save—or to bury. And we still lack stores, and stores I must have.' Then, seeing his nephew hovering by the cabin door, Don Iago de Olivarez shouted, for the first time in a courteous life: '*Alonzo!* Did you not hear me? Bring me Captain Barrameda on this instant! Unless you wish to lead the rescue yourself.'

'I? Myself? Well——'

The gallant turned and ran, before a true tempest should break, or a worse fate befall him.

X

Whatever sport the 'men in skirts' thought they might have with a foraging-party of fifteen sailors and a single boat, they must have changed their mind on sighting a hundred armed men in three craft, led by an officer who looked a shining Hercules. When Captain Barrameda returned, and made his report to Olivarez on the quarter-deck, he had nothing much to say of his landing, which had been peaceful and silent. But the tale of horror he brought back was enough to make the gorge rise.

Barrameda was a soldier magnificently tall, the finest fighting man whom the *San Vigilio* carried, and one whose regiment, the First of Biscay, followed him as others followed the Holy Father himself. Saluting Don Iago, the admired friend who was also his commander, his face beneath the gleaming casque of his helmet was stern. He reported as a soldier should, in words crisp, direct, and few.

'My lord, we found no enemy—not a hair of their cursed heads—but we found our party. They were dead to a man, the last two near the boat on the beach, as we saw. The others had been surprised and hacked to death as they were filling their water-barrels. They had also been stripped naked, with fingers cropped off for the worth of a ring or two, and rosaries taken from their necks. But their heads were taken first.'

On the silent quarter-deck, his hearers drew in their breath as Barrameda gave this last intelligence. What sort of men could these be? Olivarez, with savage thoughts of vengeance, asked:

'But you saw no one? Where were these murderers hiding?'

'I suppose in the hills, my lord. The huts, all save one, were empty. We could have spent a week chasing the villains and found none. So I wasted no time on that. We buried our men, and turned to the water-barrels and the forage. It will be worth sending again, as many times as you wish: there is good water, and a little game, and some corn of a sort. We will have to be strongly armed, but that is no matter.'

'What did you bring back?'

Barrameda had all in his head, as befitted a man with duty to discharge. 'The water—sixteen barrels full. And we killed three deer, and found ten others smoked and hanging in the huts. There was a pig or two also, and fish in plenty. And my lieutenant shot a kind of eagle.' He smiled, for the first time in a doleful tale. 'There have been times, my lord, when we could eat an eagle. . . . And we brought back one prisoner, hiding in a hut and spitting curses at us.'

Don Iago said grimly: 'We will give him more occasion to curse us—and the mother who bore him!'

'It is a woman, my lord. An old mad woman. But proud. . . . She has the look of a witch. But she might tell us something of account.'

'Where is she?'

'In the long-boat, my lord. Playing with the entrails of the deer.'

She was indeed mad, the old woman who had the look of a witch, and who was presently haled on board and set among her captors. But, as Captain Barrameda, who could tell the quality of a man at a glance, and perhaps half the quality of a woman—as Barrameda had said, she was proud. She was proud of something which they could not divine. It was only Matthew Lawe who furnished the key.

The men of the *San Vigilio*, enraged at what had happened to their comrades, had some sport with the old crone who was suddenly one of their company. They found themselves looking at a tiny woman in rags, her hands bloodied to the wrists, whose imperious eyes stared back at them as if she had the power of life or death on every soul who met her gaze.

She had a curved beak of a nose, an olive skin darker than their own, a grey crest of matted hair, and a wild freedom of movement. She hopped rather than walked, and mouthed and grimaced continually, and snapped her crooked fingers, contemptuous of this world of men who had only a chance authority.

In sum, though proud, she was ugly beyond belief; and the Spanish sailors and soldiers, after a first astonishment, brought their wit to bear on this. While Fray Bernardo, standing on a tier above among the sea-officers, listened with a stony face or affected not to hear, the jokes and the gestures grew more broad; but their intention was to one point, and addressed to one target, the boats' crew who had captured this prize: If you must bring back a woman, bring us a better one than this.

So they shouted their abuse, and jostled her, and burst out in coarse laughter when she cursed them—or seemed to curse them, for her tongue was a wild unknown. When, at the end, they went too far, she voided her water on the deck; and even this she did with pride, as if spurning a rabble beneath contempt.

For Don Iago, a liberal commander, this was the signal for a more seemly conduct of his ship. His sailing master, and then Captain Barrameda, barked out their orders, and the jostling crowd fell back. Then the officers came down to the middle deck, and took their stand in front of the old woman.

For reasons of heavenly guidance and earthly knowledge, Fray Bernardo seemed the most suited to communicate with this strange visitor. But when he had done his best, in his Latinate English, and a little French, and then—of all tongues—Spanish, he had to confess that he could make nothing of her answers.

'It is a sort of English,' he told Olivarez. 'There is a word here

and there, but twisted out of shape. . . . The curses, the male-
dictions, one may only guess at. I heard her say that she could tell
us the hour of our death, which is a matter only for God. . . . But
as to the rest——' for once he was at a loss, and took refuge in a
scholar's verdict—'I fear she lacks education.'

What else, under God's heaven, did you expect? Olivarez
thought, as he surveyed this wild, half-human, half-ape of a
woman who was their puzzlement. But he did not say it aloud.
Sane men, even sane men who had the ear of God, were only
simple creatures; the mad might have not only God's ear, but
perhaps His tongue as well. . . . Then he remembered Fray
Bernardo's disdainful phrase, 'a sort of English', and he perceived
a sort of answer, which was a sort of Englishman.

'Bring up the prisoner,' he commanded, and thus they waited
in silence, while Matthew Lawe was again lifted from his hiding-
place below, as swiftly as grumbling gaolers, with their orders
reversed at the whim of great authority, would comply.

Matthew, loosed once more into the light of day without a word
beyond the customary cursing, had thought that he was going to
his death; it would have amazed him less than the players' madhouse into which he was propelled.

Round about him was a great array of lords and officers, eyeing
this common sailor with a cutting edge of Spanish arrogance for
which not even service in the Queen's ships had prepared him.
In front was the multitude of the *San Vigilio*'s crew: turning all
their gaze upon a man whom they had heard of but never seen,
and greeting his appearance with a growling hiss which was the
word *Inglés*.

In the centre of this daunting audience, an old hag—the first
woman he had seen since Plymouth Hoe, a foul three months
ago—stood at bay: an outcast like himself, a human creature in
the great peril of utter loneliness, stripped of dignity like a hare
skinned for the pot.

The great lord who was Don Iago de Olivarez, waiting his
arrival with impatience, wasted no more time. He went straight
to command:

'We cannot understand this woman, who is a prisoner,' he told
Matthew. 'Put questions to her. Discover what she is.'

A man of Devon translating the Scottish tongue for Spanish
ears?—it was no more mad than anything else in a world turned

upside down. . . . Matthew could only answer: 'Aye, your Honour,' and bend to a strange task.

But it was more simple than he had thought. The weird old woman seemed moved by his presence, and answered him freely, even with a certain submission. Though her speech was as rough as the crags which guarded their haven, yet it was—as Fray Bernardo had judged—a sort of English.

Matthew found that he could recognize its lilt and its rasping confusion from long ago: from an usher at the Barnstaple Poor School, another sort of outcast far from his home in the north of the kingdom, a tufted hairy man who had breathed fire and brimstone upon a clutch of orphan children, but who had mourned all the time his exile from a land he called, between a sob and a belch, 'Scotland the Fair'.

The tale which the old woman told Matthew, and the tale which he told Don Iago in his turn, was not to be believed, only to be passed on to other, wiser ears, to make what they could of a riddle.

She was a witch, he reported. She was a famous witch, by the name of Morag. She had the power over humble people, and the willing ear of chiefs. She was the hereditary witch of Drumnin on the mainland, come over for the yearly fortune-telling, and caught in this chance trap by foreign tyrants who did not know the vileness of their insult—nor their own ill-luck.

She could prophesy. She could prophesy even among these barbarians, though it might fall on their ears like a plague of Egypt. She might curse them roundly, but God would curse them ten times ten, for laying hands on one of His anointed. She could prophesy. She was famed for it, like her mother before her, like her grandmother who under force had lain with a king of Scotland and woken with a dead man on her bosom. She could——

At this moment of a long and slow exchange, Fray Bernardo called a halt. Pursing his thin lips, he asked:

'If she is such a famous witch, how was it that we could take her so easily? Why did she not prophesy that, and make her escape? Why was she surprised in the hut?'

When Matthew asked the old woman this question, she replied, her eyes burning, her speech as raw as quartered flesh:

'If he lift his skirts, he will find the answer! Things shrivel when they are not put to use. For me, it is naught but my ears.'

With prudence, Matthew told Fray Bernardo: 'She says she does not know,' and when he turned back, he saw deep within the eyes of Morag the witch a gleam of laughter.

Now it was the turn of Olivarez to interest himself in their captive. A man of inquiring mind, he did not shrink from any aspect of learning; if Morag had proclaimed herself an alchemist, he would have demanded one of his brass culverins changed into gold to prove it. Since she boasted the gift of foreseeing the future, it was the future which pricked his attention.

'Let us put her to the test,' he said, to those round him. 'If she can prophesy, I would hear it.'

'Oh *Uncle*!' his nephew Alonzo exclaimed, with a delicate shudder. 'Such a dreadful woman! Did you not see what she—— Can you truly wish to hear her?'

'By all means.

But Fray Bernardo had more serious objections. He drew Don Iago aside, and told him without ceremony: 'This is sinful. It is superstition. It is not seemly for you to give it countenance.'

'Nothing is *seemly* in this accursed hell-hole,' Olivarez answered forthrightly. 'I have lost fifteen men murdered! I wish to know what is in store for the rest of us.' And to Matthew: 'Tell the old hag to prophesy. Question her.'

'Sir, what do I ask?'

'Ask, what is the fate of the Armada.'

Morag the witch was ready for this, and ready to prove her spirit. She replied: 'What Armada?'

It needed no translation. Don Iago grew irritated. 'She knows well enough, what Armada! All the might of Spain . . . What is its fate?'

'Worse than the past.' They could all hear a taunting in the old woman's voice, and there was more to come. 'And *less* than the past. They say this great might of Spain has not taken as much as one little ship of the English. . . . They have not burned one sheep-cote in all the land. . . . But they have taken two worthy prisoners. . . . A sailor who fell into their hands by chance, and an old woman. . . . Such is this great Armada! . . .' When Matthew, mortally afraid of her insolence, had passed on her words, she asked him directly: 'Do they wish more?'

'I think not.'

But Don Iago, at least, still needed to inquire. 'You put yourself in peril,' he said sternly to Morag. 'Yet we are not ashamed of our ill-luck.' Then, in a sudden bursting-out of spirit, he commanded: 'Prophesy! Prophesy! I will have you hanged for silence, not for speech!'

She gazed at him with her fiery eyes. 'Are you sure, great lord?'

'Aye.'

'Well, enough.' She mumbled to herself, and crossed her hands over her skinny bosom; then she dealt a cruel blow to the side of her own face, and seemed to fall into a trance. Her voice changed, growing deeper, yet coming from far away. She began to speak, not of 'I' but of 'we', as if crowded round about with hungry, devilish spirits. 'We see Spanish prisoners, bound one to the other with ropes, whipped homewards like little boys in disgrace.'

Matthew repeated her words in a clearer tongue. He was growing more fearful than ever, yet all he could do was obey. Olivarez, controlling his anger, asked: 'What is the fate of this ship?'

'Your ship is lying in Maclean's country. . . . We have a little chief here, Maclean of Deuart. . . . But his family will be great lords, and they will search for this ship through the ages.'

'Search? How search? *Where* search?'

'In this very place,' Morag answered—or was it the familiar spirits perched upon her shoulders?

'How can that be?'

'We tell you it is so.'

'But what is the name of it?'

The answer came out such a jumble that Matthew asked her to repeat it. 'Tobermory.'

'And they will look in Tobermory for this ship?'

'Till the end of time.'

Olivarez relaxed, with a sigh. It had no sense to it. Perhaps the old witch was only mad, after all. . . . But Fray Bernardo, who had been growing impatient at this mumbo which was also sacrilege, attacked with sudden venom.

'We have heard enough!' he cried. 'You are an evil spirit!' His fierce eyes turned to Matthew Lawe. 'Both of you are evil! And you will pay God's price for it!'

After a silence, Morag called out, in a high voice: 'He is beyond your power.'

Fray Bernardo, his throat constricted by anger, managed to say: 'My power is God's power. Nothing is beyond it!'

'We do not see with your foolish eyes. You are blind. We are not. We tell you, you will not kill this man. No mortal will kill him. Nations not yet in the womb will break their swords against the rock of his fortress.' Her voice began to fade, and her ancient body to droop, like a bird submissive to the snare. 'Hear our last word,' she said, in a cracked whisper. 'He will not die. . . . He will wander—the wild waters—until——'

'Until when?'

'Till all the seas run dry. . . .' Then, with a withered screech, the witch fell senseless to the deck; and no man on board the *San Vigilio* saw more than a crumpled bag of bones, nor heard a sound beyond the bleak mourning wind, sighing a little of life and a whole realm of death.

A grandee of Spain did not quarrel with his holy confessor; the chaplain of a great household, transported to sea, did not cross swords, nor wits, with his honoured patron. But within these formal margins of propriety, the conference of Don Iago de Olivarez and Fray Bernardo was a stormy one, and only ended in a wary truce.

Fray Bernardo, fresh from an odious scene, claimed his victim. 'You must purge your ship!' he insisted, again and again. 'The woman is nothing—perhaps she was beyond redemption, and now she must be in purgatory. But the English prisoner—cannot you see, my son, that his life is forfeit? He was condemned out of his own mouth, and the witch sealed it by her blasphemy. I pronounced him guilty of vile heresy. Now you must do your part.'

But Olivarez would have none of it. He did not know why the mad confrontation on the middle deck had moved him beyond anything in his life, but it was so. Though he could not believe the witch's words, he would not run counter to them. Nor did he wish the sacrifice of this poor runaway, whatever Holy Church decreed. Only Heaven should decide the day of such a man's death; and all that Olivarez would allow, when pressed to the limit of his patience, was to say:

'I will think on it.' And then, with all his cold authority: 'Meanwhile, I have four hundred sick—and the dead you have numbered yourself. He shall be put to work.'

Fray Bernardo, baffled, could only reply: 'I know you are wrong, my son'; and that, for a mortal season, was that.

XI

From his crag-side lair which had been a deer's covert, burrowed within the coarse heather, Matthew Lawe peeped out at the *San Vigilio*, a half-mile below him in the haven of Tobermory. He had watched her sails loosed, and her banners streamed into the frank wind. Now, as he peered, he saw a brisk muster of sailors advance to the ledge below the forecastle, to lift her anchor.

He was in wretched state, wet to the skin, shivering and starved with cold; yet the sight was the most warming of his life, and the happiest, it seemed, since their Saviour's first Christmastide. The *San Vigilio* was leaving at last, with all her company; and the number of his enemies was thus halved.

It was eight days after the evil day of Morag, during which time the galleon had been laden with water and deer's meat and all the scraps of food they could find by forage, and made ready for her sailing. But they had all been evil days for Matthew Lawe, the man with the brand of death on his brow—or, worse still, the brand of life.

Morag the witch had signed the warrant of his execution. He had suffered a great enemy before, the priest who told him he must die for his past sins; but now all the world was his enemy, on board the *San Vigilio*. To her sailors and soldiers alike, he was the object of fear and hatred; the word of his timeless salvation had gone the length and breadth of the ship, and when he was not cruelly taunted with it, he was bullied beyond endurance by men who, fearing their own miserable death, wished to share it liberally before it was too late.

Matthew had expected to die at every night-fall, and when he woke to a new day in his dungeon, it was to the same terror. What had stood in the way he could not discover, save that it was not mercy.

Meanwhile, he had been set to work, like any other prisoner who must earn his wretched bread: first at loathsome tasks in the privy daily choked with night-soil, then—in blessed relief—ashore with the forage-parties.

Here they used him to guard one of their boats, while the countryside was scoured for sustenance. At the beginning he was tied to the stern-post, so that he could not escape and, in case of surprise, would be the first to be killed; but later his captors grew lazy or contemptuous, and he was left with the other boat-keepers, to watch for six hours while the day's work was done, and to stand and stare as he wished.

On what proved to be the last morning of this easy captivity, he found that he had made one friend.

This was Juan Batista, a poor witless boy who was also the butt of much cruelty. Against the crime of life, Juan had only one defence, which was a smile; and there came a morning when he smiled on Matthew, and they drew together, and were at one. Juan, it seemed, had something to impart; and since he could not speak, nor Matthew understand, he drew pictures in the sand.

Nudging Matthew to full attention as they lay by their boats, Juan took a tide-worn splinter of driftwood, and drew the shape of a galleon. He pointed across the bay to the *San Vigilio*. Matthew nodded. So much was understood.

Juan then outlined the sun, a circle in the sky. Then he scuffed this out, and drew another, further on the larboard side. Then he smoothed away the picture of the *San Vigilio*, so that only the sun was left.

There was one other boat-keeper on the beach, a surly sottish Andalusian, and he was occupied with the nails of his toes, as a swift glance confirmed. Matthew nodded vigorously, and met the simple smile of Juan Batista with another of his own. So far, the story was clear. On the morrow, the sun would still be in the sky, but the *San Vigilio* would not be in the bay.

Juan now ceased to smile. With the splinter of wood poised, he looked at Matthew, and shook his head dolefully. Then he embarked on what was, for so simple a craftsman, his master-work.

First he drew a cross, like the cross on his breast, but tall and stark. Then he put the figure of a man against it, with small flourishes which Matthew swiftly divined were cords. Then, with much hard breathing, he told the end of his story. In the sand, wavering lines appeared, mounting upwards from the foot of the cross till they consumed the figure of the bound man, and all else beside him.

These could only be fire.

It was enough—but to make certain sure, Matthew pointed to

the picture, and then touched his own breast. Juan nodded, and laid his hand on Matthew's own. Now it was more than enough. Tonight or tomorrow, before the *San Vigilio* sailed, he was to die the cruel death which the priest had promised.

Wild thoughts assailed his brain, and then cool command, coming from God knew what remnant of spirit. Though he could not believe the mad old witch who had prophesied that he would live, he believed Juan Batista when the boy assured him of his death.

He was doomed, and he must act.

He had nothing to give this poor messenger, who might also be his saviour. Yet he had! Bending his neck, Matthew lifted from it his badge of office, the silver whistle of his command, and put it in Juan's hand.

The simple boy, who suddenly seemed to possess the last wisdom in the world, smiled his thanks. He took the silver whistle, and fingered it lovingly, and then thrust it deep into the heel of his shoe. Then—oh wise young man, oh blessed angel of light!—he lay down upon the sand, and cradled his beardless chin in his hands, and feigned sleep.

Matthew waited for a chance to seize, and it came swiftly. Above their heads, there were sudden shouts and laughter from the hillside. A musket shot rang out, and then more voices, alerting others to the chase. The sailors of the forage-party were having some sport with a cornered hind, leaving another to go free.

Matthew looked carefully about him. The man of the toe-nails was still grooming himself for Heaven. Matthew rose to his knees, and first crept, and then walked, and then ran for the shelter of the heather hills.

He reached them panting, while far away the voices still called, and not a sweet bird stirred, nor betrayed his flight.

His guards had searched for him but briefly; doubtless they had argued that it was the end of a weary working day, that there was no time remaining before they must leave, that the weather which was foul enough already might grow worse, and delay the ship's sailing—all the customary excuses of sailors who lacked their supper and would not have it declared forfeit for a little matter of a captive slipping their net.

Matthew had known that he was free, at evening when the

three laden ship-boats began to ply back to the *San Vigilio*. Now, after a night of cold and terror, he was trembling witness to the last act of all.

A trumpet sounded across the bay as the anchor broke surface and came home with a will. The galleon, skilfully handled, fell off the wind as the sails filled; then she set course for the harbour's mouth. She was a brave sight, this leviathan which might be the last ship of Spain, and a noble one—to a man no longer captive on board.

Then all melted into horror.

While the *San Vigilio* was still within the arm of the bay, a cannon-shot boomed out from her highest tier of guns. Whether this was a final thrust at the murderous foe, or a curse on Matthew Lawe, or the last gesture of bravado, only her commander could have told. But it *was* her last.

There was a huge burst of noise and smoke, and on the instant she was engulfed in fire from end to end.

A vessel in flames, whether friend or enemy, was ever a horrid sight for a sailor, and Matthew watched appalled as this great torch erupted. But the ship split asunder, and heeled and sank, so quickly that there was little time for horror; only for pity.

On the table of the sea, all the foul flotsam of disaster began to break the surface: spars which lanced themselves into the upper air as spears, and fell back like children's spent arrows: planks of wood, barrels rolling aimlessly, drenched sails which had been scorched into ribbons. Then the heads of men appeared, though only a handful of them: men screaming, and struggling against the merciless suck of the water, and swimming wildly for the shore.

Even as they did so, other upright men with swords came out of their secret hiding-places among the rocks of Tobermory, and stood waiting.

For Matthew Lawe it was the moment to move, before the lust for blood marred the difference between a Spaniard and his poor captive.

Stiffly he rose, and made for a hilltop on the further side of the bay. Just below the crest, so that he would not be outlined, he turned his face to the sun, and began to creep southwards, across the Isle of Mull towards the mainland.

The long flight had begun.

TWO

VOYAGER

1610

'The sea must needs at last have an ending.'

> *Martin Frobisher*, his constant motto,
> when searching for the North-West Passage,
> and reaching Baffin Island, 1576

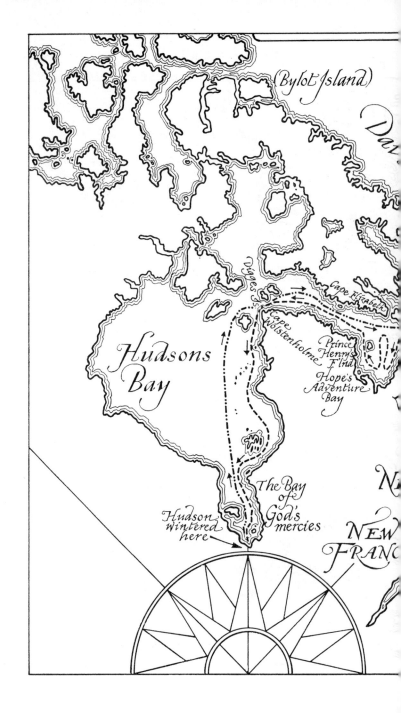

(Bylot Island)

Cape Elizabeth

Hudsons
Bay

Cape
Wolstenholme

Prince
Henry's
Flnd

Hope's
Adventure
Bay

The Bay
of
God's
mercies

Hudson
wintered
here

NEW
FRANCE

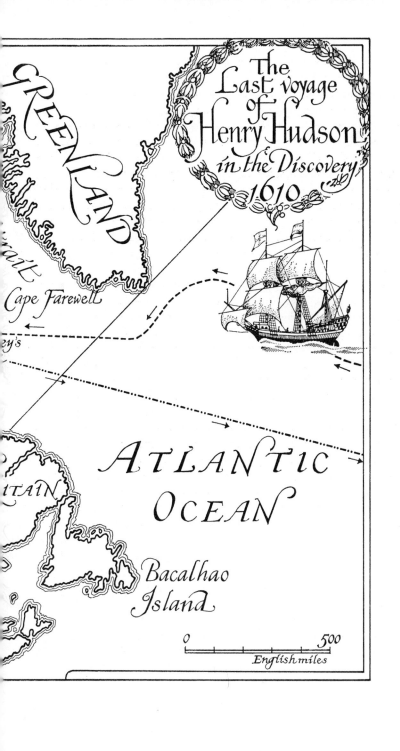

GREENLAND

The
Last voyage
of
Henry Hudson
in the Discovery
1610

Cape Farewell

ey's

ATLANTIC
OCEAN

ITAIN

Bacalhao
Island

0   500
English miles

SAVE THAT SHE WAS MORE comely than most, and young and willing, the girl in the grimy bed was like any other girl shaped for the sport of sailors. Just as there was a price for this foul room under the eaves of the inn, so there was a price on her willingness; and they must both be paid, cash on the barrel-head, before she would spread her legs, even in enticement.

Jenny the tavern-girl was not alone in this world of sinful commerce. She had her protector, a flash bully-boy with a villainous tongue, who bought and sold such meat, whether fresh or stale, from his station in the ale-bar of the Saracen's Head, and took his own pleasure when trade was slack. While Jenny and her sisters plied their craft, the rogue whore-master kept his watch below, alert for trouble.

But there would be no trouble for Matthew Lawe, on this his last night ashore in the Pool of London. He had paid his scot, from the last scourings of a shrunken purse, and thus could rule the crib and the girl till sun-up. A straw pallet, two coarse blankets, and a wanton who would keep all warm and lively, made up his royal realm.

It was enough for sailors who had run through their money.

In the space of ten days ashore, Matthew had spent all his hard-won pay on Jenny, and on this attic room, which stank like a pole-cat's rump at night and, by dawn-light, was as soiled and be-greased as a tub of swill. But the girl, praise God, was better than the room: a pretty wanton, her generous flesh formed for a lover's hands like the soft billow of a sail: merry with laughter, and her only itch a lusty hankering for the root of man.

Certainly she was enough for a sailor who must, on the morrow, rise and shine in a different endeavour.

After the first stormy bout they were wakeful, because they knew there would be more sport; a sailor on his last night was not a boy who gave all in a single spouting, nor an old man with only one spout to tap. By the light of the guttering candle, they supped their spiced ale together from the hooped leather tankard, and smiled companionably, and waited for the tide to rise again.

Meanwhile, they murmured their thoughts.

'I would I could lie here forever,' Matthew said. 'Who would choose to go to sea, when he can swim in such a sweet inlet?'

'Have you the money?' the girl asked. She knew the answer, but it was a woman's task to make all plain. Men might dream; women must be watchful.

'No.'

'Then go you must . . . Where is it you're bound, Matt?'

'Wash your ears, wench! I've told you, times enough.'

'I'll wash what parts I choose. Too many sailors tell me their tales. Can I keep score of them?'

He thrust that thought away. Tonight she was his, as on all the other nights. If the morrow was someone else's province, he did not choose to hear it. 'We go exploring. To seek the North-West Passage, may be.'

'Art afraid?'

'Nay.' It was true for this night, anyway. ''Tis a fine strong ship. The bark *Discovery*. And the captain is Henry Hudson. A great man by all account. And soft also, they say.'

Jenny smiled her lewd smile. 'Who wants a soft man?'

'Not you, and that I know! But a soft commander—that's another colour of horse. We want a captain who will victual us honestly, and not whip us all to death. If we find a way to Cathay, or the Indies, it will not be by beating and starving. A stout leader and a willing crew, treated fair—that's the berth I seek. I wish to return!'

'Will you bring back treasure?'

'Aye. I'll bring back all I have and all I find, and give it to you again.'

'Best of all, I want a jewelled stomacher!'

'You shall have one, love. . . . What is it, for the sake of God?'

'Do you know *nothing*, Matt? . . . It spreads from here to here—' she indicated her ample breasts, and then the first curves of her groin. 'It goes a-top of the skirt, it is sewn with jewels or gold rings, it sets off a gown like a rain-bow, and if I wear my buckle shoes in the same——'

'Enough, enough!' Matthew said, laughing. 'I have the likeness of it. . . . I will bring it, if I can find such a treasure. . . . Though I do not choose to see you covered from here to here. . . . Save by myself.'

'You bring it to me,' Jenny told him warmly, 'and you may cover me ten times over.'

'With my own treasures.' She had said enough, and looked enough, to set the tide mounting again. 'If they be rich enough for you.'

She lay back, as ready as a wench could be. 'There, you are the richest prince of all.'

'Not too old for your choice?'

She looked at him, surprised. 'How can a lad like you be too old? I never met such a one in my life!' Perhaps she always said this, perhaps she did not. 'Prime as the Derby Ram!'

'But how old would you say?'

'What game is this?'

'How old?'

'Five-and-twenty?'

'Near the mark . . . Now lie still, till I bury my treasure again. Then you may search it out till you have dug all dry.'

He would be four-and-forty, by his mother's oath, this Michaelmas, but he felt like five-and-twenty as he took her again.

Afterwards she sank into sleep, like a log of drift-wood lulled below the surface of life by the greater power of the sea; and Matthew was left to rule the kingdom of the land. It was no great heritage. While he swigged his cold ale in the darkness, the room itself grew colder; Jenny stole the best portion of the blankets, as the candle-wick flickered, begging piteously to be snuffed; even the ale, near the bottom of the tankard, had turned sour, needing courage in the swallowing.

He would have a fine thick head on the morrow, when he must board the *Discovery* and take up the sweat and swink of life again without the fourth part of a silver shilling to show for his pains. The sottish skull would be all. . . . While one voice told him: 'So lives every sailor, since land and water were divided,' another rebuked him with a mournful question: 'If this is the sum of a sailor's toil, is he not better dead?'

Matthew savoured, as best he could, the last of the ale, the last of his purse, and the last of the land, as the sadness of after-love made all else seem desolate. If nothing could be counted on save spent drink, spent money, and spent manhood, then what a swinish couch had Christ their Saviour spread for the limbs of man!

Yet it was a bed he must lie on. Though he knew it by rote, he

could not change it. No single man of the company of sailors could
alter such laws, set as they were by God, or by greater folk, or by
the foolishness of mariners themselves. He was doomed to a
squalid calendar of life: doomed to make a harsh voyage, to come
ashore rich (by the humble standards of his trade), and then to
drink his fill, lay claim to a wench, spend on her a year's earnings in
a few days or weeks, and ship out again without a groat in his pocket.

This wheel of fortune, immemorial as sun-rise and set, furnished
girls with trinkets, tavern-keepers with money, men with the pox
and a splitting head, and ships with crews—and that was all, and
ever would be.

Yet it was an honest web of life, for all its brutish strands. He
had grown used to it, and it had served him well enough, apart
from the darkest thoughts of all. Though Drake their famous
Captain now slept in his leaden hammock, fathoms down in
Nombre Dios Bay off Panama, the brave line continued; victory
over the Great Armada had flavoured a whole nation with pride
in its new destiny, which they would never yield, and sailors with a
sort of repute, of which they could not be robbed.

Matthew tipped the tankard for its last bitter dregs. Sailors
could be robbed of all else. . . . And pride, in truth, was only for
the great commanders: pride did not bake bread, nor buy beef nor
ale nor wenches hot and cold. Pride, especially, was not for himself. He had not earned it: he had earned only the ordured side of
its coin. So he must go on and on, until—and now those dark
thoughts came flooding in and drowned everything else in a swift
nightmare of doubt and fear.

He could not readily believe his doom, as proclaimed by the
mad old Witch of Drumnin so long ago; but by now he had begun
to live with such fancies. He was four-and-forty, yet his body had
not moved with the years, nor had his face altered by a line, nor
his hair by a single lock. *He was the same man.* Could such fantastic
tales of cock and bull be true?

The chill wind sighed in the eaves; behind the wainscot the
rats scuttled and slid. No comfort there save cold and corruption.
He lay back, and by long habit plucked at a string of courage.
Even these fearful thoughts were part of his calendar, part of
every last night ashore; and what he must do now, to defeat them,
took hold on him again. Some divine or devil's spark came to his
aid, reminding him of the sovereign cure for such despair.

Tomorrow he sailed, but tomorrow was not yet. Tonight was still with him, and indeed still lay at his side. Tonight was still asleep, but not for long. . . .

He turned towards Jenny, and pressed and stroked her flank until she awoke. She grumbled briefly of sailors' wicks unsnuffable, and their ways with a poor girl who had served full measure already; but she proved as prime as ever when she had warmed to it.

He murmured: ''Tis the last time,' and she answered: ''Fore God, I hope so! I will need crutches after this!' and then she made a sturdy crutch of the best timber to hand.

On a whispered command, she mounted him. It was the way of love he liked best, a true sailor's holiday: above him the fine sails swelled and dipped and tugged, while he lay aback and kept the main-mast tall and trim. To rise like a sword into those willing thighs, to watch her jaunty bubs tossing as she swung to and fro, to see her face in the last of the candle-light glow and grow wild, to hear her gasp 'Thrust up, sailor! *Now!*' as the twin tides of love crested and broke—this was the answer to all!

What a wench it was! She was *not* just any dock-side drab. She loved her work, but she loved him best of all. Truly loved him! She had told him so!

When he slept at last, sweatily, snoring in unison with a jaded bed-fellow, he dreamed of apes and peacocks, roaring seas and soft strands, and mermaids with gleaming breasts and tails which happily parted. He awoke by dawn-light, to see Jenny out of his bed, and astride the stone pot in the corner.

The spell of love was over, and it was time to leave.

Swallowing against a foul tongue, he dressed and stowed his gear. When Jenny, snug beneath the blankets again, received her farewell, a fond slap on the rump, she did not stir. Matthew hoisted his sea-chest on his shoulder, descended by the creaking stairway, and crossed the evil-smelling tap-room.

The surly pimp at his post nodded, letting him go without argument. Empty sailors were no loss.

Out in the pale sunlight, Matthew tautened certain idle sinews and began to stride across the cobble-stones towards St Katherine's Pool. It was the morning of Low Sunday, and the April of 1610.

II

Though the *Discovery* was already astir when Matthew Lawe found her berth, she had the look of a ship which would never sail, not in a month of Low Sundays. All ships lost virtue by contact with the land—as did all sailors—and this one was no exception. She was in the labour of storing, and all was in chaos, with each thing on top of another, like a muddled shopman's wares. The broad deck could scarcely be glimpsed, through its burden of barrels and chests, spars and tumbled sails, all lying as if a mountain-side of gear had descended upon her, and been left to moulder.

Among this gross huddle men moved sluggishly, like ants whose hill had been disturbed: men crouching under heavy loads, men labouring to heap one great chest upon another, men toiling down ladders step by step, like children learning: while yet other men marched on board with more fuel for the confusion.

Matthew stopped to stare, and found that what a sailor's eye could see, beneath the turmoil, was more heartening. The *Discovery* was a bark, of three masts and perhaps sixty tons lading. She was small—smaller than many a ship he had sailed in; but her stout timbers and tall masts had the look of quality. It was easy to imagine her sailing westwards, or northwards, or wherever else was fated, without fear or wavering: breasting her natural element, the boundless sea, and making a handsome task of it.

Though she was dirty now—land-soiled, as sailors termed it— yet she would soon be sluiced clean; and when honest salt brine took the place of that spillage from the great dung-hill of the land, it could only be to her profit.

If all else matched this ship, then he might count himself fortunate to have found such a berth.

Matthew was about to pace out her length, which could not be more than sixty feet, when his dreaming was rudely interrupted. It came as a roar, an ancient screech, as if from old Stentor the herald of Troy himself.

'*Matthew Lawe!*'

He turned, abashed, to meet the burning gaze of an old man, stationed on top of a corded chest on the deck of the *Discovery*. He had a fist raised, ready to shake against Heaven if need be, and certainly at any man who failed his duty.

Matthew knew him instantly; it was Robert Juet, mate of this ship, who had engaged him as common seaman the morning before, and witnessed the deed with a crabbed hand: Robert Juet, who had seemed no more than a kindly grandfather when their bargain was struck, but who now stared at him with eyes as sour as last night's ale, prepared to spew out such a bargain already.

Matthew made as if to answer, but Robert Juet broke in, not so loudly but with no lack of spite:

'Did I not sign you as seaman, yester morn?'

'Aye, sir.'

'To stand and stare, like Tom Fool at the fair?'

'No, sir.'

'Then come on board!' The men on deck near Juet were grinning at Matthew's discomfort, as the mate continued: 'Stow your gear, and get to work! Do you think we carry gentlemen of quality here, and pay them for the voyage?'

Matthew shouldered his sea-chest, and walked across the plank from the dock-side to the ship. He passed one of his smirking shipmates, who muttered: 'You made a fine beginning, Tom Fool!' before turning back to his work. Then Matthew was face-to-face with authority, standing ten feet tall on its lofty pulpit.

Robert Juet's look had settled into sulky discontent. 'Stow your gear,' he repeated. 'And make haste! We carry no idle hands.'

Matthew looked round him at the cumbered hull, and then towards the nearest ladder leading below. 'Where, sir? Between decks?'

'Who said aught of "between decks"?' Robert Juet rounded on him. 'You will be in Master Hudson's great cabin next!' He swung about, and pointed at the ship's long boat behind him. 'In the shallop. Where else? You are last a-board, and must take what is left.'

Matthew looked at the shallop, a rough-hewn craft of some twelve feet, half-covered by a sheet of ragged sail-cloth. 'This is my berth for the voyage?'

'Aye. We strike it below when the weather grows foul. You will be warm enough.'

Matthew had the spirit to turn stubborn. 'What keeps me warm, sir?'

'Six other men,' the mate answered, with a smile as wintry as the wind.

'What—*seven* in this cockle-back?'

Robert Juet had had his fill of question and answer. 'You will not sleep all the time,' he said, with careful spite. 'Nor half the time, nor one quarter, if I can devise something to keep you wakeful. . . . Now stow your gear, and the quicker you are, the better I'll be pleased, and——' he drew a hissing breath through his mean lips, 'if you please me, you may still live to see the Pool of London again!'

Perhaps it was not such a fortunate berth after all.

But all on board the *Discovery* was not meanness or spite. Like the newest orphan at the Poor School, Matthew Lawe found the beginning strange, and certain of his masters enjoying good sport in cruelty; but carrying stores into a ship, and seeing them safely stowed, was the most familiar work of his life. It was also the best cure for shyness, since his strong arms and willing back soon earned him a place among his shipmates. This was the same for any sailor, the world over; let him board a new ship, show himself lively, and he found more friends than foes.

The men who worked at the loading, and Matthew not least among them, were full of curiosity at what they carried on board: there was much furtive peeping and prying into sacks and chests, much sniffing at barrels and reading of signs and letters. On such information, their peace of mind depended; and on its reality, their comfort and perhaps their lives.

Making a pool of their knowledge, they tallied good supplies of hard biscuit, salt beef, and pickled pork: an abundance of dried peas and beans: cheeses of several kinds, and small beer to wash all down.

There were also some rare blessings, delicacies not to be met with save in a sailor's dream—lemons, sweet oil, sugar, cloves, and raisins—which made one of the ship's boys exclaim: 'We shall all be living like kings!' But he was swiftly put down by Michael Perse, the seaman who had first jested with Matthew as he came on board with his gear:

'You will see little enough of these, lad. If they wander further than the great cabin, or Juet's private snug, I will eat the foremast. Should you fall sick to dying, they might push the half of a lemon-skin between your teeth, to stop you babbling. But you must earn it!'

Then the boatswain's whistle called, for a spell below and a bite to eat; and Matthew was free to discover more of his new ship-mates.

Some half dozen of them retreated one deck below, away from the turmoil overhead; and there, in cramped peace, they took their ease in a haven lined with triple-tiered cots, of the kind which Matthew Lawe had failed to win. They found themselves in a pleasant twilight, full—since the bilges were not yet fouled—of the tarry smell of cordage and new sail-cloth; and having worked like dogs for four long hours, they flung themselves down, wordless, and waited for the ship's boys to bring them food for revival.

It came speedily; and, still in silence, they wolfed down cuts of crusty bread—the last they might see for many months—and sturdy carvings from a round of fresh West Country cheese, sharp and satisfying. Then, their pangs stayed, they fell at last to talking, while the *Discovery* rocked gently to the lift of the tide, and all was tranquil.

It was Michael Perse, his first friend—though he had proved a mocking one—who turned to the new-comer Matthew, and sought to soften his mockery.

'You should pay no heed to Robert Juet,' he counselled Matthew. He was a strong man with a strong voice, and he spoke with friendly authority. 'He barks well enough: so does any dog, even an old one. But he is past biting. . . . Juet is jealous, that is all: he has made three voyages with Master Hudson, and he wants to rule, instead of standing in his shadow.'

'God help us all if he did!' another strong voice proclaimed. This was Philip Staffe, the carpenter, whom Matthew would come to know as a man devoted to Henry Hudson, and ready to prove it. 'We would be fast on the rocks, and he swearing it was all our fault! Do you mind that day when we found the mouth of the great river, and he shouted out: "This is the awful hole in the middle of the world! Bear away, bear away!"'

'I mind it well,' a third voice confirmed. It was a mincing voice, and the man who owned it, Arnold Lodlo, was a mincing man, a little pink fellow who would have seemed more at home with a fiddle under his chin than a rope in his hands. But at least, Matthew Lawe thought, this one has made a voyage with the great Henry Hudson, and still lives, and is ready for another. . . . 'I mind it well,' Lodlo went on, 'because I have been loose in the bowels ever since. The mate screeched out: "Bear away, bear away!", and then he called: "The other tack, you witless fool!" and the next we knew, we were fast by the tail on a bank of sand, with painted savages ready to leap on board!'

'He mistook the land?' Matthew asked, amid the laughter which greeted Lodlo's story.

'Mistook the sun for the moon, more like,' Lodlo answered, with a delicate shudder. 'All I can tell you is, *I* was the witless fool he cursed—in front of all! I was ready to die! It needed no *savages*!'

Philip Staffe, of sterner mind than could be suited by a licensed droll like Arnold Lodlo, interrupted:

'Henry Hudson is our man, and we should not forget it! Juet could not take his place in a privy!'

'Oh, I think he could do *that*,' Lodlo said, in such a way that they all laughed again.

It was his last jest, so far as the honest carpenter was concerned. 'Think what the man has done,' Staffe said. 'Only three voyages, and already the greatest sailor that England can boast!'

Matthew Lawe, his knife busy with the last of the cheese, felt ready to have his say.

'But what quality of great sailor?' he asked, echoing the words of Philip Staffe. 'A great commander? How does he rule a ship? I heard tell that he was a soft man.'

The carpenter looked at Matthew as if gauging whether friend or foe, but prepared for either. He was prepared also to give an honest judgement. 'He is just, and fair. He will listen. He can be persuaded, which is more than can be said for others who would slander him. Of course he grows old. They say he is a grandfather already. But I would rather follow him. . . . Three great voyages already. . . . No soft man could match that.'

Michael Perse, who was also a lover of fresh cheese and had long been silent, now made his voice heard.

'He was no soft man on our last voyage. He led us into that hole in the middle of the world,' he went on, scornfully repeating Arnold Lodlo's words, 'and it was no more than a river, but the fairest I ever saw in my life. A broad river leading north through the New World! We broached it more than a hundred miles, till we were checked by rocks that crowded together into a wall. It was not savages that stopped Master Hudson. It was the end of the sea, that no ship could pass. But there was clear water ahead—I saw it from the top-mast with my own eyes—and one day, one day, men will hop over it, and they will find——' he paused, and finished lamely, as if his dream was not strong enough for such mysteries, 'they will find whatever lies on the other side.'

'Is that where we are bound?' Matthew asked.

'Who can tell?'

'Aye, who can tell?' Another voice broke its silence, that of Sylvanus Bond the cooper, a work-mate of Philip Staffe: a wizened sad-eyed man who seemed to look at the world only to discover whence the next buffeting would come. 'God knows where we might find ourselves, half a year from now! Who confides in common sailors? We are provisioned for six months, that is all we know, and after that we might starve, or eat the seaweed off the rocks, or devour each other, for all we can tell!'

'Cheerly, man!' Arnold Lodlo told him. 'We will not eat you, till you put on some flesh.'

'You can jest,' Sylvanus Bond grumbled. 'But is it enough to be commanded, "Set sail!", and then for Master Hudson to tell us where we are when we reach it?'

'Aye,' Philip Staffe said sternly.

'What do you mean, aye?'

'I meant what the word means. He is our captain, and a great man, famous for his voyaging. To be told, at the last, "We are in the Indies", or "We are in Cathay", is enough for me. It is all I need to hear from him.'

'I would rather hear, "We are *going* to Cathay",' the little cooper persevered, to make his meaning clear. 'Then I would know where the captain was leading us, and I could tell myself, each day, "the land that lies ahead is Cathay".'

'Belike he does not know that himself,' Matthew Lawe said.

'Aye!' Sylvanus Bond was glad of a sort of ally, in a world of unknown terrors. 'And there is always the day when he might over-reach himself, and not return. And us lost with him!'

Matthew Lawe was bound to answer: 'I have heard of such captains.'

But he had fastened attention on himself, perhaps too early; and Arnold Lodlo, a butterfly among men, was the first to be drawn to a new colour. He flung away the rind of his cheese, and stretched his little legs towards a shaft on sunlight which penetrated the opening in the deck above. Then he spoke:

'Did I not see you at the Saracen's Head, these last few days?'

'Aye.'

'Fresh from a voyage, yet ready for another?'

'Aye.'

Michael Perse asked: 'What voyage was that? And what ship?'

Matthew did not want any such questions; he wished only to be a sailor on board the *Discovery*, in the here and the now, without past and without future. But he must tell some story, and it had best be a true one. Sailors were the first to discover a liar in their trade.

'To the Middle Sea,' he answered. 'Aboard the *Judith Mary*. We touched last at the Bay of Bantry in Ireland. . . . We brought back horses. . . . For King James, it was said, but I know not the truth of that.'

Michael Perse knew a little more. 'They say you touched at the coast of Italy also.'

'Aye. . . . We were gone for eight months or more, trading where we could. In Italy it was wine—foul stuff! In Morocco——' he essayed a laugh, 'God knows what!'

Arnold Lodlo took up the questioning, with a certain lickerish relish in his voice, as if he knew a secret.

'But you touched last at the old Saracen?'

'For sure. Where else would a sailor go, for a flagon to drink?'

'And went in hiding?' Arnold Lodlo asked slyly.

'In hiding? What is this?' Matthew could not be baited, even by such a lap-dog as Lodlo, without protest. If it was dangerous to answer back, then he would face such danger. There was more peril in becoming the butt of such plaguing, in a new ship at the beginning of a long voyage.

'In hiding under the eaves, with a fine bed-fellow.'

Matthew expelled the breath of relief. So that was all. . . . He made the easiest answer he could. 'Who would be alone, under the eaves, on a cold night?'

Lodlo was still intent to pursue his jest. 'Tell us, how did you find her?'

'Who?'

'Jenny Wallop.'

'How did I find her? I looked under the blanket, and there she lay!'

There was an honest roar of laughter at this sally, which changed all, from conflict to good fellowship.

Michael Perse exclaimed: 'Jenny Wallop? You must have come ashore rich, Matthew! I'll warrant, that wench cost you a gold sovereign, each time you set foot on the stair. She's as bright as a breast of bacon!'

They were all attention, as sailors were when such talk caught their ears.

'I can warrant *you*,' Matthew answered, 'I am not rich now. I spent all, under the eaves.'

'But worth the price?'

'Aye.'

The talk soon became bawdy.

'There's some fine breast of bacon there.'

'Some fine bacon, too!'

'Aye—what a haunch! A man could lose himself, before he reached private parts.'

'Who said her parts were private?'

A voice coming from the outer darkness suddenly roared out: '*Filthy whoredom!*'

Startled, they turned towards the intrusion. A man towered at the entrance door, a tall man clad in stark black broadcloth, his face gaunt and fierce. He looked as strange as a raven among a flock of gulls, a very bird of ill-omen. But when he had glared at them all, one by one, he disappeared, like a ghost from the battlements.

Only unfeeling laughter was suitable to their relief—and only Matthew Lawe remained startled.

'Who was he, for the love of God?'

'Abacuck Prickett,' Michael Perse told him. 'But do not fret. He plagues us all the time, and we are still no nearer to salvation.'

'But has he no taste for these—privities?'

There was further laughter as Michael Perse answered: 'He would have to be instructed where and what they were, first.'

Arnold Lodlo added: 'He would run a mile if he caught a glimpse through the fog! But he can thump a Bible till it bursts!'

'Then what is his station?' Matthew persisted, still in confusion.

'A landsman,' Sylvanus Bond answered scornfully. 'What else? He is Sir Dudley Digges' manservant, set to spy on us. He will watch all, and report.'

'There is no certainty of that,' said Philip Staffe.

'There is enough for me! I can smell a Company spy, from here to Grave's-End! By the time he has preached a bushel of hell and damnation, and tied us all in knots, he will be able to tell his master, *if need be,* "The crew of the *Discovery* was not worth your expense." Then, if we succeed, he will beat his black breast, and declare: "I, Abacuck Prickett, showed them the right path!", and if we fail, he will ask: "What could a righteous man do? I wrestled with these devils till the air was blue; and all to no purpose! Alas, alas, and lack-a-day!" '

He said these last words with so desolate a mourning air, coupled with his own natural dejection, that they all fell to laughing again.

Matthew had one more question. 'But who is his master? Who is this Sir Dudley Digges?'

'One of our patrons,' Philip Staffe answered, wishing for a more seemly discussion. 'A rich man in the Kentish country, and joined with two others of the same repute. . . . So you may see,' he finished severely, 'Master Henry Hudson has great friends who trust him, and will lay out gold to prove it.'

There was that in his voice which seemed ready to add: 'And I could wish that all doubters and runagates were of the same mind!'

Before they had time to swallow this rebuke, a shrill whistle called from above, and the harsh voice of Francis Clemens the boatswain ordered: 'Watch below! Turn to!' And then, even louder and harsher: 'Side-boys! Attend the captain!'

When Matthew Lawe, eager and lively up the ladder, reached the deck, he was in time to see Henry Hudson come on board his ship. He saw a man of middle age, grey and slow-moving, climbing up the boarding-plank against the slope of the high tide: he saw a man gravely studious, because he had much on his mind, and yet smiling confidently as he reached his journey's end, since this was his own ship, his private ground, and as he stepped upon the deck he stepped once more into a longed-for dream of voyaging.

Henry Hudson paused as he reached the deck, and looked about him with a captain's eye, first at the cluttered waist of the ship, and then aloft at the spars and rigging which would bear them away. He spoke a word to Boatswain Clemens, and Clemens went running to set something to rights. Then Hudson turned landwards again, and noted one of the ship's boys, the slimmest and the youngest, making as if to hoist a heavy chest on his shoulders.

He called out: 'Leave it there, John. It is too heavy for you.

You lack a year's growth still!' Then he looked round for a sturdier man, and found him. 'You there!' It was the voice of command, but not harsh. 'What is your name?'

'Matthew Lawe, your Honour.'

'Take up that chest, Matthew Lawe,' Hudson ordered him. 'And have a care! It holds all my instruments, and the best of my charts as well. If you let it fall,' he ended, with a grave smile, 'you may let fall our voyage.'

Matthew, proud to do this bidding, strode across the plank, shouldered young John aside, and hoisted the sea-chest on his back. It was not till a day later that he discovered that this bright-eyed 'John', the ship's boy, whom later he chided for some careless-ness, was named John Hudson, and was the captain's own son.

But in the space of moments, much had changed for Matthew Lawe. He found that he owned, already, some respect for Henry Hudson, and some strong liking also. This was a captain to trust: no tyrant was here, but one who spoke to his men as if he thought them better than bags of lumpish brawn, and a commander too full of years and wisdom to carry such a crew into danger.

When he followed Henry Hudson down to the great cabin, and towed the chest as carefully as if he bore his own first-born in a cradle, he had begun to count himself a loyal man among the ship's company.

Many men had taken this liking to Henry Hudson; many had followed him; some had failed his trust, and had been forgiven. As a mariner, his star had topped the horizon three short years ago, in three sea-quests which had proved his daring and his skill at navigation, in equal proportion. He had made two voyages for the Muscovy Company, a guild of English merchant venturers who founded their first fortunes upon trade with that monstrous Duke of Muscovy who went by the apter name of Ivan the Terrible; and then a third voyage for the Dutch East India Company which placed the crown upon his repute.

His first essay was in the *Hopewell*, of 80 tons' burden, with a mere eleven men as crew, and its purpose, to find a way to Japan or China by vaulting over the North Pole. Though Hudson reached the island of Spitzbergen, within the bitter circle of the Arctic, his way was barred by ice, and he was forced to put back.

His second voyage, with the same purpose and in the same ship, took him far to the north-east, along the snowy pinnacle of Russia, to the island of Novaya Zemlya, in the same icy latitude. But when neither ship nor men could endure more from the wild weather, their captain brought them home again, their purpose unfulfilled. It needed a third voyage to earn him true fame, in a wondrous corner of the world.

His new masters, the Dutch East India Company, had lured him from the service of England, for one reason alone: because this was the quality of man who could bring to flower their own hope of a North-East Passage, through the ice and snow at the top of the globe. But something far different was the outcome.

His ship was named the *Half Moon*, and her crew was half English, half Dutch; and the Dutchmen, used to service among the warm and fragrant spice islands on the other side of the world, proved half-hearted. They did not like the cold. After much muttering and head-shaking, sulking and shirking, they refused to take the *Half Moon* beyond the North Cape which was the crown of Lapland.

It could be called a mutiny, and Henry Hudson did not know how to master it. He could govern a ship with rare skill, but he could not persuade a company of men whose square heads, he thought, were too thick to perceive his dream. Yet it must not be the end of the voyage. His Dutchmen had no more wish to return home so soon, in clear disgrace, than Henry Hudson himself.

He told them straitly: 'If you are cold, we will go where it is warmer. If you will not go north, we will go west. But go we must, and you know it.' The *Half Moon* then crossed three thousand miles of rolling Atlantic, turned south down the coast of the New World, discovered the entrance to the 'Great River' which seemed ready to pierce a continent, and voyaged nearly one hundred and fifty miles on its broad bosom, before rocks and rapids barred their way, and the wild Indians grew so bold and murderous that prudence, at last, must rule over valour.

Hudson came home full of honours, but under the displeasure of his king—for a reason which was no disgrace. He was of too much value to serve any nation but his own. The secrets of that 'Great River', with all its promise of profit, were not to be shared with anyone before they must. Nor could Henry Hudson be allowed further liberty of this kind. Such a drain-away of brain and skill must be staunched.

By Order in Council, he was censured for 'voyaging to the detriment of his country', and forbidden to undertake further foreign service of any kind.

But this was no more than a pause, one missed step in ten thousand miles. There were other men besides his sovereign who were jealous of Hudson's skill and daring—but for more loving reasons: they wished such a man to bring honour and profit where they truly belonged. A new company of merchant venturers was formed by Sir Thomas Smith and Sir Dudley Digges, both men of Kent, and by John Wolstenholme of Yorkshire. They had money, and their own brand of daring, and that great curiosity about the world which often seizes men of generous mind.

Thus they had furnished Henry Hudson with this ship, the *Discovery*, and were ready to send him voyaging again. Their sole direction was the measure of their trust. They told him only: 'Do not try for a North-East Passage again. Such vaulting of the Pole is for flying monsters, not ships. Go west, and then north, and try another passage—and God speed!'

Within four days their storing was completed, and the *Discovery*, as laden as a ship could be, and all well bestowed, dropped down the river on the ebb tide. With her crew of twenty-one men, she touched at Grave's-End; and here they took on board a twenty-second man, who remained a mystery to all until he became a manifest villain.

His name was set down in the ship's book as Henry Greene. Though he was greeted in the most friendly fashion by Henry Hudson, this name was not among the list of those approved by the owners of the *Discovery*, and the manner of his boarding had in it all the smell of deceit. Why was he lifted on board so secretly, when it was too late for the owners to question it? The ship was a-buzz with rumour, which the man's appearance did nothing to allay.

Henry Greene was young, strong, and well-made, though the marks of dissipation in his face told another sort of story—and indeed, he was in liquor when he made his first appearance, and never flagged in this regard. In clothing, he was as bright as a jay: in manner, impudent as the devil; and hail-fellow with all, until he was crossed, when an evil-furrowed brow banished all trace of good humour.

Who was he, and what, and why? Philip Staffe thought him a

runaway murderer; Michael Perse suspected a bastard son, secretly given a father's favour; Sylvanus Bond pronounced him yet another spy for the Company, to see that their money was well spent.

Matthew Lawe, after many covert glances at this strange new shipmate, guessed him to be a young gentleman of fashion who, falling to lechery and deeply mired in it, had been shipped off to sea for his own good—and perhaps his own health.

Abacuck Prickett, that sour sniffer-out of evils great and small, lost no time in agreement—and improvement.

'Such filthiness!' he croaked. 'If he prefers the company of bawds and trollops to the consort of the godly——'

'Who does not?' asked Michael Perse, and laughter turned a fiery sermon into a cinder.

But something let fall by Prickett later, when he had nosed his fill in this noisome bucket, placed this judgement nearest to the truth. It seemed that Henry Greene was a dissipated young rake-hell of good family, whom his father had disowned, and who was now befriended by Henry Hudson, for a reason unknown. His mother had secretly furnished five pounds for his outfitting. It was her last expense upon a worthless son.

Not Henry Hudson, nor any of his crew, were to know that they had embarked, to their misfortune, the biggest scoundrel who ever joined an honest ship; and that thus the rampant seeds of treachery were sown already, in what should have been a happy virgin field.

III

Yet it all began handsomely enough. These were the matchless days of an English spring, when April turned her back on winter and gave a loving welcome to May; and the *Discovery*, with fair winds, voyaged up the east coast of England and Scotland, past Harwich and Flamborough Head, and the Orkneys and Shetlands, with never a cross word from the elements until she reached the oldest enemy of all, the Atlantic Ocean.

Though this was the route of the ruined Armada, fleeing in despair and shame, yet no trace of this ancient torment was left for native sailors to endure. Matthew Lawe, ploughing the same furrow as had beset a wretched Spanish keel when he was a prisoner within, could only relish the badge of English blood, and a stout English ship to match it.

Of necessity, it was hard lying on board the *Discovery*. A small ship bred discomfort, in any circumstance; a small ship crammed for a long voyage, and manned on a measure of one sailor for every three feet of length, could match a spital-house at a time of plague.

There was no one on board, from Henry Hudson the master and Robert Juet the mate, past Edward Wilson the surgeon and Thomas Wydowse the man of mathematics (for they were prepared for everything, whether reaching for the stars or just short of the grave), down to John Williams the seaman-gunner and John Hudson the boy, who did not suffer from the close confinement, when one human trampled on another for space to live, and the stink of a man's breeches could destroy a day's happiness and poison a night's sleep. But it was the common sailor, as ever, who suffered more, and was expected to speak less.

For Matthew Lawe and other unfortunates, there was the business of living in the shallop, secured on deck between the mainmast and the poop. Though it was May on the voyage up the east coast of England, and May still as they turned westwards past the Scottish isles, the nights were cold, and the morning dew (which a landsman would have termed a bone-chilling fog) was enough to bring a shivering sailor a great bite nearer to mortality. But when Matthew, elected spokesman for the shallop-dwellers because 'he knew his rights better than most, and could find the words', broached the matter with Robert Juet, the mate gave him a sour reply.

'What did you look for when you came aboard?' he asked, in that ancient whining tone which could set an angel's teeth on edge. 'Cushions and velvet hangings? Log fires and mulled ale? . . . I marked you out as a trouble-maker from the start, and by God's blood I made no mistake! Did I not say, you would be in the master's cabin, given the first chance? You want to improve your station? Then work and sweat and earn it! In the shallop you are, and in the shallop you shall stay, and the shallop remains on deck, till I give the word!'

'Sir,' Matthew Lawe answered, waiting for a pause, not daring to stem the onrush of such vehemence, 'all I say is, it is bitter cold on deck, and when we rouse out for the dawn watch, we are half-dead already. We can never be warm! Would you have us fall from the mast-head, to prove it?'

'Yes,' Robert Juet answered readily. 'If it will rid me of all this whining and complaining.'

'But we cannot do our work.'

'I will see that you do, like a man—or starve like a dog!'

Thus it remained, until the weather turned suddenly harsh, between the Faeroes and Iceland, and even a man with a heart of rusted nail-heads, or a spite of the purest venom, must have relented. Then, with much grumbling and cursing for ingratitude, the shallop was struck down and placed two decks below, just above the bilges.

These were already growing foul, since that moment when Arnold Lodlo had tossed the first rind of cheese downwards, and began to make, of a clean ship, a midden afloat. But at least they were warm.

They were warm from the efforts of Bennett Matheus the cook, a small round dumpling of a man (as befitted his trade), and a determined landsman who carried his ideas of comfort with him where'er he went, not to be altered because he found himself half-way between the arch of Heaven and the pit of the sea, in a tossing tumult as far from one Godless coast as another.

Just beside the space made for the shallop, Bennett Matheus had fashioned an oven which would have earned praise from his own mother: a shapely hearth of brick, laid upon a bed of sand resting on their bottom timbers. Within this glowing cavern he could make miracles—if he was in happy mood; if not, all others save himself lived on hope and cold porridge.

It was the crew's task to keep him happy, and secretly to fawn upon him; to be the cook's favourite was an early taste of paradise, and to be in his bad books, a glimpse of another place. So, without ceasing, they fetched and carried for this capricious tyrant, whether it was brushwood for fuel, or water for the pot, or stores which he needed for his confections. Then they waited, slavering their desire for extra portions, or the best cut near the bone, and outbid each other to win his favours.

At times it seemed that they did not work for Master Hudson at all, but for the True Almighty, the Master Cook, the Royal Pharaoh of a succulent pyramid. A man must live.

The living grew harder, as the *Discovery* surged and stumbled deeper into seas icy cold, gales more bitter than the savage heart of man, and ice itself which tumbled about the ship as if a swimming giant, just below the surface, was seeking to sink their vessel by raw bombardment. Past Iceland, such living finally proved intolerable, to ship and men alike; and Henry Hudson, after taking counsel and listening to sailors at the edge of exhaustion, put back into shelter.

Their shelter was on the west coast of Iceland, in a haven which the men disdainfully christened Lousy Bay, but which proved as merciful as a woman's gentle bosom. There were hot springs ashore, coming like magic out of a freezing land, which warmed the air; and the captain of this turbulence, the volcano beneath Mount Hekla, sometimes darkened the sky with black snow, drifting down the wind till it fell below the eastern horizon.

Here they put ashore, to forage, to climb hills and see for fifty miles around; to drink deep, to fish, to hunt the wildfowl which abounded, and to swim in sheltered water. But here, being only men, vile men, they fell to quarrelling.

It was the swaggering Henry Greene who was the spark of this ungrateful villainy. He fell into conflict with the ship's surgeon, Edward Wilson, for no more reason than that Wilson, a young man of twenty-two fresh from university, was of superior mind, and showed it in his speech. Greene saw fit to mimic this, and then to say he doubted if such pale flummery could take the place of manhood.

But when they came to hard words, the surgeon proved to have a tongue as sharp as a lash, and more talent to quicken it than (as he said with relish) a hundred young ne'er-do-wells whose principal brains were in their private parts.

The taunt stung, as it was intended (since Greene was thought to have trouble in that quarter), and on the instant they came disgracefully to blows. It was witnessed by the crew that Henry Greene, who had travelled to Holland, cursed the surgeon in Dutch and then beat him in English.

A clamour for justice came from the bleeding mouth of Edward Wilson, who had the backing of the majority. But when the two men took their quarrel to Henry Hudson, the captain made nothing of it. He seemed to support his favourite, Greene: he said that there were rights and wrongs on both sides; and then he bade them forget the matter. By this feebleness he opened wounds more sore than any suffered by the surgeon; and straightway they began to bleed their infection.

When old Robert Juet, whose sense of authority was bound by the book, sought to intervene and to have this verdict changed, with punishment for Greene, Henry Hudson—for the first time in their voyage—flew into a rage.

'Keep your fingers out of these embers!' he thundered at the old man, 'or by Christ's blood they will be too hot for you! What matters Henry Greene to you?'

'It matters if he is quarrelsome,' Juet said manfully. 'It matters if he strikes an officer of the ship. It matters,' he went on with ancient craftiness, 'if he is only come aboard to spy on us, and crack our credit if we displease you.'

The old grey grandfather who was their gentle captain was now beside himself with anger.

'Spy on you!' he roared again. 'Say one word more, and I will put you ashore, and you can wait for the fishing fleet to take you home! Spy on you? Crack your credit? When I *spy* you sitting on the coals of that volcano, with your backside *cracking* instead of your credit, then I shall be happy. And you know whom I would put in your place of mate? *Think on it!* . . . Do not try me, Master Juet, or I will show you who rules this ship, and who is the judge of all these matters, and who——' here he borrowed all the majesty and wrath of Holy Writ, as handily as Abacuck Prickett, 'who says Go, and thou goest!'

At that, Robert quailed, and fled, and was no more seen that day. But for all its violence, this was only the first ripple in a most troubled pool.

Henry Greene, smirking in victory, now set no bounds to his arrogance. Edward Wilson remained in fury, and smarting as only a young man who knew himself wronged by authority could feel. Robert Juet, who had taken one step forward, and two steps back, in his pursuit of justice, and been made to look a fool for his pains, was in a smoulder of hate. To be threatened with putting ashore! . . . (Better for all, Matthew came to think, as odious time went on, if this threat had been enforced.) But it was the effect upon the crew which was the most wounding.

Seasoned from the very first to obey the bidding of the next man above, so that Juet said 'Aye, sir' to Henry Hudson, and Boatswain Clemens to Juet, and Robert Bylot their leading seaman to Clemens, and so on down to the ship's rats, which scurried at the command of Nicholas Syms the smallest boy—brought up to this strict rule, the greater part of them were enraged that the surgeon, a ship's officer, could be insulted and beaten by Henry Greene, a volunteer without licence or rank.

If there was no punishment for this, then there should be no punishment for a common sailor who called Henry Hudson a fool to his face.

Bitter words were said, bitter threats made, bitter quarrels hatched, and all were remembered later, like the wounds of war which came to throbbing pain at a change of weather. A seaman with the strange name of Syracke Fanner, who had been mocked for it by Greene, made dark vows of vengeance—to be answered by Arnold Lodlo, who admired Greene and thought him 'a real man, not a bag of wind', and inquired of Syracke how he would wreak vengeance upon a young gentleman who topped him by twenty inches or more, and whose fist was bigger than Fanner's head.

Michael Perse said this question had nothing of right or wrong in it: Henry Greene was villainously *wrong*, and would have been punished for it if he were not perhaps the captain's bastard—did they not both bear the name of Henry? . . . Bennett Matheus the cook, in sulks for some earlier slight from Greene, vowed that he would rather douse the fire than warm another morsel for such a reptile.

Sylvanus Bond was full of woe, and promised 'a voyage to ruin, and all because of one foul rogue and one captain who swallows his spirit when they come face to face'.

Only Philip Staffe the carpenter was truly loyal.

'The master must be the judge,' he told them. 'That is his station, and you had best remember it.'

'If he were a true *judge*,' Michael Perse answered him warmly, 'he would have judged Greene guilty—and hanged him, for all I care!'

'He *did* make his judgement,' Staffe insisted, 'and found them both at fault, and so did nothing more. I say that he was right—but even if he was wrong, it is not our business. He is the captain.'

At which they all rounded on Philip Staffe, and called him a lick-spittle—which he was not—and a Company man who jumped where'er he was bidden—which, in all honour, he was.

In the outer ring of all this turmoil, sour Abacuck Prickett did pucker and pout his lips as he silently searched the Scriptures. When asked which side he stood for, he groaned out: 'The side of

God our Father!'—which placed him in lofty company, but made none the wiser.

Matthew Lawe ventured nothing, whether sitting with his back against the shallop in the darkness below, or aloft and stretching a new sail on the yard, to catch the sun. He could not form a true picture of Henry Hudson so well as others who had sailed with him; and, being the newest-joined among the crew, he still hung back when it came to dispute.

He felt at one with Philip Staffe—that all eyes should go up to the master of the ship, and when this man had said his say, there was the end of it. Yet he thought Henry Greene a true villain who should have been put down for insolence, before he fed so fat on liberty that he left none for others.

Above all else, he was filled with amazement that Henry Hudson, a wise captain of great repute, should have assembled such a picture gallery of jealous rogues, malcontents, and men who sought their own interest—and be-damned-to-the-rest! If this was the future of their voyage, if this was to be the spirit of *Discovery*, then the sea would have them all before the Pool of London saw them again!

In such a mood of discontent did the ship, replenished with fish, and fresh meat, and fat Arctic geese waylaid on their flight to the far North, leave Iceland and take up her voyage into the unknown.

In the clear cold of a June morning, with *Discovery* creeping ever westwards before a light breeze, there were only two men at work on the poop-deck: Matthew Lawe at the helm, and Henry Hudson bending over the table on which his charts were spread.

For the rest, the watch on deck were idling in the waist, waiting for any call to tend the sails; Boatswain Clemens was slowly fashioning a wedge of wood for the foot of the mainmast, which sometimes groaned more than a mainmast should; and Nick Syms, the boy who imagined himself the first discoverer of the world, gazed steadfastly forwards as if, without his efforts, they might miss Cathay and end their voyage at some limbo on the border of hell.

In such a sailor's perfect peace, Matthew had time to look about him. Steering the ship was nothing at all: one finger could take care of it: their creaming wake spread behind them as straight and honest as an arrow. But his eyes were constantly drawn to the land which lay on their beam: a land of green-black mountains,

seen across a shire of tumbling, growling ice which no ship could
penetrate: a land mysterious and proud, beckoning and threatening with the same careless hand, as if to say: 'You are a ship. I am a
land. Try me if you dare—and welcome!'

But what land? Matthew waited for the chance to ask his
question. He knew already that Henry Hudson, though captain of
them all, would never be too proud or too cold in authority to
answer.

Since every skilled seaman took his turn at the steerage, Matthew
Lawe—once he had proved himself and secured the grudging nod
of old Juet—had spent many hours at the task. He had relished it,
for seven hundred miles or more, by day and by night, ever since
they had left the haven of Lousy Bay and taken to the ocean; and
most of all because it brought him closer to Henry Hudson, a man
more to be admired the more he was known.

After his one furious outburst, Hudson had returned to the
world of shipboard as a centre of calm judgement, the father of
them all, the peg of sober sense where it was most needed. Though
the murmuring and grumbling on board continued, he overrode it
all, not by harsh command but by a certain sweet reason which was
far more difficult to withstand.

His message, whether in word or in silence, by look or by look-
away, was simple: all was going well with their voyage, under a
commander whom they could trust. As long as this good fortune
held, it needed a brave man to deny it outright, and a fool to say
that he could do better.

Now Henry Hudson, satisfied with calculation, straightened his
back and turned from the thick curling leaves spread across his
chart-table. He cast an eye to the sails, which were gently filled,
and another, more searching, to the land gliding by on the northern
horizon. Then he took a turn upon the poop-deck, and stationed
himself by Matthew's side. The favourable moment had arrived.

'She steers easy?' the captain asked him.

'Aye, sir. A touch now and then, no more. . . . What is that land
we pass by?'

Henry Hudson, who perhaps had been a schoolmaster before he
took to the sea, answered in his own fashion:

'What would you say?'

'I know not.'

'But at a guess? On the whole surface of the globe?'

'The New World?'

'Not yet. It is Greenland.'

Matthew Lawe had heard tell of it, but nothing more. Even today, with all discoveries brought to the very margin of perfection, there were certain names which must be taken on trust. Greenland was one.

'Shall we put in there?'

'No. We are warned away.'

Hudson spoke like an old prophet—or a boy with a secret. He was wearing, as always at sea, what the crew called his motley gown: a rusty black cloak, plain and gaunt, but trimmed with fur such as a merchant of the Levant might affect: strips of Arctic fox circling the neck, and broad bands of beaver skin to adorn the skirts.

He had once said, in the hearing of Arnold Lodlo (whose ears could catch a whisper down a mile of drain-pipe): 'These are all the riches I brought back from Russia, and from the New World. Do you wonder that I still go a-sailing?'

It was a good jest for sailors, so long as a search for a third choice of pelt did not skin them of their lives.

But Matthew Lawe was more intent on learning. 'Sir, how are we warned away?'

'Can you read, Matthew?'

'Aye, sir. Enough.'

'No man can read *enough*. . . .' But Henry Hudson was smiling. He crossed a few paces to the table where he had been working, and returned with an ancient tattered map, which he thrust under Matthew's nose. His thumb marked a great spur of land. 'What does it say there?'

Matthew spelled out slowly: 'Groenlandia. Here the Ice Reigneth.'

'Warning enough,' said Hudson. 'And writ by the man who discovered it for himself.'

'Who was that, sir?'

'John Davis, the friend of Sir Walter Raleigh. He voyaged in these waters twenty-five years ago, looking for the Passage. Once he sighted a whirlpool moving to the west, with furious overfalls, but dared not penetrate. That is the opening I seek.'

'Where is his Honour now?' Matthew asked, after time to digest this foreboding news.

Once again Henry Hudson smiled his gentle smile. 'If you search diligently the other side of the world, you might find his bones among the Spice Islands. . . . He was killed by Japanese pirates. . . . Better for him if he had stayed on this quiet coast!'

'Then we sail on?'

'We sail on till Hell itself freezes over.' A sail fluttered at the foremast, the *Discovery* yawed off the wind, and Henry Hudson swiftly became a working captain again. 'Watch your steering, lad! We would sail on the sea, not on our hands and knees across the ice!'

It was the end of their talking watch, but not of their converse, which often continued. Henry Hudson was full of travellers' tales, and never grudging when he came to speak of brother sailors who had ploughed these same furrows, or others unknown, and had perished in lonely misery, or lived to plant a banner on a far-off strand, and come home laden with honours, to make their children's eyes sparkle with their stories, and to die by their own fireside.

Sometimes he talked of Sir Humphrey Gilbert, half-brother to old Sir Walter. 'By God, there was a man!' he would say, with fire in his eyes. 'And died like one, too! He perished in the *Squirrel*, between the New World and the old. Near the Azores, on passage back from taking possession of Newfoundland in the name of the Queen. And what would you say were his last words?'

Matthew waited.

'He picked up his speaking trumpet, in the midst of the greatest storm of his life, and shouted to his other consort, the bark *Golden Hind*, which was trying to counsel prudence: "Cheer up, lads! We are as near Heaven at sea as on land!" Then a squall of rain blotted him out, and when it cleared, the *Squirrel* had vanished. Ten tons' burden. Ten tons! . . . A flea on the ocean, but a flea with a lion's heart! They teach them more than Greek and hand-ball at Eton College!'

Then he would come nearer to Matthew's own heart. 'But old Drake was the man for my money. The greatest captain we have ever bred. All the way round the globe, in a ship no bigger than this one. You know of whom I speak?'

'Aye, sir.' It was the most that Matthew could admit. 'Yet they say he was over-rash.'

'Rash?' Henry Hudson could not allow the word. 'Boys' tales! You were a boy when he lived—if you were alive at all! He was a man of his time. He saw his dream, and he made it sing true, and if any thing or any man stood in his way, that man or thing must go down. That is not rashness.' Hudson paused, as if he had some secret to impart, but all he could summon, from an inmost yearning, was: 'It is mastery.'

Once he drew them both back into the true mist of time, but made it seem living and breathing, as if they had been there together. It was on a night watch, but with a fine midsummer moon overhead, like a lantern showing all; and as Hudson spoke of the first known voyagers, the Norsemen, the broad deck of the *Discovery* seemed to change its shape in the shadows, and become a much older ship, long and lean like a Crusader's hound.

'They were open ships, like skiffs,' Henry Hudson told him. 'There was not a plank of cover within. To keep the waves from curling on board, the warriors held their shields half in and half out of the vessel, making a wall against the sea. A leaking wall, by God! And their only roof, winter and summer, was the sky. But in a ship such as that, sometimes rowing, sometimes sailing down the wind, they crossed this whole Atlantic, and found the New World, centuries of years before we did it for ourselves.'

Matthew said: 'I thought we were the first.'

'I like to think so still! But it is not true. What they found, they called Vinland in their language, and a Viking sailor—Leif Ericsson—landed there six hundred years ago. He was not the first in these waters, either! His own father, Eric the Red, placed colonies on that land we passed, two weeks back. In Greenland itself! But he had fair reason to travel from Norway. He was exiled for murder.'

'But which then is Vinland?'

'Perhaps Newfoundland, perhaps further southwards where we voyaged in the *Half Moon*. It means "Wine Land", but why it was so called I know not, for not a drop did we see. We might have stayed longer else! But the wonder of it is, they voyaged all these three thousand miles, in slim open shells such as I would not trust between London Bridge and Greenwich! With scarce an instrument of navigation worth calling so. They reached perhaps the very deeps where we are now——' and he motioned with his hand towards the black water, sparkling with starlight, running past the *Discovery*'s hull, 'but while we have the finest instruments to give us aid, they came here by guess and by God!'

Matthew, standing with spread legs on the sturdy deck, could almost feel himself a Viking, lost, lonely, yet valiant, in a ship surging across a dark waste of water with the waves pounding against the warriors' wall of shields like mailed fists, hungry to take their hull.

'But did they have nothing?'

He could hear the smile in the captain's voice as Hudson answered: 'Well, they had ravens.'

'Ravens!'

Henry Hudson, as sometimes happened when he was imparting knowledge, seemed to steer a round-about course.

'Have you read your Bible, Matthew?'

'Aye, sir.' Matthew could answer with honest certainty. For the orphans beleaguered within the Poor School at Barnstaple, no volume in the history of scholarship had been driven home with more lusty strokes, many of them on the backside, than that most holy book.

'Then you know the story of the Ark.'

'Aye.'

'When Noah thought there might be some land uncovered after the Great Flood, he released a dove from the Ark. But the dove returned, with nowhere to roost—and so did another, seven days later, but with an olive branch in her beak. There must be tree-tops showing! Then a third was let go, and sped away, and never returned. What did that mean?'

'That there was land to settle on, at last.'

'I'll warrant you! And so it was with the ravens carried by the Norsemen, though they added more skill to it. When they wanted to divine where was the nearest land, or the new land they sought, they released their shore-sighting birds. Sometimes these flew homewards, sometimes they returned to the ship in puzzlement, sometimes they set off on a new course—north, or west, or south—and never returned. So that way lay the land.'

'But what land?'

'What land indeed! The sailors must travel there to find out, and thus was the world enlarged. Sometimes, when all else failed, and they were truly lost, they followed the great skeins of geese, winging and barking all the way from Ireland where they spent the winter, to Iceland or Greenland which was their summer breeding-ground.'

'But was it only birds they followed? What else besides?'

'Little else. They had the lead-line, as we have today, armed with a lump of tallow to pick up sand, or stone, or grey mud. So they could tell when the sea-bed grew shallow, and what sort of coast it might be. They learned to love a handful of stars, like the Stella Maris——' he pointed heavenwards, to the Pole, 'the one still above us, and still our guide. And the Great Dog which we call Sirius, and Orion with his belt, and the little Pleiades. They knew enough to keep these on the left hand, or the right, as they voyaged. They could mark their height against a knot in the rigging, and note when it changed. But that was all. *No true measurement!* No dipping needle. No compass. No clock, nor even a sand-glass. No astrolabe.'

There was no astrolabe for Matthew Lawe, either, and he harked back to his principal doubt.

'But how do the birds know where to fly, when men do not?'

Henry Hudson sighed, as if suddenly out of spirits. 'If we knew that, we would know every secret on the face of the waters.'

Word must have travelled of this long talk, and perhaps been added to other earlier whispers, for when Matthew awoke next morning, and climbed out of his angled corner of the shallop, and sought to set up his spirits for a new day with that chaw-bacon's delight, a knuckle-end of salted hog, he was refused this titbit by Bennett Matheus the cook, who gave him a heel of mouldy beef instead.

When he protested, though in good humour, Matheus told him with a snarl: 'Take your complaints to Master Hudson! Perhaps he will give you pressed pheasant in the great cabin!' As others within hearing joined in this baiting, Matthew knew that suddenly he was in poor favour.

Robert Juet, sour and sarcastic, inquired if he wished to be mate of the ship—for if so, he was moving too fast for his own good. Sylvanus Bond the cooper made a sad jest of what might happen to the ship if he, Matthew Lawe, took control of their course. Arnold Lodlo, as peevish as a woman with piles, said that the *Discovery* was in bad enough case already, steering on and on away from home, without Matthew making worse of it.

'Do not,' he sneered, 'encourage our Lord Master Hudson to go further. He may love you, but we shall not—and we are twenty to one!'

Shocked beyond measure, red-faced with shame, Matthew could only stand and gape. It recalled to him the orphans' school of long ago, when a boy could be the darling of a warm world at one moment, and then an outcast, fit only to be pinched and cuffed, at the next—and all by a change of tide, a breath of poisonous air. He answered nothing, but slunk away on deck, and ate his foul breakfast in solitude.

There was worse to come. Presently Matthew was sent up to the main yard to make repair of a rope-end which was frayed by chafing. It was work he enjoyed, and he did not hurry the time spent on it. From his perch he could see Henry Hudson, back on his station at the chart-table, this time with Robert Bylot, the young man who was called their leading seaman but who had such skill in navigation that he might better be named the pilot.

Suddenly Matthew was roughly called down by Boatswain Clemens, who now seemed another enemy, and sent on an errand. It was nothing more than to fetch the captain his fur bonnet from the great cabin, and take it up to the poop-deck. This was boy's work, and a foolish thing for which to take a seaman off the yard and make him climb up and down and round-about, like a monkey on his master's leash.

Yet the weather was against him, and he obeyed. Waiting by the table to take the captain's attention, he heard Hudson say to Bylot, with his fingers on the spread chart:

'By the cold in the air, we should be near a main of ice. Or more close to a coast than we think. When you sighted the star at dawn, were you sure of it?'

Robert Bylot, as usual, thought with care before he answered. He was a grave young man, studious and set apart; his strength was in the head, and his passion the casting-up of sums. He was happiest when such a sum had a measure of doubt in it, and might be resolved in two ways, or even more. He was happy now, though he was not the young man to show it.

Finally he said: 'I was sure of it. Perhaps the ice is further to the south than formerly. It varies by the year, as we know.'

'Well, God knows I trust you,' Henry Hudson answered, without delay. He turned from the table, and saw Matthew Lawe proffering the cap. 'You hear that, Matthew? I would trust him with this ship, just as I trust you with my best fur bonnet!'

It was not the finest praise in the world, but Matthew smiled none the less. Robert Bylot, seeming to know his own worth, answered nothing. Exclaiming against the sudden cold, Hudson put on his cap, while Matthew lingered, despite his better judgement, and looked at the table, spread with charts and a great array of instruments.

Henry Hudson, noting his glance, asked:

'What do you know of navigation, lad?'

'Nothing, sir.'

'It is the skill of all skills! But think what aid we need to have, compared with those shore-sighting birds we spoke of last night.' He motioned with his hand across the table, and then to a wooden trestle-bench beneath it. 'Besides these charts, the best of the age, we have a fine floating compass, that has never failed me yet. There is this traverse board,' he pointed, 'on which I peg off the length of each course, with an hour-glass to keep tally of the time. . . . A clock as well—London made—though it does not do so well at sea as in my lodging.'

He was full of happy pride, like a woman showing her jewels, or a boy his treasured toys. 'Here is our true anchor—our anchor in the sky—the cross-staff, to measure the height of sun and stars.' The cross-staff was six feet tall, splendidly carved, mysteriously shaped: Matthew would not have known whether to balance it on his head or hold it between his legs. 'Here is John Davis' pet—his back-staff, to view the sun in reflection, so to soften the glare and be gentle to old eyes.' He turned aside. 'My first star-tables, that I have used these five years. Thirty-two fixed stars, all named and measured! But the Bible of all is this new Waggoner: the nautical almanac with the declination of one hundred stars, and all the tables to make measuring easy. Dutch brains and English printing—I would rather lose the wits from my head than this!'

Matthew himself felt as witless as if he were a dove newly sprung from the Ark. 'How is it worked?' was all he could ask.

Henry Hudson smiled. 'You must ask Robert Bylot, when he is not so busy. He knows more than myself!'

They both laughed together; and even Bylot smiled, though primly, as if this might not be such a great jest after all. Then Matthew, dismissed, turned to leave the poop-deck.

He had to pass close to the helmsman to do so, and the man, who was Arnold Lodlo, turned and gave him a look of such contempt that Matthew almost quailed. The glance said, as loudly and venomously as any words:

'Captain's toad!'

It was as if some of his men could not bear Henry Hudson to have another friend.

A strong contender might have ignored it, and even found amusement in such antics, which proved nothing save that when little dogs rose on their hind legs, they showed more parts than pith. But the morning had not armed Matthew Lawe with strength, and he found that he was open to wounds which could bleed him of the wish to strike back. This contagion of turmoil grew until it was no longer a matter of himself and Lodlo, himself and Juet, himself and those odds of twenty men to one.

The hateful stare of Arnold Lodlo had served as a signal, warning of a great plague in the air, which might wither all good fortune and outrun anything so pettish as conflict between one man and another. With it, a cloudy curse threatened to settle on the *Discovery*, marking this out as their last day of ease.

So it proved.

At noon, as the bitter cold increased, they sailed without warning into a bank of fog, so thick that between one moment and the next, not only was the sea and then the sky blotted out, but the whole forepart of the *Discovery* as well. Matthew, returned to his post at the helm, was ordered: 'Set her head to the north-west, and keep it so!' But the command served for little. The wind had died, and when it returned, it blew this way and that, like a swordsman probing where next to strike.

Sometimes the sails filled, sometimes they slatted sullenly against the masts, with the crack of a whip-lash. When the ship was taken aback, Matthew was roundly cursed for his steering; but ten men with twenty eyes could not have done better. The *Discovery* drifted on, rolling drunk, as blind as a bat in this noon-time gloom: with half of the ship in hazy outline, and the rest vanished as if by a wiping-cloth.

An unearthly pallor overlaid every corner of the decks. Men's beards dripped with moisture, and their faces ran glistening. Voices came out of nowhere. Even more daunting were the threats and curses to be heard, clamouring against all those who had brought them to this misfortune. A speaker who could be recognized as Henry Greene made a jest of 'the blind leading the blind', and mocking laughter came from the shrouded foredeck—but it was the laughter more of demons than of men.

Henry Hudson stood steadfast, nosing this malevolent plague of blindness, while all around him the spirits of men crumpled.

With evening and night, the fog only grew darker, and then fell black as a pall upon a coffin. So it continued for hour after hour; there was not a star to be seen, not a glimmer of the lantern at their mast-head, nor even the whiteness of their nearest sail. The *Discovery* might have been coffined in her own grave, and when the grudging dawn came, it only illumined the fog again, turning a black blindfold to one of mourning grey, icy and dreadful to all.

So they groped forward until the second noon, when a brisk wind came up from the east. They ran before it, seeming to carry their own blanket of nothingness with them, and to have no other choice; soundings with the lead-line came up with naught save idle strands of weed from the stilly depths, and no more bottom than the invisible heavens had top.

Then a piercing hail sounded from the mast-head, and from the mouth of a boy who could not even be seen from the deck. It was young Nick Syms, and he had news for the blinded mariners below, as if he were gifted with eyes set on stalks—which, fancifully, he was.

'Land-ho!' was the call, and when Mate Juet bawled back: 'Where away?' the answer came straight-way: 'Starboard bow! A league or more!' and then, more astonishing than the news itself: 'I am in sunlight up here!' Sunlight—they had almost forgot the meaning of the word. . . . But they remembered soon enough, when the fog cleared without warning, and they sailed out into seas visible, and a bright blue heaven overhead.

Henry Hudson, and Bylot beside him, stared amazed at what they saw. Close by, on the starboard side, was a grey headland which shone and glittered in the sun; and then, far away to the south, there was another range of land, whether main coast or island could not be judged. In between them was restless ocean which heaved and swirled and twisted, as if in a cauldron stirred by a mighty spoon.

Hudson, needing time and sea-room for thought, called to the helmsman, who was Michael Perse: 'Down helm! Steer south!' and as the bows of the *Discovery* turned away from the nearest land, Boatswain Clemens and his watch sprang forward to trim the yards. Sharper on the wind, the ship gathered way and began to cross the troubled water.

As if he had discovered the universe to be in error, rather than himself, Robert Bylot said: 'We have been moved more than we should, during the night of fog.'

Henry Hudson growled assent, but he was watching rather than listening. His eyes were intent, not on the land nearest to hand, but on the distant coast at the further side of the great tumbling bay. Presently he called back to Michael Perse: 'Bring round to larboard! Steer south-east!' and then to Francis Clemens: 'Brace back the yards again! As near to the wind as we can point.'

After it was done, and the new course set, Hudson kept silence, and watched the southern point of land ahead. He had eyes also for the sea over which they sailed, mysteriously disturbed by something more than the wind, beset with whirlpools of ice like the wicked jewels of sin. Yet he was the first to notice that the *Discovery*, for all her sharp sailing, was making more progress sideways than ahead.

They had become the prisoners of a racing tide. Their ship would not clear the arm of land to the south; and if she turned and ran back, she could not escape the cliffs they had first seen. They were being drawn steadily westwards, and all choice was already gone.

Hudson's decision was enforced, even as he faced it. A puny man in the surging welter of the sea, he could only give it voice. He looked towards Michael Perse again, and described a wide sweeping arc with his hand, from left to right.

'Come round westwards,' he commanded. 'But gently, while we ease the braces.'

All this westering had been his intent, or his hope—but not so swiftly, nor so far out of his power. Too early, they must have been snared and sucked into those 'furious overfalls' which John Davis had marked on his chart, and had shunned. But the *Discovery* could not shun them. Betrayed by her day and night of blindness, she could only submit.

The broad entrance was carpeted with the same swirling, foaming water, as far as the eye could reach; and across this they began to run on, ever westwards, ever faster. Might it be, Matthew thought, staring in wonder and fear, the lip of that awful hole in the middle of the globe which had so stricken Robert Juet? Or was it the true passage west to China and the Indies, and would they live when they had run its gauntlet?

Soon they were slipping into an unknown world, past tall and distant cliffs which drew together and then parted; through tormented, engulfing waters whose giant strength could not be denied.

The answer to all questions was mortally at hand.

IV

Wrapped in his motley gown, ignoring bitter cold, lumpish malevolent seas, and the gnawing doubt of a captain taking his ship beyond the guidance of any chart yet known to the world of sailors, Henry Hudson remained on deck for hour after hour, with a fortitude which would have shamed his own grandson of twenty years. He was warmed by his dream, and by nothing else: sustained by hope, and pricked by a true lust to see what lay across the next horizon, and round the next corner of this savage maze.

He lived triumphantly from moment to moment, and as these sped by, unfolding picture after picture, he could exclaim, within his very soul: 'By God's blood, we may be the first to see it!' But he was also a prudent navigator, and he was involved as closely as Robert Bylot, the mouse in their sea-borne counting-house, with the mathematics of their advance.

His principal thought was directed to this question of a great tide. If it was truly a tide which was carrying the *Discovery* along so swiftly, and not a river of salt running backwards out of the Atlantic Ocean, then it must soon come to a stop. No tide in the world ebbed or flowed for more than six hours—the space of twelve turns of his newest half-hour-glass. If this tide ran on, without check, then it ran to Cathay.

Hudson waited, stamping the deck-timbers against the cold, with perhaps the only patience and confidence in his ship, and certainly the greatest resolution; and at length he was rewarded—with an answer, if not with triumphant news. After four hours, the violent overfall slackened to a stop; and after an hour when the sea quarrelled with itself, in a duel wild and tumultuous beyond description, the great tide turned, and began to run with equal fury to the eastward.

The *Discovery*, even with a fair wind astern, started to slip backwards, and the precious cliff-points which she had gained were swallowed one by one, in the greedy maw of God's command.

Swept back again, compelled into flight, he turned and ran—having no choice, and needing space to solve this puzzle of threat and promise, open door and barred entrance, which could tease a cold man and turn the tail of a hot. He touched briefly on the north shore of this mysterious strait, but found it as blank and bare as a closet full of ice. But when he turned aside, and then to the south, he discovered the entrance of a broad bay. Here the water was more still, free of that racing tide, and so he ventured on.

Hudson was not downcast. If the water to the north always ebbed and flowed as wildly, yet took him nowhere, then this calm corner, so majestic and wide, might be the true entrance of the North-West Passage, which was his goal.

He had begun his close search, when the *Discovery* was first swept to the westward, at the end of June; and before many more days it was clear that his ship was no longer voyaging onwards, but was caught in a backwater. His fine compass had lost its property, and swung and spun with the malice of the mad, as if some demon hand, of monstrous strength, lurked underneath their keel; and wherever he went he was faced with the most mournful landscape, the most desolate sea, that any searching sailor could endure. No twist or turn would release them.

'Here's misery!' Sylvanus Bond was wont to croak, to anyone who would listen, a dozen times a day; and the sad-eyed little cooper was right. They spent much of their time under a canopy of fog, and when they emerged, it was to see nothing more comforting than bare mountains, snow-filled valleys, and perilous slopes of ice—an encircling prison as dreary and disagreeable as any in the world.

When any life moved, it was strange and fearful. They saw a host of white bears which could dive and swim, and horses like stags, with spreading antlers and gross noses. It was enough to rouse a superstitious dread, in a crew not bound together by the same hope as their captain. Their spirits faltered; and when they cast further to the south, seeking an outlet, and the calm water boiled again, and the *Discovery* was battered by waves and loose ice, courage itself was on the ebb-tide.

It seemed safer to Henry Hudson, alone in resolution, that he should be bold, and thrust his ship into the thickest part of the ice, where the water might be calmer. But on a day in the middle of

July their advance ended in a blank, forbidding wall of rock. After pressing more than two hundred miles southwards, they had voyaged nowhere.

Hudson, sick at heart, turned his prow north again; and there, at an island so bare and miserable that it might have been the last sign-post of their journey, he received his first taste of the true spirit of his crew. The men who had muttered long and grumbled bitterly now refused their duty.

They were bound by the ice in a cold hell of wind and wave, snow and frozen spray—and it was the middle of July! They would go no further, save homewards.

Their leader in this disgrace was the mate himself, Robert Juet, who thus dishonoured his position out of spite and malice. But behind him stood the young rogue Henry Greene, who spent more time among the crew than any man of quality should, and never ceased to sow discord.

Where two or three were gathered together, as the Holy Bible had it, he was there in the midst of them; but 'he' was no merciful Christ, but a villain with a lickerish love of mischief greater than any other love in his life. He always started well, with a worn jest which never lost its power.

'Here's misery!' he would say, aping the wretched croak of Sylvanus Bond. It did not lack a laugh from those who warmed to the condescension of the swaggering great; but Greene intended more than laughter, and went on to prove it. Whatever he said, or left unsaid, or hinted at, was always a reminder, day by day, that there *was* misery, and cold, and despair, all marrowed deep into their very bones; and that they were only being led on to worse affliction.

'Henry Hudson is lost, and he knows it!' was the text of the message of this evil man. And when Robert Juet found it in his bitter spirit to mock their captain publicly, saying that he was pursuing nothing but shadows, and dragging men after him in a Tom Fool dream of reaching Cathay by Candlemas, it was enough for a crew which, looking no further than the next hour, could only hate what they saw.

On the eve of their mutiny, there was a meeting deep below decks, in the shadow of the shallop which was still Matthew's only home. But who could tell what friends or foes were there? While, outside, the ice cracked and scraped against their hull, men moved in and out of the gloom, listening, or having their say, and then withdrawing, as if in shame or doubt.

Robert Juet was always present, and Sylvanus Bond the prophet of doom. Abacuck Prickett listened, but said no more than that whatever happened was all to the glory of the Lord. Henry Greene was ever in attendance, and ready with a jest which held more poison than wit. Philip Staffe, when he could, spoke out boldly for duty and obedience, but he was already cursed with the name of captain's favourite. Matthew, in this same low company, did not speak.

He also knew his duty, but duty on this darkling night was out of the fashion.

On one matter which they all shared, they were all agreed. It was the cold, which gripped the *Discovery* like the embrace of the dead. For though the cook's fire had fallen low, he would not mend it.

'Fetch me firing!' Bennett Matheus snapped, whenever a man complained. 'You wish to be warm? Take an axe and cut down a tree! Until then, the wood is for my cooking, and you may all die of cold before I put on another log.'

In their barren wilderness, to say 'Cut down a tree' was to advise: 'Go climb the steps of Heaven.' It was one more denial, in a world of harsh imprisonment.

A voice which was Greene's came from the darkness: 'You can have a cord of the finest apple-wood, Bennett, at any time you wish!'

'Where, in God's name?'

'In Devon.'

They laughed in company, but then they thought, and dreamed as well. Devon. . . . Against the grinding of the ice, it sounded the dearest place on earth, and the most distant. Devon. . . . Apple cider. The sunshine of July in the West Country. Comfort and safety. Home and a warm hearth. Devon. . . .

'We'll not see Devon,' Sylvanus Bond prophesied, 'on this side of the grave.'

'We can see it when we choose to.'

The voices became confused. Sometimes a name could be attached, and sometimes a stranger spoke, or an angel, or a demon, from the bleak and howling limbo outside.

'How can we choose to see Devon, when the master has the command?'

'We must take the command from him.'

'You are mad!' This was Philip Staffe. 'Without the captain, you would not see tomorrow!'

'Maybe we shall not see tomorrow, whether he is captain or not.'

''Tis the Lord's will,' said Abacuck Prickett. 'Search the Scriptures!'

'I would rather search the great cabin.'

After a long silence: 'What would you find there?'

'Better food than we find in this damned hole. And a chart to guide us home.'

Philip Staffe again: 'We have signed for this voyage. Not even three months have passed, and you are all turned cowards! And what do you know of a chart? You would use it to wipe your nose, before you recognized it!'

Arnold Lodlo, the little sneak: 'Tell us how you wipe his tail, Philip. And lick it too, for good measure!'

Laughter, from this small pit of hell. It was foul beyond measure, but foulness was gaining the advantage. Robert Juet gave it more voice:

'Surely we signed for a voyage. *I* signed you. But I did not sign you for a burial.'

'What do we do, then?'

'Refuse your duty. Tell him straight that we will go no further.'

'Suppose he says nay to that.'

A new voice: 'He will not. He is not the man.'

Philip Staffe, in a fury: 'He is *ten* men, set alongside a paltry moon-calf such as you!' Through the dark he peered helplessly. 'Who are you?'

'Will-o'-the-Wisp!'

'Jack-in-a-box!'

'Jenny Wallop!'

But Robert Juet would not allow laughter to weaken their temper. 'Are you resolved to go no further?'

'Aye.'

'No,' from Philip Staffe.

'Aye!' in full-throated chorus.

'Then we put it plain to the captain, at first light tomorrow?'

'Aye.'

'I will lead you, and speak out. Then we shall see who is the master here.'

And Prickett's voice gave the Amen: 'God prosper us, every one!'

It was the most shameful, and also, for Matthew Lawe, the saddest turn of any voyage. He thought of Henry Hudson as a friend. He *knew* him as a friend. But though, this their captain, could fire one or two with his dream, he remained no more than a dear man trapped between his precious vision and those fainthearts who alone could make it possible.

On the morrow, it seemed, he must be vanquished.

On the morrow Henry Hudson was not vanquished at all, nor even held to a halt; indeed, after the night of boastful words and dark threats, he carried the day so swiftly that the malcontents were still looking one at the other when the *Discovery* had taken up her voyage again. It was the triumph both of a stout spirit and a warm heart; and if it did not make him friends, it added no single enemy either.

When Henry Hudson stepped down to the main deck to meet his rebellious crew, he had at his back young John Hudson his son, Robert Bylot the navigator, Edward Wilson the surgeon, and Boatswain Clemens. He was fortunate in the weather of that morning; it had turned warm, as the biting wind dropped, and the jostling knives of ice no longer threatened their timbers.

There was not a sailor in the world who could not be cheered by such a respite, after the turmoil which had made their lives a misery. The men of the *Discovery* had woken to a day which seemed to promise them peace and hope. Hudson had only to persuade them that this was so.

It was remarked that Henry Greene, who had been everywhere in the fierce whispering of the night, did not station himself behind his captain. But he was not among the men, either. Suddenly, at an early sniff of the climate, he made himself a bystander. The man of straw was the first straw in the wind.

Henry Hudson, though apprised of the demands of his crew, did not burst out into anger and damn their eyes, as he had done with Juet on an earlier day. He rebuked them straightway for their wrens' hearts, but he was more a sorrowing father than a captain faced with mutiny. His message was simple, and enshrined in a single sentence:

'Do not waste the toil we have already spent on a noble venture.'

Robert Juet the mate set himself to speak his brief, though his sneering tone, by contrast with the captain's calm appeal, placed him in the situation of a traitor.

'Fine words!' he declared. 'But what is the *noble venture* which has put us in this miserable corner? How do we know where we are, and where you would take us?'

'I will show you.'

Show them he did, in a manner which perhaps, even so early, decided the outcome of the dispute. Robert Bylot had a chart at hand, and Hudson placed it on the deck, and bade the men gather round, and took his long cross-staff for a pointer, and made all as plain as he could. The men were constrained to kneel down in a circle on the deck, to see matters clearly, while the captain stood. It was a kind of submission, though Henry Hudson gave it enough comradeship to make it seem innocent.

'You may be proud,' he told them presently, 'that this new picture is the only one in the world to be made of these waters. We lie there——' he tapped the cross-staff on the chart, 'but it is a mere back-water, with no passage through—and we are the first to discover it, and to warn those who come after! So we go north again, and then press on westwards. You know now what we seek, and we may be nearer to it than we think.'

'But how do we *know*?' This was Arnold Lodlo, whining and questioning, but more like a puppet of the mate than a separate man. 'It is all writ on a piece of paper. But how can you swear it is true? What is writ on the water? What lies ahead?'

Suddenly Philip Staffe the carpenter—whom Henry Hudson was later to call, in affection, 'my staffe'—spoke out: 'That is why we sail on! To find out! We will discover nothing if we sit here like mangy dogs, biting at every hand that reaches out. This may be the North-West Passage! Who would not sail the *Discovery* onwards, to prove it?'

'I would not.' But from Lodlo this did not sound bold; it was more a petulant child, which would not put on its boots. He took refuge in pedlar's talk. 'If I had a hundred pounds, I would give ninety of it to be safe and sound at home.'

'I would not give ten!' said Philip Staffe—and the difference between man and mouse was suddenly made plain.

Though Juet tried to rally his faction, the heart had gone out of it. By ten o'clock, there was no more mutiny; by noon, the crew had returned to duty and were working their ship clear of the ice.

Once again in open water, their course was north, and then west; and no man on that day spoke another word of home.

Thus did the first axe-blow fall, and was thus turned aside. The wound was not mortal; but it sapped the tree, and laid it open to the next.

It was Matthew Lawe's great fortune to be at the helm of the *Discovery* when she burst out of the straits on the roaring westerly tide, and into an inland ocean, broad beyond limit and belief, which could be called the Middle Sea of the New World.

It had been won in hard fashion, dictated by the first rule in the sailors' primer—never to challenge the might of the sea, which was stronger than any ship that could be built, but to use it cunningly for their own purposes. So the *Discovery* had nursed her progress westwards, for seventeen days and more than three hundred miles; swimming with the great tide, anchoring under a lee shore when it turned against, and picking up the journey as soon as the fierce current relented.

The result was with them now, as they ran through the last of their ordeal, between a pair of great cliffs two miles apart and two thousand feet high, and made their entrance upon the noblest stage which any had ever seen. The whole crew was on deck—malcontent, coward, honest worker, and dreamer alike—to cheer the moment, and to exclaim at the vast prospect of an open sea, extending as far as a feeble eye could reach, and sparkling like the jewel of discovery which it was.

It was the second day of August, and Henry Hudson could record it as a triumph, though not the last one for which he hoped. He looked for this expanse to be the very doorway of the Pacific Ocean, which old Drake had conquered, and thus the new road to the Indies. For him it was no more than a single giant step—but at least he had found out whence the salt stream came and went!

He stood by Matthew's side, with Robert Bylot and Henry Greene at his back, and exclaimed like a boy at what had suddenly blessed their eyes.

'It is the gateway through,' he told them, 'and it explains all! Here are two levels of water—the east and the west. First the Atlantic tide leans this way, and breaks through. Then it ebbs, and the western sea surges back again. But what a moment of the world it must have been!'

Henry Greene, for once in awe, and ready to learn rather than to instruct or to put down, asked:

'What moment was that?'

'When it pierced through for the first time.' Hudson was impatient with anyone who could not comprehend his dream at the mere word of it. 'This is the source of what John Davis named his furious overfalls, and they are with us yet! But there was a day when this tide first breached the land, like a roaring hammer-blow. When did that happen before, in the Atlantic Ocean?'

Robert Bylot, careful and concerned with fact, gave the answer: 'With the Middle Sea.'

'Aye! There the tide burst through, between the Pillars of Hercules, between Europe and Africa, and opened up a whole ocean. Here we have it again. But this time it is ours!'

'How do we use it?' Henry Greene asked.

'We move on. But first we anchor, and return thanks, and make a landing, and give to all this——' his hand swept round in possession and pride, as if he were the Creator Himself, 'its first christening.'

The names came easily, in honour of the patrons of their voyage. On their larboard hand, new born, there now stood Cape Wolstenholme; and on the other, Digges Island, to salute Abacuck Prickett's master, and theirs. Henry Greene, who claimed wide acquaintance with the great, instructed all who would listen: 'No matter how it is writ, Wolstenholme should be said as Worsnam.' They thanked him kindly, being too content at that moment to quarrel with any man of such refined scholarship.

Intent on making an exact chart of all his new discoveries, from the very beginning, Henry Hudson sent a party ashore to explore Digges Island. They were Henry Greene, Bylot, and Prickett, and after some difficulty in landing against the high cliffs, they found a hopeful scene. There were deer moving about, and fresh green grass in abundance, and birds like the fat ptarmigan of Scotland roosting, and an air of simple prosperity such as they had never hoped to see in a wilderness so far from home.

Abacuck Prickett was much affected. 'If Sir Dudley Digges could see this paradise,' he declared, 'he would praise the Lord on his knees, as I do!'—and, as when Henry Greene declared that 'Worsnam' was the only word, there was no one to gainsay him.

They made one more mysterious discovery: certain heaps of

stones like hay-cocks which, when prised open, were found to be full of fowls hanging by the neck. These were hailed by Prickett, a nose-poking manservant, with the greatest delight. A larder in this waste-land! What manna from Heaven! To Bylot, it seemed of more account to find out whose larder it was; and for Greene, to rifle it.

When they returned on board, and gave the story of their great exploration, all who heard them wanted to stay in this agreeable haven. But Henry Hudson would have none of it. He had one ambition, and one only: to push onwards, taking advantage of the brief summer, in quest of the western passage. He felt in his heart that the *Discovery* must be on the very threshold. Thus he sailed southwards, with scarcely a day's delay—or a day's rest for a hard-worked crew.

Naming capes and islands as he went, he voyaged on: two hundred miles, four hundred miles, six hundred miles, *seven hundred miles*: casting to this side and to that like a questing hound, searching for the doorway out of the huge inland sea and into a new world. When the crew murmured, he would not listen. When their food fell short, he bade them hope—and make do with less.

But there came a day when the ship reached a last wall of rock and ice, deep to the south, impassable, and could not move except to retrace her steps. It was clear to all that even this ocean was no more than a vast backwater, like the last one.

Discontent blazed up, and came to the bursting-point. Once more, the crew made plain their intent. They did not know their captain's mind, because he would not share it. But they knew their own! Once more, duty was refused and a sullen disobedience ruled the *Discovery*.

It was the second blow from the same axe of mutiny. It was turned aside, as had happened before. But this time the battle ran a more fiery course.

Robert Juet was yet again the appointed leader. But hardly had he begun to set out his complaints, and make his demands, all with the whining insolence of the slave who would be king, than it became plain to the men behind him that the time of sweet reason was over. Henry Hudson flew into a passion, denounced the mate as a mutinous dog, and ordered a public trial of his conduct.

The captain's towering rage, and Juet's reputation, now made for a turn-about of a most curious sort, so that the great dispute became a mad mummers' play, with the prince to lose his bride and couple with a goat instead. For the crew, remembering how the mate's earlier failure had given them the look of fools, or because they *were* fools, or for any other reason which could make, of a man, a lion one day and a fawning dog the next, now united to overthrow their leader.

They began to testify against Robert Juet, with savage appetite. He was proved to have been heard uttering threats of 'bloody manslaughter' and 'vowing to turn the ship's head homewards', as far away as Iceland. Both Arnold Lodlo and Philip Staffe, the strangest bedfellows since elephant joined viper in the Ark, swore that Robert Juet had warned them to keep their swords and muskets handy, because they would be needed before the voyage was out.

Henry Greene, the most brazen weathercock of all, once more kept his silence.

Before the morning passed, Robert Juet had been voted guilty by his shipmates, and his punishment at Hudson's hands was swift. He was dismissed from his office of mate, and Robert Bylot was put in his place. For good measure, Boatswain Francis Clemens, who alone had spoken up for him, was also dismissed, and a common sailor of no skill or repute, William Wilson, was raised in his stead.

Wilson was a fearsome choice for the office: one might as well, thought Matthew Lawe, have named the first sea-bird that dropped its filth on the deck. He had the foulest tongue in the ship, and the quickest fist. Matthew had steered a wide course of him, from the first day of the voyage.

It was therefore no blessing from Heaven when Matthew Lawe himself, to his astonishment, was named as boatswain's mate.

With this good and bad day's work done, Henry Hudson began to press forward again. It was the middle of September, bitter cold already, and the promise of time was running against him. But— as it seemed to his men—he continued to sail without purpose: now going westwards, now north again, now south to the barred door of rock.

There was little profit in this wandering, and then none. They found a foot-print ashore, but no man to match it. They ran upon a rock, and were stranded like a blown whale for twelve hours. It did little for the captain's credit, and once more the doleful murmuring of 'We are lost' began to be heard on board.

Hudson was not lost: on any day or night he could put his finger on the chart, and say with certainty: 'We are there.' But he was searching against time, and losing the race. He knew already that he must winter here, but in common prudence he kept his counsel. All he could do was to set his men to fish, and hunt for deer and game-birds ashore, and build up some stock of food and fuel for a long, cruel siege.

As October turned to November, and the nights grew long and the snow fell thick, he looked about for a place for wintering, and found it, in the very south-east corner of their giant bay. In this desolate lair they anchored, and hauled their ship aground, and made what provision they could for a life half ashore and half afloat, which might endure for six months.

By the tenth day of November, in a howling blizzard of snow which drove splinters of ice into their lungs, the *Discovery* was frozen in, and a vile northern winter took merciless command.

V

The blind months fell one by one, like withered leaves from the barren tree of the world. The sun had deserted them for an endless season, and when men dared to face the biting cold they found their lonely ship mostly wrapped in a shroud of fog. When this was blown aside by the wind, their view was dismal: only a haggard landscape of mile after mile of snow, a shore-line like the snarling teeth of a skull, and a sea frozen dead.

The wicked north wind had stunted everything that grew; and if a man lingered too long on the open deck, or while hunting ashore, it was likely to stunt him as well.

But though they lived in desolation, life was never silent. The frame of their beleaguered ship creaked and groaned constantly, protesting its torture. Booted foot-falls were magnified to thunder. The bitter wind screamed like a demon when it did not sigh like a woman going mad, and at night the packs of ravening timber-wolves crept close, and howled for blood—blood of any kind.

No band of men could have been trapped in a deeper, colder pit. At times it seemed easier for them to fold their hands, and to die. But since the human spirit always chose life, however vile, rather than the dreaded unknown of death, and since there was not enough food to last the winter months, much less to see them home at the end, they must needs drag their limbs upright, and chafe them into a thin warmth, and go look for something to support their living.

They were set to hunt in pairs, for the safety of each and perhaps to make sure that there was no secret hoarding. Rewards were offered for any beast or bird brought back on board. When the game failed, they lived on the mouldering scraps, nine or ten months old, of what they had loaded in the Pool of London. When fortune smiled, they ate the ptarmigan they could stalk and kill, and any wandering swan or goose or teal-duck which put down near the ship.

At the worst, frogs caught unaware in copulation were not to be despised, nor the rats which had been their shipmates for an anniversary year, and then the very moss scraped off the feet of trees, and boiled to make the thinnest soup that one orphan pauper could set before another.

As the months crept by, they grew diseased, and feeble, and sick to death. The principal plague was scurvy, which made their limbs swell monstrously but left their blackened gums to rot about their teeth, so that they could neither work nor eat. Edward Wilson the surgeon, who was a very angel when set beside his namesake, William Wilson the raw-fisted boatswain, worked faithfully to ease their misery. From his skill he conceived a brew of pine-buds, collected one by one and crushed for their drops of oil, which did something to unloose the knotted cramps of an empty stomach.

None the less, men began to fail one by one, as at the first hiss of the scythe of Death. Abacuck Prickett took to his bed, swearing by Almighty God that he was as lame as a horse with its leg in a bear-trap. For once, he spoke true. . . . Francis Clemens, having lost his rank from cold dis-favour, lost his toe-nails by frostbite. Syracke Fanner was a mere cripple. ('Mortally sick of his name,' Henry Greene jested unfeelingly.) John Williams, the seaman-gunner, died in his cot—safe below decks, but he died of the cold shivers even in this protection. Other infected men became foul to the sight, and then to the nose.

Even the strongest among them grew, or were shrunk, into gaunt spectres of men, with only gaunt thoughts to sustain them. Their yearning hope was for the spring, and their only thought was the cry of a child in captivity: When that spring comes—*if* that spring comes—shall we go home again?

They had grown suspicious about the division of the food, and whether Master Henry Hudson had favourites, who crept into his cabin for a favourite's bite, while others—the forlorn moss-suckers at the starveling foot of the table—were denied this shy teat of privilege. But they had no strength to insist.

It all returned to this one thought: would Henry Hudson take them home? By April they had been a year absent. There could be no more of these fearful seasons! Not another one! And not another sea-mile to the west!

The captain on whom their hopes and fears and doubts were all centred was also yearning, but not for home. Immured in his great cabin, as hungry and cold as any of his men, this old lion in winter had lost none of his shining spirit. He knew himself to be solitary, and near to friendless, among scowling ruffians who wished only to be quit of the voyage. He knew, better than all, that the *Discovery*'s situation was desperate—until the spring, and perhaps for all time afterwards.

But against this he also placed one single thought and purpose. He had a voyage to make, and a goal to find. Since southwards was not the path to it, then they must go west. As soon as they were free, he would sail on—with his back to home, and his face to the hidden half of the world; and not a man was to stand in his way!

To this dauntless dreamer, it mattered not that he could count his friends on the pinched fingers of one hand. Henry Greene, a blustering rogue, was still a favourite. Robert Bylot had the same thirst for discovery as his captain, and would never desert him. John Hudson, his youngest son—of the same age as his grandson, he sometimes boasted—was as stalwart and true as a grown tree. Matthew Lawe was simple and faithful. William Wilson owed him loyalty for his promotion to boatswain. Philip Staffe—no, he could no longer number 'his staffe' among his friends.

The fault was his own, and Hudson mourned it even as he persisted in the stiff-necked temper which had brought about their quarrel. When the *Discovery* had been two months frozen in, he had ordered the carpenter to build a hut ashore. Philip Staffe, who had volunteered for this duty as long ago as October, and had been roughly dismissed, now refused in his turn.

His argument was clear, and his resolution also. It was too late, Staffe declared with great firmness. They should have built earlier. Now the planks were frozen to the ground, and a man holding nails in his mouth, as a carpenter did, would have his lips skinned as he drew them out. He did not choose to frost his face for another's delay. He was a shipwright, and no shore-carpenter, in any case.

Henry Hudson, when he heard of this refusal, flew into another great rage. He ferreted Philip Staffe out of his berth, struck him about the head, abused him foully, and threatened to hang him. But Staffe was a man of equal spirit, and as hot a temper when roused. He said, with the scorn of an honest man under bitter insult, that he knew his business better than did the captain. He was no house-builder, and no slave either. The answer was No, and be damned to all tyrants!

Thus was Hudson's most loyal friend put into disgrace, and thus he remained.

For weeks, and then months, Henry Hudson dreamed on in his cabin, waiting for the axle of the world to turn and the sun to set them free. He did not show himself. There was no pleasure or profit in facing this surly crew—selfish, quarrelsome, not to be governed. There was only poison, which he need not swallow.

But it would all be different in the spring. Then there would be warmth, and full bellies, and the magic of a kindly season, and he would talk to his friends, and they would spread the word—the word of courage which would take their ship onwards.

He knew, or thought he knew, his friends. But he did not yet know his enemies.

Certain of the children of captivity were already old in sin, and, though working secretly, bore a brand of ill-fame which could be discerned by any honest man, being as stark as the Mark of the Beast. Matthew Lawe, returning from the great cabin, which he had been setting to rights under Henry Hudson's listless eye, found his way barred by one of these knaves. It was Boatswain William Wilson, to whom Matthew was the appointed mate—though to 'mate' with this reptile would have dishonoured the meanest child of God.

The boatswain, whose bulk filled the passage-way like a gross bung in a pot-bellied cask, confronted Matthew with his accustomed glare.

'Did you do as I bid you?' he demanded. 'Make all straight?'

'Aye.'

'And is the master satisfied?'

'Aye.'

'I'll warrant he is!' Boatswain Wilson had two faces only: one showed rage, the other contempt. It was the turn for contempt. 'To do a dog's work, I send the most faithful dog of all! . . . Did he tell you his plans for the ship?'

'No.'

'How do I know the truth of that?'

'Why should I lie?'

'Do not give me insolence!' William Wilson roared, and drew back his knotted fist. '*I* put the questions! You answer them, or by Christ's wounds I will deal you wounds to match them!'

Matthew, who had felt the weight of the boatswain's arm more times than he could number, kept his silence. A soft answer provoked a spluttering wrath, and a spirited one, the swiftest punishment since Cain settled the world's first score with his brother. Thus he had learned submission, vile though it was. He wished to live, like every other soul on board.

'Remember that!' Wilson snarled, seeing that even this small rebellion was at an end. 'Or I will fix it in your skull with a nail, and use this——' he bunched his monstrous fist again, 'to drive it home!'

They were joined now by a third man, drawn by the dispute. Matthew, with a sinking heart, saw that it was Robert Juet, the deposed mate, who fed on quarrels like a crab on a corpse. Was there to be no end to goading and tyranny, on a day which was miserable enough already? He turned, as the old man shuffled towards them.

It was clear that Juet had been listening. His rheumy old eyes glistened as he asked:

'What did he give you to eat, Matthew Lawe?'

'Nothing.'

'What? No fine slice of meat?' The whining voice was like a worn-out, gap-toothed saw. 'No soft loaves? No stoup of wine? No cuts of cheese from his private store?'

'He has no store, that I know of.'

'What has he, then, that keeps up his strength?' Wilson demanded.

'He has nothing, like ourselves.' Matthew could not betray the captain on so false a charge. 'His strength is the same as our own—starved to skin and bone. And his food also. And his empty belly.'

'Do you defend him?'

'He is the master.'

'He is the master *today*,' said Juet, with evil meaning.

But this was dangerous ground for William Wilson, who was thought to maintain that loyalty to their captain, of which Juet had been robbed. Yet he had his own question to put.

'Did you see aught of the secret scuttle, the opening between the great cabin and the storage hold?'

'No.'

'I told you to search for it!'

'There is no such opening, that I could see.'

'Then how does he live, while we starve? Whence come all those fine suppers for his favourites?'

It was an old question, an old suspicion, an old rumour, and a thought more foolish than any that sped about the ship. Matthew did not answer.

'If I see as much as a crumb on your mouth,' Wilson roared out, 'I will drive it down with the shaft of an oar!'

He might well do so.

Presently they let him go, with a last oath from the boatswain and a warning from Juet that he should 'keep his fingers out of such pilfering'. He took a gulp of the foul ship-board air as he turned away, knowing well enough that he had been speaking in close encounter with two of the three wicked men, as he called them in his heart. Any freedom from such infection came as a relief.

The three were Henry Greene, William Wilson, and Robert Juet, and the merest fool could number them. This triple of knaves would be the hard heart of mutiny, if mutiny it was to be. Greene was a natural villain: Juet a malcontent who in disgrace had become an implacable enemy; and Wilson a masterless rogue, whose cursing, cruel as the wind, concealed a viper's treachery.

While they went about their work with slow cunning, never showing an open hand yet losing no chance of sly insult, or doubt, or derision towards their captain, they were making all ready for the spring. Henry Hudson did nothing to counter them. Could he not see? Or did he see, and not care? Or had he lost his wits, while others kept theirs as sharp as daggers?

It was the spring, in the year of Our Lord, 1611. It was the spring!

The darling season had come slowly, and late: the sun climbed painfully above a bleak horizon, like an old man lifting himself from bed, and not till the last of May did it show its strength, and true compassion for the chained prisoners. But then the claws of winter fell from power, loosing the ice, warming all their thin bones, and sending clouds of biting black-fly as warrantors of hope.

Their ship was free at last, and all her children with her.

As soon as men could move about instead of crouching like cripples under the cold wind, they pushed and pulled their ship till her keel was clear of the sea-bottom, and anchored her in liberal water. The *Discovery* began to rock, and then to reek. Spring was warmth, and warmth was the melting of good elements as well as bad. Foul bilges lost their frozen innocence; crusted turds became noisome floating filth.

Within a few days she stank to Heaven. Men grew restive, even in their new freedom. The winter had been the worst of their lives. Was spring to be the same? As the blood rose, so did hope—but it was hope for one single end. Their captain *must* take them home!

He must do so, they reckoned, to let them stay alive. Their food-store was still desperately bare. Though they managed a little fowl-hunting, the birds were shy, and thin as the men who sought them. The first spring fishing won a miraculous draught of five hundred trout, and other fish as big as herrings. But after that one happy day, they caught scarcely a minnow. Perhaps fish were wiser than sailors. At the least, they had the freedom to swim elsewhere.

The ship, alive with bugs, and ravenous infant rats, and flies with teeth, and loathsome smells, was rampant with rumour as well. When Henry Hudson began to show himself on deck, and to look about him, and to talk with Bylot the new mate or Edward Wilson the surgeon, and to examine sails and spars for winter damage, his face and bearing were eagerly watched, and every chance word overheard was borne below decks, to be turned this way and that in the search for meaning.

To guess their future course was becoming a torture. Food was their aching lust. The two themes were married: both Juet and Abacuck Prickett, still a-bed with his lameness, preached a mathematical surety of starvation, if they did not return to Digges Island, where the meagre 'larders' had been robbed and where the fowl bred, and take on every scrap of game they could snare or shoot, and sail straight for home.

'If not,' Prickett declared from his gloomy couch, 'we will starve where we lie.'

Henry Greene was with him, and said: 'It would be best if we were fewer and stronger.'

Prickett, his eyes burning with more than fever, answered: 'None can tell who is among God's chosen!'

'Would you be among them?'

'Aye. With God's help.'

'And mine,' said Henry Greene.

There came a day when the captain, and Mate Bylot, were talking together on the poop-deck, with charts spread out before them. Matthew Lawe chanced to be near them, putting a wrapping on the worn back-stay of the mast; and the fact was noted by his desperate shipmates. When he came below, he was surrounded by questioners.

'What did you hear?' Robert Juet demanded.

'Nothing much,' Matthew answered, which was the truth.

'But what was the talk?'

'Of where we go, I suppose. They were looking at the charts and pointing this way and that. You should ask the mate himself.'

'Bylot will not talk,' Michael Perse said roughly. 'He is the captain's man now. . . . What charts were these?'

'The ones we have used before.'

'For the love of Christ!' Arnold Lodlo exclaimed. 'Are you another captain's man? *What was said?*'

They were all thronging round him, and William Wilson was gazing at him as if he saw his next meal.

'I *think*,' Matthew answered, 'that they plan to sail some miles north, till we can find food, and then voyage onwards.'

'To England?'

'It was not said. I think westwards.'

The discomfort and rage in all their faces, the muttering and threats to be heard on all sides, increased day by day as the *Discovery* was made ready to sail. But before she did so, on the twelfth day of June, Henry Hudson took three courses which confirmed the number of his enemies, and added to them.

He called his crew together, and declared that he would make a division of the food. ''Tis all I have to give you,' he said as he had the stores brought up. 'So each man must make it last.'

What each man received served only to increase their forebodings and their suspicious fears. Hudson himself was near to weeping as he gave his sailors a single pound of bread each—'And it is *all*!'— and three and a half pounds of cheese to last for seven days. Some of them ate the greater part immediately, being sure that there must be more hidden. William Wilson, an arrant hog, ate all his share in a single day, and then took to his bed for the next three.

The crew did not trust their master—there was new talk of that secret scuttle between the great cabin and the barred storage-hold, whereby he must be increasing his own share—and it became clear that Hudson did not trust them either. When the murmuring reached his ears, he waited his chance and then sent the boy, Nicholas Syms, to search the men's sea-chests.

Nick Syms returned with a hoard of thirty bread-cakes, and (having been discovered) many a flea in his ear. To have their private chests searched by the cabin-boy! . . . The crew was thrown into the utmost fury; and Henry Hudson's next stroke, of a foolishness to match this search, completed a course which seemed to prove that the captain might have lost his senses.

On the eve of sailing he picked a quarrel with Robert Bylot, his skilled and loyal mate, dismissed him out of hand, and replaced him with one John King, an ignorant, choleric quartermaster who could neither read nor write. The choice was first disbelieved, then ridiculed, and Hudson's credit was utterly destroyed.

It was in an evil humour that the *Discovery*, now a nest of hateful intrigue, sailed on the twelfth of June. Matthew Lawe was secretly appalled at their position, and the captain's folly, and the men's spite. It was madness beyond belief!

This was not some great carrack of Spain, with 800 men and room for warring factions within. It was a little bark of England, carrying a score of men who should have been united in their ordeal; and—with scarcely any food, and no prospect of more until they reached Digges Island, 700 miles to the north—they were being led to ruin, with half of them at each other's throats and the rest sick and starving.

Eleven days later, with a miserable hundred miles behind them, they came to a fatal stop. They were becalmed in thick ice, in the lee of a bare island. They were twenty-two men, with hardly food enough for half that number to face the shortest voyage.

It was the eve of Midsummer's Day, the twenty-third of June.

VI

Though midnight was near, there was still dim lantern-light in the cabin where sat a black-browed villain, Henry Greene, with an old malcontent, Robert Juet, at his elbow, and a coarse ruffian, William Wilson, a-squat a barrel-top and keeping watchful guard over the doorway. They shared a flagon of sour wine, though Henry Greene took a greater portion than any, since it came from his own hoard.

Their talk was not new; it was now a matter of certain last strands which must fall into place, to weave the most wicked plot that ever fouled the sea. They were gathered for this purpose alone; and the table at which Henry Greene sat, and the paper that lay upon it, and the pen in his hand, were all the weapons needed until next day dawned.

Juet said: 'I have searched about, and talked to all I can trust. We have enough stores for three days at the most. Then there is what the captain keeps hidden. But it cannot amount to more than six days altogether.'

'Six days' food for all on board,' Henry Greene answered. 'It is not enough, and we know it.' The lantern flickered on his face, in which self-will and resolution were joined in a strong draught. 'We can be sure of twelve days or more, if we carry fewer mouths. But to make certain that we live, the weakest must go. And anyone else who is not of our mind.'

They had known this for many days, but it had not been so starkly said.

'How do we decide?' old Juet asked.

'We decide now.' Already Greene had assumed captain's airs, and as he drew himself to the table, and took a swig of the wine, and prepared to write, he had a look of command which the strongest man on board would not have dared to challenge. 'Now keep silence for a space, while I set it all out. Then we will talk. Then we will act.'

Thus, in an infamous hour, on the stroke of midnight, he drew up an infamous muster-roll. It made only two divisions—of those who would stay on board, and those who would not: a list of the quick and the dead.

'THOSE WE KEEP', he set out at the head of the paper, in a broad hand. 'Myself' was the next word, and then, to show at once his strength and his evil humour, he noted: 'With a sword, two pistols, and a dagger!' Then he wrote on, with a word here and there to show his mind:

Robert Juet, Mate. 'In his rightful place again.'

William Wilson, Boatswain. 'Has proved himself.'

Robert Bylot, Leading Seaman. 'Our pilot, needed to take us home, and the only one who can.'

Edward Wilson, Surgeon. 'We must keep a doctor.'

Abacuck Prickett, Landsman. 'Will procure pardon from Sir Dudley Digges.'

Bennett Matheus, Cook. 'A cook we must have!'

Philip Staffe, Carpenter. 'A known friend, and needed.'

Sylvanus Bond, Cooper. 'To work with Staffe. And he will keep us merry!'

TO WORK THE SHIP: John Thomas, Seaman. 'Strong and well.'

Francis Clemens. 'The same.'

Matthew Lawe. 'The same.'

Michael Perse. 'The same.'

Nicholas Syms, Boy. 'Useful. Eats small. And we cannot take the other.'

So ended the tally of the blessed elect. Henry Greene now came to the remnant, which was expendable flesh.

'THOSE WE PUT AWAY', he wrote large at the head of the next paper, and swigged his wine again. It could be remarked that he now showed more appetite even than before.

Henry Hudson, Master. 'Our great captain!'

John Hudson, Boy. 'Of the same brood.'

John King, new Mate. 'We have our mate.'

Thomas Wydowse, Scholar. 'He has no friends.'

Arnold Lodlo, Seaman. 'That little reptile curls my soul!'

Michael Butt, Seaman. 'Sick.'

Adam Moore, Seaman. 'Sick.'

Syracke Fanner, Seaman. 'Sick—of his name!'

When all was done, Henry Greene sanded his script, and shook it dry, and perused it for himself. Then, in a low voice which lacked nothing of resolute malice, he read it to his fellow-plotters.

'If you have questions,' he said at the end, 'voice them now. But we have talked much already, and there can be little to say.'

Old Robert Juet, a man of poor spirit when it came to turning word into action, asked timorously:

'What did you intend, when you wrote that they are to be *put away*? Is it to be murder after all?'

'No, no! They shall go into the shallop, to fend for themselves.'

'You swear it?'

'Aye, if you wish.'

William Wilson, not so nice in his scruples, had more particular questions.

'Why do we keep Matthew Lawe? He is a captain's man!'

'He is one of those we need to work the ship,' Greene answered. 'If we do not take enough of the strongest, we shall never reach England. And I do not think Matthew will be such a captain's man when he sees where it will lead him.'

Juet asked: 'Can we be sure of Bylot? He has been another standing close to the master.'

'He is out of that humour now. And I want his brains, more than his little squeaks of loyalty. There is none else on board who can guide us home.'

The boatswain had another particular objection. 'I mislike Francis Clemens,' he growled. 'Since I took his post when he was put down, I cannot trust him.'

Henry Greene looked from face to face in the gloomy cabin, and said with threatening care:

'If we cast off all we did not trust, the shallop would be full and the ship empty.' He waited for this shaft to find its mark; then he said heartily: 'Enough, enough! The lists are made. If alteration is needed, I will trim the man, not his name!' He brought his hand down flat on the oaken table-top. 'An end to talking! Now we move.'

'What comes first?' William Wilson, a man more for action than words, asked.

'We go quietly to old Prickett's cabin, to make sure of a pious scoundrel.'

The pious scoundrel, a-bed in a black night-cap which had the look of a bishop's mitre that had lost its faith, looked up fearfully as the conspirators stole into his room. He knew their purpose, and approved it, so long as it would not rasp his own repute; his sole concern was to combine treachery and Holy Writ in a dish which man could justify and God would swallow—in short, to keep his long nose clean.

It turned upon a midnight oath. Whatever they did, they must swear not to spill blood. Setting sick men loose in an open boat, foodless and forlorn, to make what they could of an Arctic wilderness, was not murder—not for certain. Who could tell what God's mercy might not provide? But they must all swear that the castaways would leave the *Discovery* with a whole skin, and their blood within it.

They took such an oath, as sanctimonious, perjured, and false as any in the world; and honest Abacuck intoned: 'The will of God be done!' at the end.

As other men came in secret to swear on Prickett's own Bible, there was much creeping about, much furtive movement along the dark alley-ways of the ship. After Greene, Wilson, and Juet, John Thomas swore, and Michael Perse, and Bennett Matheus. Bylot was already thought to be secure. The surgeon would look up when all was over. The morning would unfold the rest.

Amid the whispering and the shuffling foot-falls, Matthew Lawe lay low. His cold cradle of the shallop was gone; at the first of spring it had been lifted above decks again, and set up by Philip Staffe, and it now swung to its mooring alongside their hull.

Matthew was curled up in a corner of the lower deck, feigning sleep, closing his ears to the creeping and the talking, trembling with more than the cold. Like old Abacuck Prickett, he knew what was a-foot, and he could not bring his mind to it with any surety.

Mutiny was vile and disgraceful—a hanging matter, if there was any justice in the world. Yet Henry Hudson had behaved with such rashness, if nothing worse, that he scarcely deserved command, and had begun to forfeit loyalty.

There were sailors, adventurers, who were gripped by fear too early—these were the cowards; against them were those who grew afraid too late—the forlorn dead, without heritage or headstone. In between were a few living heroes, coldly rash, prudently aware. Matthew could not trust their captain to be among this number.

Hudson seemed to bear the brand of failure—his quest too great for his ability, his dream too wild for sharing. He seemed destined to lead them all into ruin. Thus, the choice for his crew might lie between death by hanging, and death from starvation. At any time within the hour, there might be a third extremity—a dagger in the throat from a desperate traitor.

He, Matthew Lawe, wanted only to close his eyes and ears, and, when it was all over, somehow to live.

Between two o'clock and break-of-day, the ship grew silent as the grave which it might become. All were at last a-bed. But wickedness did not sleep. Waiting for that midsummer morn in that year of their forgotten Lord, Henry Greene conceived a small stroke which would have done honour to Judas Iscariot. He pretended sleeplessness and, in all good comradeship, passed the night in Henry Hudson's cabin, talking and keeping watch.

Nothing must warn his captain that the last axe-blow was poised; one which would prove mortal to all, whether in honour or in blood.

It was all done with the swiftness of evil.

At first full daylight Bennett Matheus the cook went forward to fill his kettle at the water-butt. Passing the shallop, innocently tethered alongside after its winter sleep, he began to whistle cheerfully. He might only have been greeting a fine June morning. But this was the signal for treachery.

As the minutes passed, certain men took up their stations in certain places, and waited. Henry Greene walked close by a just-awakened Philip Staffe on the poop-deck, and held him talking. Though the mutineers wished to keep Staffe on board, they feared what he might do when he saw the fate which would now overtake his captain.

John King, the new mate, was lured down to the hold on a pretext, and the bolt shot fast on him. Abacuck Prickett, who might still have holy qualms, was warned by Juet to keep himself well— in his own cabin. Then the conspirators waited for their main prize.

At sun-up, Henry Hudson came out of his cabin, and saluted a fine Midsummer's Day, and a Sunday, with a smile. It was his last. On another signal he was seized and pinioned by three men: Bennett Matheus, John Thomas, and Boatswain Wilson, who bound his hands behind his back. Though Hudson put up an old man's struggle, it was mere moments until, surrounded by faces with no trace of pity or remorse in them, he was thrust downwards into the shallop.

His son John was the next to follow, with Matheus and Thomas set to guard them. The search then began for the rest of the doomed.

There were loud cries, and shouts for help, and a confusion in this harsh shepherding which might have occasioned laughter on any stage save this bloody circus. Only Henry Greene, sword in hand, seemingly consumed by a black rage, knew for sure who was to go and who to stay. Juet kept aloof, and William Wilson, whose best brains were in his fists, kept losing his way in memory. At one moment Matheus and John Thomas were both in danger of being left in the shallop, until the true victims, Arnold Lodlo and Michael Butt, were found to replace them.

John King, released from his prison below in order to be sent to his doom, put up a stout struggle. He attacked old Juet with a sword, and might have killed him; but he too was presently overpowered, and thrown into the boat as roughly as a sack. Wydowse, the mathematician of Cambridge whom no one knew, had sufficient scholarship to see his mortal danger and to beg for his life at any cost. They might have his keys, and all his goods. . . . Down he went.

The rest of the named-and-damned were easier meat for sacrifice. They were taken by surprise, when half asleep or weak from sickness, and made little resistance as they were dragged along the deck and dropped over the side.

Within an hour, they were all sat in the tiny cockle-back which was now their flag-ship. All but one man, that is, who now put the vile world of treason to shame by declaring that he wished to join them.

Philip Staffe the carpenter had fought with his good conscience, and conquered his bad. Now he faced the mutineers, and cursed them for scum, lumps of foulness, thieves, murderers, eaters of their mother's dung. He would have nothing of this mutiny, nor of the dogs who had made it.

'Do you choose to be hanged when you reach home?' he shouted in their faces. 'Well, I do not! And I will not stay except by force!'

'There is no force to make you stay,' Henry Greene told him, between rage and shame. 'Go if you will. But you may be climbing down to your death.'

'Then give me my chest of tools, and be damned to you! I would rather trust to God's mercy, and go down into the shallop for love of the master, than take a better chance with villains such as you!'

Sullenly they let him by, and added to the carpenter's chest—from their own noisome consciences—a musket, an iron cooking-pot, and a little pease-meal. Shouldering these burdens, Philip

Staffe, the single hero, stepped down to his doom without a word or a glance.

This was the moment when Matthew Lawe, lurking below for fear of his own expulsion, judged that the search must now be over and came up on deck. What he saw when he looked down at the shallop was the most pitiful sight of his life, and the most terrifying.

The little boat was full of its misery. Henry Hudson sat in the stern, in his motley gown which had once been a badge of honour. His son's arm was clasped round his shoulder. The pale sick lay huddled on the floorboards. Entreaties rose from any who could speak, and were ignored. There was no store of food to be seen, no drink, no firing, no clothes. Philip Staffe sat astride his chest of tools, and looked up at the mutineers as if he saw a pack of foul dogs eating, because they could not mount, a dead bitch.

'*Well!*' a voice at Matthew's shoulder suddenly roared.

He turned in alarm, to meet another sort of gaze: that of an armed man close enough to murder him if he chose. William Wilson, with a butcher's carcass-cleaver in his hand, was staring furiously down at him, like a mad serving-man who would know his wishes.

'What is it?' Matthew asked fearfully.

Wilson shook his gleaming weapon, as if in a frenzy. 'It is for you to tell! Would you go with them?'

Matthew glanced at the laden shallop, and back again. 'There is no room for me.'

'We will make room, if you choose.'

'I do not choose.'

'Well enough.' With a last glare, William Wilson stood back. 'Get to the helm—and if you move false, I will split your skull and toss it down to them! They will need their supper!' Then he turned, and roared out his first order as a pirate-boatswain: 'Up anchor! Make sail!'

The *Discovery*, with her little doomed chicken in tow, gathered way and moved clear of the thick of the ice. Then, at a signal from Henry Greene, the rope was cut, and the shallop dropped away astern.

Matthew Lawe, from his high station on the poop, turned to look back as the gulf of water widened. Hudson still sat tall among his tumble of wretches, staring ahead as if frozen in a nightmare.

His son's arm still cradled a father's shoulder, though who the comforter, and who the comforted, was in God's mind alone.

But it was young John Hudson's eyes which lifted towards the ship, and a boy's voice, borne on a wailing wind, which gave to the only man he could see the last salute of the condemned:

'Matthew! *Why?*'

THREE

PIRATE

1670

'Narratives of cruelty and bloodshed hardly to be surpassed in the annals of crime.'

Chambers's Encyclopaedia, Vol. II, under 'Buccaneers'

THE FORLORN CLUSTER of men still continued to wave as the gap widened between themselves and their ship, seeming to believe, even at this late hour, that far-off gestures of entreaty might soften a ferocious will. When they were no more than dots, mere atomies of men marooned on a Carib rock under a burning sun, these tentacles of prayer still wavered and shimmered round about them, as if such tiny wisps might suck pity from the pitiless.

Ben Pannikin, the water-boy of the *Cambridge* brigantine, had not seen such a sight before, in all his eighteen years. Matthew Lawe had. Though they were watching the same scene, under the same bright sun, a chasm of feeling gaped between them. While young Ben drank it in with the thirsty amazement of youth, Matthew, having spread this murderous net before, could have spat it all out with no second thought.

It was only punishment, and all that mattered was to be the man who served it, not the wretch who felt the whistling lash. If a pinched old face, swallowed into fog and cold misery, was now succeeded by lusty young desperadoes, screaming curses at their executioners until they woke to their fate and begged for mercy, it was no more than a change of sacrifice.

The waste of water widened in the same way, whether bitter-grey with ice, or magically blue and pink, veined with coral milk to soften all save the heart of man. The barbarous treachery was one and the same. All that it taught was the first or last lesson of life: to belong to the victors.

Ben Pannikin, who should have been helping him to secure the long-boat after it had run its cruel errand, and returned with five men fewer than it had carried ashore—Ben Pannikin was still staring sternwards, at what was now only a jagged rock on the horizon. It was as if he could not bear to break the link with the marooned men, his shipmates in the *Cambridge* brigantine until two hours ago.

He gazed so long that Matthew, making all allowance for a tender spirit, recalled him roughly to his duty. He was the mate, not the wet-nurse; and the boy must learn, if he was ever to be made a man fit for their wolfish trade.

'Bear a hand here!' he shouted. 'And have done with staring! You'll not see them again till you are all haled to the Judgement Seat. And better have your knife sharp on that day!'

Ben came shambling forward to the waist of the ship. He had grown beyond his strength during the past year, and his arms and legs were like the parts of an ill-made puppet, going this way and that with no law to rule them. Only the look in his face was stead-fast—steadfast in pity.

'But what will become of them?' he asked, as if he could not believe the truth of what they had done. 'We put them off without so much as a sup to drink.'

'Serve them right,' Matthew said harshly. 'They knew the penalty. We would have taken that Dutch pinnace, if they had not faltered and turned tail.'

'The odds were too great.'

'So they were cowards! They crossed Henry Morgan. Do you think a man lives, who does that? This is the brotherhood of the Coast, not a whining girls' school! You know the watchword— "No prey, no pay". Morgan lusted for that cargo of silver. So did I, for my share! But the fools hung back, when they were in the very rigging, and lost it. They were lucky not to be cut up for fish-bait!'

'I'm sorry, Matthew,' Ben Pannikin said, humbled before this wrath. 'I never saw a marooning till today.'

'You'll see more before you shed your last tooth.'

'But what will become of them?' Ben asked again.

'They will live on gull's eggs, and seaweed, and fish if they can lure them within reach. If it rains, they will drink.' Matthew cocked an eye at the cloudless Carib sky. 'If not, they will thirst. Perhaps they will fall to fighting among themselves. Perhaps they will eat each other at the end, when the sea-birds are frightened off and the fish turn wily. Then the last man will starve, and all will bleach—and in fifty years another ship will wander by, and mark her chart with "Many Bones Here", and sail away again, with none the wiser, nor the sadder either.'

'God have mercy on them all!' Ben muttered fearfully.

'And He alone . . . Now coil down that rope, before I coil it round your backside.'

Presently, after some minutes of quiet work, Ben asked: 'When did you see it before?'

'Only once. Long ago.'

'Tell me, Matthew.' Ben Pannikin was always saying this, and always forgetting the answer, and so asking again. His head was like a corn-sieve, and as soft as his heart. 'What happened? Where was it?'

'Far to the north.'

'But how many?'

'Eight. An old man, and a boy like you. . . . And six others.'

'Had they disobeyed the captain?'

'Questions, questions. . . . No, it was a sort of rebellion. So we cast them off, and went on our way.'

'Did they die?'

'They must be dead by now.'

'And were you punished?'

'Aye. Well, some of us. Some were killed, or died. But some were even honoured.'

'And you?'

'I walked away. . . . Now *you* walk away, Ben Pannikin, and fetch me a cup of water. This noonday sun is a furnace!'

It was already a lifetime away, yet, in the demented numbering of his days, only yesterday. Their punishment had begun even as the shallop was left behind. In a small Indian war, waged by revengeful men whose larders had been robbed once, *but never again*, four of the mutineers met a cruel end.

John Thomas spilt his bowels on Digges Island, from an upthrust Indian knife. Michael Perse fared the same. William Wilson died, cursing horribly, from an arrow wound in the breast. Henry Greene, the chief monster, closed his evil score by dying bravely, crying '*Coraggio!*' as if to show the travelled man, even in death.

Their ship was clawed out of the roaring straits only by the skill of Robert Bylot, who matched it with his courage till the end of their voyage homewards. But the ship fared worse and worse, whether becalmed, or thrust back by gales, or lost in a blind welter.

At the end they were starving, on seaweed and fowls' bones fried in tallow, while old Robert Juet—whom no one troubled to feed—perished of mere want in mid-Atlantic. On the ninetieth miserable day, their frayed and tattered sails brought them at last into Bantry Bay, where smiling Irish eyes gave them so warm a welcome that they must pawn their anchor chain, link by link, for water and a crust to take them onwards.

With their spectre of a ship come at last into the Pool of London, one year and five months from their departure, the brains of this infamy—Abacuck Prickett, Robert Bylot, and Edward Wilson the surgeon—marched up the hill to justify their murder before authority and, of course, before God.

Left alone to guard their dishonoured pest-house, Matthew Lawe had taken a last look down at Bennett Matheus, lying sick beside his cold hearth-stone, and slipped ashore, and melted into the crowd of the curious. He had Henry Hudson's pair of pistols, and a necklace carved from walrus tusks, and that was all to show, for a voyage of such hardship and shame that it deserved a knighthood—or a noose.

While Robert Bylot, too valued a man to suffer punishment for a murder not proved beyond doubt, lived on in profitable Crown employ, and the island named for him was now upon all the charts of the north-land, saluting his honourable death, Matthew had merely lived. He lived still; and thus, in the vast peep-show of the world, all the pictures changed, and one damned man did not.

Ben Pannikin returned with his laden water-bucket, and began to pass round a dole of water, first to Mate Matthew Lawe and then to other thirsty men gathered in the waist of the brigantine. It was a time for holiday. They were set fair on their voyage again, running north from the rock of doom whose nearest habitant land, far below the horizon, was the Dutch lair of Curaçao.

The *Cambridge*, a sturdy two-master with gun-ports bristling along her sides, was more serviceable than handsome; as buccaneers, her crew wasted no muscle on clean decks or polished brasswork—it was enough for them, and for their master Henry Morgan, that sails and rigging were sound, guns quickly served, and the rack of cutlasses at the foot of each mast gleaming with sharp steel.

If the decks had been scoured clean of blood, they would have felt far from home. . . . The burnished sun bore down on bellying sails and tarry cordage; the sky was blue as a bonnet; the fair breeze, which fat and prosperous merchants in the counting-

houses of Bristol or Cadiz or Amsterdam were beginning to call the
'Trade Winds', blew steady as the compass-point itself from the
north-east.

After a month of saucy poaching in Dutch waters, they had
taken treasure enough; and all would share in it, according to the
law of the Coast, except those five hapless cowards who had not
fought their best, and must therefore pay the forfeit. It was, or
should have been, a moment of ease.

Yet among the men gathered in the waist, there were more
grave faces than merry. Blood-letting was their common sport:
ships of trade, whether Dutchman or Spaniard or French moun-
seer, were their natural prey. But to condemn five of their own
shipmates to a miserable end was enough to give pause. It might
be a deserved death, however it befell; but the empty berths below,
the comrades missing, could not be forgotten between one watch
and the next.

Young Ben had said his soft-hearted say, and been rebuked for
it, and he spoke no more. It was left to men more seasoned in
cruelty to give their verdict, and among them a trusted friend of
Matthew's—in so far as a man had friends to be trusted in this
trade—laid down the law as he saw it.

He was John Flowerdewe, and one must beware of making a
jest of his name, since he was a small, broad, villainous fellow
who would fight any man with any choice of weapons—pistols,
cutlass, dirk, ox-whip, barrel-stave, or bare hands—for the price
of the next stoup of wine.

He was reputed once to have torn out a man's throat with his
teeth, in some tavern brawl when he had lost all other means of
persuasion; and whether true or false, the tale was enough to give
him a courteous hearing whene'er he spoke.

'Waste no tears on Parrish and the others,' he declared, when
someone seemed to mourn the men they had marooned. 'To
board a prize, and then to fall back, when a few fat Dutchmen
stand in your path—is that the way to grow rich? I would
have nailed them by the ears to the cross-tree, had I been
Morgan!'

There was no doubt among his hearers that Flowerdewe would
not have done less. But there was still space in their minds for
argument.

'There were fierce odds against,' said Berryman the cook, who was himself a better fighter than a cook—and that was no miracle, so the men who ate his slops were ready to swear. 'Those Dutchmen must have been ten to one.'

'They were landsmen,' Flowerdewe said scornfully. 'Or fair weather sailors. Show them the sharp end of a cutlass, with proper spirit, and they would have run from here to Amsterdam. We have fought better than ten to one, a hundred times in the past year. To send a long-boat from the *Cambridge*, and have it turned back? —there's a fine tale to tell in Port Royal!'

'Who will tell it?' another man asked.

'Not the men who showed their yellow breeches,' said Flowerdewe coarsely. 'They can tell it to each other, until their dying day, and no one to hear but the fish.'

'And all in the dark,' a voice said suddenly.

It was young Ben Pannikin, come to life at last. He was standing alone, looking down at his feet, and what could be seen of his face was dejected. They rounded on him mockingly, John Flowerdewe to the fore.

'What was that, lad? Who is in the dark?'

'The men marooned on the rock. It will be dark tonight.'

'Wonder of wonders!'

'I mean, it will be their first night there. It will be desperate lonely.'

'It will not be Port Royal on New Year's night, with the taverns bulging,' Berryman the cook agreed sarcastically. 'But it is their new home, and they must set it to rights. Tell us—do you think they have earned a palace?'

'Well——' Ben Pannikin began, in sad confusion.

' "Well" is right,' Matthew Lawe interrupted, coming to the rescue, though with no great comfort to help the boy. 'Parrish and the rest were put ashore by the captain, and when Morgan says it is right, so it is. If you plan differently, you must wait for twenty years, till you are the most famous captain in the brotherhood of the Coast. Then you may do it in your own fashion.'

The rebuke silenced all, as Matthew had intended. Ben Pannikin hung his head and said no more, while the brigantine rolled to the easterly swell, and the only sound was the creak of seasoned timbers and the wind sighing in the sails. Suddenly, from above, came a harsh dry cough which all of them knew. They turned, to find their captain leaning over the rail of the poop-deck, staring down with fierce eyes which seemed to nail each several man to the spot where he stood.

Choosing its own time, the Welsh voice which could lilt as well as snarl in the same moment commanded them:

'No long faces there! If you mourn, you are not too late to join the burial!'

He was jesting, in his own fashion, but his repute was enough to make them tremble. One could never tell, with Henry Morgan; when he was grave, he might be planning to give them all a holiday, with a handful of guineas to sweeten it; and when he smiled, he could only be considering the best sauce with which to eat their hearts. While they kept their prudent silence, their captain stepped down the short poop-ladder, and was among them.

His splendid appearance never ceased to awe his own crew; and what it did for strangers was enough to make them doubly proud. A lean black Welshman, all fire and sinew, he dressed in a fashion never seen on the Coast before—and only to be copied, by pirate or nobleman alike, at their immediate peril.

Some years earlier Morgan had chanced upon a portrait of his new monarch, Charles II, now restored to his rightful throne after the rule of the drab Puritans; and nothing would satisfy him save to adopt the likeness of his king. It might be said that Morgan outdid the royal pattern, by a long sea-mile.

His curled black wig, falling to his shoulders, had a gleaming magnificence: silk and lace and ribboned love-knots adorned his body from throat to foot: his breeches shone, his hose glistened in the dullest light, his shoes were afire with monstrous silver buckles. A golden cross, fit to daze a bishop, on a golden chain fit to lift the anchor, clung fondly to his breast. When he wore a hat, it was curved like a cathedral and feathered like a peacock.

He carried other ornaments which Charles his king need not affect: a pair of pistols at his belt, a stabbing knife tucked into his garter, and a cutlass in a jewelled scabbard. He also had a mind of such abominable cruelty that no lawful monarch would have dared to match it. Many pirates bore private descriptions, some of them shameful. The name of 'Morgan the Terrible' had been well earned, rarely challenged, and never lost.

No one knew his story until his story was fully fashioned, written in blood and printed on a hundred human skins. It was said that he had been kidnapped as a lad, and sent to sea; or sold as an apprentice cabin-boy to a Bristol merchant; or had run away from a girl with a full belly and a father with a loaded musket, and stowed himself in the first ship he could find.

It was certain that he had been put ashore, or shipwrecked, in the Barbados, and had suffered from some Spanish mis-treatment, and had sworn to match them both in seamanship and cruelty, as soon as he grew to manhood. It was beyond doubt that among the pirate captains of the coast, the boy from Glamorgan had risen like a flaming star and now, at thirty-five, outshone them all.

His hatred of the Spaniard, as hot as old Francis Drake's, was also beyond doubt. Matthew Lawe had heard him tell a story, which must have been burned into his spirit while it was still tender, of a girl whom he had seen tortured in foul fashion by her Spanish captors; and of certain shipmates having their hands, feet, noses, and ears cut off by these same villains, who then tied them to trees, smeared them with honey, and left them to the ants.

It had planted within his breast a lifelong lust for Spanish blood and treasure. He saw the Dons both as the public enemies of England, which he loved, and the private foes of his own advancement; the duty of a patriot thus became the pleasure of a man. If he had lived in the time of Sir Francis Drake, he might have been a royal hero; one hundred years after such robustious times, he was no more than an outlaw.

But even outlaws could become allies of the men of government, if there were pockets to be lined—as Henry Morgan was to prove before his murderous course was run.

A fine seaman and a ferocious blade, who would let no man stand in his way while trickery, treachery, a river of blood, false friendship, theft, or gross dishonour could serve to remove him, Morgan was now a very admiral among pirates. He had ten ships, and five hundred men, under his close command; in Port Royal, Jamaica, his 'fleet' maintained an armoured lair as strong as any in the western world. There he kept his vast treasure-house; and from thence he sailed out, as the mood took him, to increase a staggering wealth.

He treated prizes and prisoners like slaughter-house offal. He practised, year after year, one trick with a singular badge of infamy, which made his name reviled and his reputation feared far beyond the Devil's. To signal a victory, he would hoist to the yard-arm, as a battle trophy, a decapitated corpse dressed in a gory suit of muslin and silk ('Morgan's mummery', such fearful gear was called), with the severed head lashed to the feet.

But his own men, so long as they were loyal, he cherished, treating them as necessary friends. Every ounce of treasure which was captured, down to the meanest fistful of coin, was brought into the common pool, and there divided with great scruple, as if they all partook of some evil Sacrament. So much went to the captain, so much to his officers, so much to the surgeon's kist which might save their very lives, so much to each man on board; and so much to certain venal gentlemen ashore in Jamaica whose blind eyes were worth the blindfold.

To this holy communion Morgan added a system of insurance, so that any man wounded or maimed in a pirate venture would be recompensed. For a right arm lost, this most-respected fund paid 600 pieces of eight, or six slaves: for a left arm, 500 pieces or five slaves: for a right leg, the same: for a left leg, 400 pieces: for an eye, one hundred pieces, and for a finger, the same.

Answering a man who once asked Henry Morgan: 'What is my reward if I lose my member?', the captain replied most civilly: 'For that I will be generous. Five *female* slaves!' It was the kind of jest which, along with all the rest, bound him firmly to his men, in a manner not to be broken save by two mortal events: death, or cowardice.

As he stood among his crew, Henry Morgan's nimble mind was already busy on other matters, besides that morning's harsh work. He had all the spirit of a great commander, whether high officer-of-state or cruel rogue; and as he talked to his men, it shone out.

'Look you!' he began—and when Morgan said 'Look you', with that hint of Welsh song in his captain's voice, no listener allowed his wits to wander. 'We have done well on this our little voyage—not so well as I hoped, but even I cannot foresee that men sent out on a simple capture will start to bleed before they feel a wound. . . . No matter—the rest of us have earned our sweat, with three prizes taken, and ten chests of Dutch guilders, and some prisoners who might pay a ransom as fat as their paunches. But we will fish no more in these waters. There is too much stir now; and it is the Spaniard who is our enemy, not the Dutch—they are for sport. I have a better plan, the biggest plan of my life. So we go home to Port Royal.'

They raised a cheer at that. Port Royal was indeed home: it was the vision they carried with them on all their voyages, with its promise of drink and women and roistering song, in the town which they commanded, by right of conquest: where no one dared say them nay, for fear of a broken head or a spilled gut. . . . What Morgan's 'biggest plan' might be, they did not care. If it was big enough for him, then they were big enough for it.

'So, to Port Royal,' Henry Morgan went on. Then he turned swiftly to Matthew Lawe, with one of those darting questions which any man who served him closely—and was paid more for the privilege—must be able to answer, with an equal swiftness. 'Matthew—how far and how long, from sundown today?'

'Six hundred miles, less a score,' Matthew answered promptly. The great pool of the Caribbean Sea was firmly fixed in both their minds: they knew it as well as a farmer would know the exact limit of his pasture: the question was asked, and the answer given, for a spreading nosegay of reasons—because listening men must be taught, because Morgan must be seen to trust his mate, because Matthew Lawe must be seen to deserve it, because it was a fine warm day on a homeward voyage: for all the reasons, indeed, which made it more pleasurable for men to speak like angels than to grunt like apes. 'With this steady wind abeam,' Matthew went on, 'and the weather set fair, four days from now should see us berthed in Port Royal.'

'Make it so,' said Henry Morgan. 'And Port Royal will be welcome, even to your continent captain.' There was a laugh at this: Henry Morgan was not continent: he loved as he drank, as ready to top one shapely mould as to drain another. 'We lack good liquor, we lack the company of females—save for the tubs of swill-meat we have got by capture. And by God!' he said suddenly, coming nearer to the common mind, 'we would have more pleasure in selling these Dutch cows by the stone, than in——' he had no need to finish. Henry Morgan could catch a mood with one sniff of the ambient air; his crew had already surveyed their new cargo, and found them more fit for the kitchen scales than the cot. 'So—when we reach home we will play for three days, and then I will summon the fleet and make known my plan.'

John Flowerdewe, whose fierce repute gave him a certain licence, made bold to ask:

'After we play, captain—do we fight?'

'Until we drop,' Henry Morgan answered. 'Or till we all march ashore, each with a chest of gold, to set up as English gentlemen, with all the quality begging for our favours, and all the wenches as hot as monkeys. . . .' In a handful of words, he had drawn aside the curtain of each man's secret dream; and then, to finish the matter, he plucked two more notes which showed his particular spirit. One was of friendship, the other of fear.

'I tell you this now, privately, because you are my own crew—flag-ship's men! But if any one of you says a word of any great plan, before I do so myself, I will slit his tongue, Berryman will cook it, and *he* will eat it!'

Here and there, a licking of dry lips from within a dryer mouth showed that he had thrust home. Without soiling a stitch of his finery, Morgan would do all that he promised.

Then: 'Matthew!'

'Sir?'

'Rout out the blue foresail from Genoa. Let us voyage home in style—and swiftly.'

Four days later, to the very hour, the *Cambridge* berthed within the haven of Port Royal; and the cry of 'Morgan is returned' swept through all the four corners of this, the most wicked city in the world.

II

Cromwell, the late Lord Protector of England, whom Henry Morgan, as a king's man, despised, had taken Jamaica fifteen years before. It was a time when determined men, and kings, and lords-protector, took swift possession of vacant territory before any greedy neighbour might forestall them. The Spaniards were active in these waters, and the French, and the Dutch; who could challenge England's right to win a base of power, and put it to use? And who could deny it, when it was done?

Earlier, England had seized Tortuga from the French, and expelled its pirates, and made it a sort of colony; but Tortuga—which drew its name from the Spanish word for a turtle, since it looked like a monstrous sea-turtle floating on the surface of the waves—Tortuga proved too mean a conquest, and not enough to serve as a fulcrum for an ocean of commerce. So Cromwell, whose motives were to gain a trading interest and to overcome Popish idolatry at the same time, cast eyes upon Jamaica, and took it from the Spaniards with 8,000 men.

Jamaica was one hundred and fifty miles across, 7,000 feet high—and no floating turtle. Its fine harbours could serve many purposes. When the pirates expelled from Tortuga began to move in its direction, England installed a strong man as governor, to see all in proper form. He was Sir Thomas Modyford, a planter from the Barbados, and assumed in faraway England—since he had the style of Baronet—to be a reputable man for a worthy task.

His principal instructions, inscribed for him by earnest, un-worldly clerks in the shadows of Whitehall, were to keep good terms with the Spaniard, for reasons of high policy, and thus to suppress privateering and piracy, of which Spanish commerce was the main victim. But he found it prudent to undertake this policy 'by degrees', and then by degrees wonderfully slow, and then not at all.

For as the years passed, Governor Modyford found himself with some strange subjects in his island kingdom, and some stranger friends. Jamaica had become a very soil-pit for all the refuse of English prisons; it was more cheap to transport them away, once and for all, than to keep them in gaol upon the public purse. They could not all be hanged, and a man far away was a man out of mind, and harmless by four thousand miles.

Thus Jamaica, and Port Royal principally, was peopled by vagabonds, petty thieves, mutinous soldiers, Irish pirates, vagrants, lewd and dangerous persons, runaway apprentices, idlers as shy of work as a maid of a pair of dropped breeches: rebellious politickals, men in hopeless debt, men who must emigrate or lose their heads.

Here they found a life of ease, under a heavenly sun, and no laws which they did not choose for themselves; and for their privy needs, contractors supplied parcels of fifty women at a time—for marriage perhaps, for use certainly. Such was the breeding-ground of this vile city; and to it had lately been added more pirates driven out of Tortuga, who would not change their ways, and were intent only to teach them to others.

Sir Thomas Modyford did his best to suppress the Trade, with-out that excess of zeal which might have spurred a less indolent man. When such buccaneers were caught, he deprived them of their rudders and sails—thus establishing a market for rudders and sails, at second-hand, which curiously never lacked supplies. But even this faint-hearted show of government made him unpopular.

The citizens of Port Royal had firm affection for their pirates. The merchants grew fat by buying pirate plunder, and then recovering their outlay by pandering to their vices. The common people loved them as heroes, and for the sea-protection which such a forceful fleet gave to their island, at no increase to their own taxes. Shipwrights and rope-makers were in full employ, tavern land-lords could scarce keep pace with the custom, and harlots, when they slept at all, dreamed that the base metal of their aching rumps had turned, by lascivious alchemy, to gold—and woke to find it true.

Thus there was profit in it for all. There was even—and this came louder than a whisper—profit in it for Sir Thomas Mody-ford. . . . As privateering multiplied, and the pirates swarmed down upon Jamaica like locusts, from every corner of the sea-world, the Governor, standing now on one foot, now on the other, came to the resolve that what was best for so many must be best for all.

By the fifth year of his tenure, he was so far in bondage as to grant licences for the buccaneers to proceed against the com-merce of Spain, with a *per centum* profit for his public and private coffers. To protests from England, echoing the grief of the Spanish Ambassador, he replied with dispatches so long, so tortuous, and so disarming that by the time they were deciphered another season had passed, and his offence was half forgotten—or overtaken by another.

The city to which Henry Morgan now returned in his fine brigantine was a pirate empire, and nothing less; and it was Morgan's own stronghold, above all else. A most capable rogue, whose only tenderness was towards women and lap-dogs, whose fearful cruelty was a weapon of policy, whose hatred of the Spanish had blackened his blood, Henry Morgan had swiftly climbed to the very throne of power.

Now he ruled absolutely, while Governor Modyford nodded and took his honorarium. As the world of Spain fumed at these affronts, Port Royal in Jamaica became Port Pirate, where insolence of power, corruption of greed, and contempt for all save the strong, had turned a city into a shameless stew of blood, lust, and guzzling madness—a Vanity Fair in Hell.

Under the flaring torch set in a rusty gibbet above the sign of the Boucan Ear, on the seawall of Morgan's Harbour, the three men looked and lolled like expelled devils. Matthew Lawe, John Flowerdewe, and young Ben Pannikin had exhausted all that the flesh might aspire to, and the purse endure. Now they were three spent sailors, to whom the merciful fumes of wine still allowed an hour of that amiable glow preceding the fall of man.

Their clothing was stained, their hair like stacks of hay, their faces running with the sweat of a hot night. Matthew had lost the heel of his shoe, John Flowerdewe a trusted stabbing-knife (''Twas a present from my mother!' he roared, and they must all believe him), and Ben Pannikin his fringed leather purse—warranted to be full, known to be empty. They sat with their backs against the still-warm coral stone of the seawall, while the earth heaved beneath them and the stars whirled overhead. They were tipsy, and hoarse with singing, and slack-limbed like babes—and most happy.

Young Ben Pannikin, who had drunk the least and squandered his wits the most, was full of boasting. He had drunk ten flagons of ale! He had sat upright as their mainmast, while all around him men were snoring under the table! He had made a French pirate, a huge fellow with two pistols, back away and give him the wall! And as for girls—one after the other they had fallen back, snoring also, while he went on to the next! They had thought him a giant!

The other two said 'Aye' to Ben when the occasion demanded, and 'Is it so?' and 'I'll warrant you did!' They need not warrant anything, but a boy was a boy, and an allowance of faith between friends cost nothing.

Presently Ben, on whom the fresh air was making certain necessary improvements, took notice of the tavern sign across the cobbled street. Whatever wandering painter had designed the emblem of the Boucan Ear, he must have been a man of hot imagination—or an honest fellow who gave his money's worth, whether a king's ransom or the price of a pint.

The single monstrous ear, roasting on a spit, glowed as if in the fires of Hades: parts of it were burned black, parts still bled scarlet from their removal, parts were outlined by the licking flames. No human flesh could have surrendered more movingly.

'What does it mean, Matthew?' Ben Pannikin asked, after long surveyance.

'What does what?' Matthew answered. It was not the sharpest

question in the world, and he tried to improve it. 'What does what mean?'

'That burning thing,' Ben answered, and pointed. 'The tavern sign.'

''Tis an ear.'

'But why? What does it mean? And why are they roasting it?'

Matthew looked owlishly at Ben, thinking him to be jesting, but the lad was truly puzzled. He gathered his scattered wits, and put his own question:

'What are we, Ben?'

'You mean, our names?'

'No—our trade.'

'Why, pirates, of course!'

'True. But the sign does not say Pirate Ear. Think on.'

'Privateers,' Ben suggested, after long thought.

John Flowerdewe, whom they had thought asleep, gave a bellow of laughter. 'If your private had an ear, the girls would truly marvel!'

'It could listen for the babby,' said Matthew.

'God forbid!'

But Ben Pannikin was more adrift than ever. 'I cannot make head nor tail. . . .' He would be in tears the next moment. Matthew his mentor came to his aid again.

'Think of one more name for our trade.'

'Well, there is buccaneer . . .' Suddenly Ben's face lighted up, as bright as the torch overhead, and he pointed to the tavern sign in wonder. 'Buccaneer! Boucan Ear! 'Fore God, that is a neat jest!' Then his castle collapsed once more. 'But what is the sense in it, after all? What is boucan?'

John Flowerdewe joined in again. ''Twas a sort of oven, lad. For making smoked meat. I never saw it done, but it was the old trade of the natives hereabouts.'

'I watched them at it once,' Matthew said. 'In Hispaniola. Flesh-hunters, they called themselves. Filthy fellows in rags, sotted with blood and grease. I would as soon have put *them* on the roasting fire as the meat they sold. But now the name has fallen to us. Boucaniers—buccaneers. Is it clear, Ben?'

'But we do not roast meat.'

'Live long enough,' Flowerdewe growled, 'and you will see all happen on the coast.'

Matthew said nothing. Already he had witnessed Henry Morgan roasting such meat, once: the flesh of an old woman who would not tell her wealth, when they took Porto Bello on the coast of Panama. Morgan had set her bare upon a baking stove, and persisted in his cooking till she was past telling anything—and all they had from such foul kitchen work was the smell of it. But he spared Ben the story.

Instead, he rose clumsily to his feet, and stood rocking upon his heels like a ship-boat in a turbulent tide. Then he said: 'We lack spirit—and so early! Who will dance?

Since no pirate who would call himself a man ever sat when he could stand, nor kept still when he could cut a caper, the others joined him. They tried a hornpipe, but it proved too lively, and when it came to the moment when they must pantomime the hauling of a rope, their circle fell to pieces and they were more giddy goats than men.

They fared no better with a morris, which might indeed have been Moorish, like its ancestor, so outlandish was the measure. It was Ben Pannikin, tumbling about with laughter, who suddenly grew grave, and asked:

'Do you know the pavan?'

'Oh yes!' John Flowerdewe answered, sounding almost angry. 'I know the pavan as I know the palm of my hand. . . . What is it, for the blood of God?—the name of a girl, or something to eat?'

''Tis a Spanish dance, and slow—which will suit us better.'

'If the Dons can dance it, so can we!' Flowerdewe made a courtly bow which came near to splitting his breeches. 'Mistress Pannikin—pray lead us in the pavan!'

Ben, with much instruction, set them to the stately *pavana* of Spain, counting one-two-three-hup! like a Seville dancing-master whose mother had submitted briefly to the Cornish hug. Clod-hoppers three, they circled the rough stones of the quay, giving of their best, bowing like courtiers with the ague, curtsying like wenches with wooden legs. Passers-by paused to watch, and to laugh, and to persuade them to more extravagance.

When Flowerdewe called out 'Music! We must have music! 'Tis my turn to play the *señora*!' a beggar with a flute came out of the shadows of the Boucan Ear, and gave them a trill on his shabby pipe. But he was a beggar who knew the price of accomplishment.

'A guinea, masters,' he whined. 'A guinea for such slow, lordly music.'

Without a word, Matthew thrust a hand into his pocket, and threw the man a pair of gold pieces. Pirate spending, like pirate quarrelling, did not wait on argument. Picking them up with a grin, the beggar began to play 'My Lady Hawksbee's Round'— but slowly, as if my Lady Hawksbee's joints were responding neither to music nor to medical skill. Under the flaring torchlight, they took up their *pavana* again.

Matthew's two golden guineas, flung away carelessly for the sake of a measure on the flute, were among the last remaining. Each man had come ashore from the *Cambridge* with two hundred Spanish *reals*-of-eight, changed by the sharks to fifty pounds English. It had gone the way of all, in three days.

This wanton foolishness, and worse, was their common bond with Port Royal, where the lust for blood and treasure at sea was topped by the lust to toss away all that had been gained by their cruel valour, as swiftly as a man could curl his hands round the twin goods-of-sale, liquor and women. The younger men took the women first, like bulls in pasture, then soused themselves in fiery drink, then returned to the rut again; while the older lingered more at wine before spending, by necessity, longer time atop the girl.

But it was all one. Long years ago the Spaniards had set loose in Jamaica herds of pigs and horses, to ensure meat for their ships and draught-animals for their carriage. Now, grown wild, these coursed about the island, copulating with the brazen freedom of rabbits. The pirates took their sweet tune from this example; and if one of them could not wait, but must stand and couple on the street corner, few paid any heed—though a good friend might hold his cutlass while he set to with his dirk.

Thus, for a short storm of devilry, men spent in a few hours all the earnings of weeks or months of bloody work and sweat. For Matthew Lawe, it was a pattern of seagoing which he knew well enough; but the pirates' appetite for it stupefied by its folly.

Within this roaring swing-boat which never stopped, which could be boarded as soon as a man came back to Port Royal, by day or by night: within this wicked nest, men ripped and tore at their dream of ease, whether it was drink or girls, guzzling or gaming, buying trinkets for a whore or liquor for a tavernful—and shooting anyone who would not finish his flagon. They slept in foulness, woke in gross appetite, and turned to sport again until their last stiver was spent.

They were cheated vilely at every turn; and tipsy men could be robbed outright in the first shadows of an alley-way outside the tavern—or their breeches rifled as they took their breech-less pleasure upstairs. Yet it did not happen to excess. The pirates knew their own madness, and gloried in it; but they could exact a terrible revenge if they thought that the price had been hoist too high.

Bare-faced robbery was for pirates alone. Let greedy landsmen beware!

A luckless handful of such landsmen had been taught this lesson, in the week before the *Cambridge* had sailed for Dutch waters off Curaçao.

The Phoenix, a tavern on the edge of Spanish Town, where the sinuous Creole girls spread their wares as wide as St Peter's net, had long incurred wrathful muttering among the pirates. Not content with the roaring trade of every tavern in the town, the Phoenix had hatched its own greedy brood of vultures: to their lawful profit they added short measures of wine, drugged food to twist a man's gut, a scheme of cut-throat robbery on their own front doorstep, and gaming as crooked as a shepherd's staff.

When a victim protested, he might find himself flung outside, or knocked on the head, or worse, as soon as he could be lured into solitude. A young fellow from the brig *Clementine*, a favourite with his shipmates, had ended with his throat slit as deep as the gutter where he lay, because he had dared to question his score; and the foul slut who had brought him to this fate was heard to boast that she had earned a guinea for the task, without the trouble of lifting her skirts.

As if a Port Royal tavern did not make enough profit already! . . . It was held that the Phoenix had now gone beyond the mark. A society of revenge was swiftly formed, with a foolish title to suit its resolve—the Pirates' Benevolent Fire Company. Its benevolence came to a halt in the navel of its name.

On a night agreed, ten men each from five of Morgan's ships went into handy action. Armed guards emptied the Phoenix tavern of all save the two brothers who owned it, their servants and 'troopers', their treacherous relatives, and their known hangers-on.

Then the doors were locked and barred, and each window beset by sentries with muskets. Then, from doorstep to hanging eaves, the house of ill-fame was put to the torch.

Around this merry blaze a hundred pirates danced, while hideous screams supplied their music, and frenzied hammering on the doors and walls beat a funeral drum-roll. Wretches leaping from the upper rooms and the roof were shot in mid-air, or hacked as they ran. Then, when silence fell, and crackling flames took over the requiem, the pirates began to toss handfuls of guineas to the mob, in a golden shower which matched the fiery sparks of the blaze.

Each coin seemed to bear its own message, clear and loud: 'We are not bad men. We are simply just. *Remember this!*'

In the morning the Phoenix was gone, with not a singed feather to show. Its smouldering timbers were raked over, and the bones made up into fifty bundles, and buried in the common trench. Carrying their custom elsewhere, the pirates could warrant that it would be many moons before a bird of the same quality rose from the ashes to take its place.

Governor Thomas Modyford's answer to this outrage, as had been feared, was stern and swift:

'Each tavern shall keep two fire-buckets, filled with water, in all its public rooms. *Penalty, Five Shillings!*'

'By God!' said Henry Morgan, straight-faced: 'This fellow will *kill* us with his harshness!'

Now the *pavana* of Spain was beginning to falter, less through weariness than the tedium of repetition: when three untutored men trod the same measure for upwards of twenty times, the first wild spring in the blood must slacken. Presently the flute-player himself grew silent, between one trill and another; and when they looked for him, they saw only a glow of torchlight and an empty space. The crowd drifted away, leaving three dancers whose occupation was gone.

'Where is the damned villain?' John Flowerdewe demanded, with that swift progress from good humour to murderous spite which distinguished a pirate from a more sober citizen. 'We paid him gold! . . . By God's blood, if I catch him I will put his flute where it will play only bottom notes!'

He sounded as fierce as a man boarding a prize; but when his companions laughed, he forgot what it was that had angered him, and remembered only his jest. He repeated 'Bottom notes!' and slapped his thigh heartily. Then he said: 'What next? Another stoup of wine to pay our labours?'

The movement of the dance, and the cool night breeze which was the sea's gift to the burning land, had revived them all. But they were no longer free-spending men with the world their oyster.

'Aye, if we can raise the wind,' Matthew Lawe answered. 'But those two guineas were the last in my pocket. Who has gold?' He shrugged. 'Who has silver? Who has a string of savages' sea-shells?'

'Not I,' said Ben Pannikin. 'And to think I had twenty pounds in my purse when it was filched!'

Flowerdewe thrust his hand into his breeches pocket. 'I have a guinea,' he reported. 'But I was saving it for a new knife. . . . No matter! I will make do with tooth and nail until I win another.' He looked up at the broad front of the Boucan Ear, where the lower windows glowed cheerfully, and the babble of voices promised good fellowship within. 'And when that last guinea is spent, we will see what friends we have left in the world.'

They found friends as thick as fleas within the Boucan Ear, but scarce one who was richer than themselves: the men of the *Cambridge*, and of other ships which had lain in harbour for less than a week, were now enjoying as onlookers what they had once owned as proud proprietors. They sat slack-handed in a corner of the taproom, nursing the farewell inches of their once foaming pots of ale, while near the open hearth, under the best of the lantern light, the men still with money to spend lorded it among their fellows.

Though there were girls aplenty, they did not sit on poor men's knees, nor stroke a pair of breeches from which the golden thread had been drawn. Girls had sharp noses, as well as all the rest. The *Cambridge* crew were out of their turn already.

Thus it would be, until their ship sailed again, and came home again, and her men had gold pieces to buy tinsel favours; and thus did Matthew Lawe, and Ben Pannikin, and John Flowerdewe, and Berryman the cook, and Jackman and Pardew the waisters, and Bracegirdle the captain's coxswain, and Oliphant the renowned hand-strangler, and Brant the wicked main-yard marksman, and half a dozen other honest fellows—thus did they sit and stare, and forlornly wish for yesterday, or for tomorrow.

At this moment, they had something most teasing to stare at. Berryman nudged Matthew Lawe, and pointed:

''Fore God, they have got Hole Mary for their sport!'

'What it is to be rich!'

It was indeed only yesterday that they themselves had been feeding Hole Mary her chosen fodder. Now, like poor church mice, they must watch from afar, as she showed her skill for other lordly patrons.

Hole Mary—her name, blasphemous to the Spanish ear, had been Henry Morgan's own quip, when he heard of her tricks— Hole Mary was a high-yellow girl, a mulatto as big and strong and self-determined as that Spanish mule from which her breed-name was drawn. She was a working whore with one particular antic; and this had taken the public imagination, just as a singer might briefly catch the public ear for a profitable season, or a virulent plague could stock the graveyards.

She could pick up a gold piece—nothing less was acceptable— from a table, without recourse to hands, feet, fingers, toes, lips, or any other claw save the one remaining.

She was doing so now, under the flaring lamplight, to the joy of her proprietor and the lickerish delight of all others in the tap-room. Clad in her cover-slut, the rough apron which all the serving-girls wore, but with nothing else either to hinder or to cloak her talents, she advanced towards the corner-edge of a table, on which the gold piece had been placed.

She lifted her apron stylishly, and covered the coin, while her mouth split into a great grin of satisfaction which carried a message to all the watchers. There was a moment of waiting, while her clutching body worked its mysterious grasp. Then she called 'Hiyo!' on a note of triumph, while Snap! went that other generous mouth which only the favoured few could see.

Then she jumped back, and the cover-slut dropped again. But the coin had vanished, and all that was left was a prancing brown body, a sigh of wonder from the throng, and then a storm of cheers to salute her. Hole Mary, a private investor, had banked her wages.

The men of the *Cambridge* were affected in their various ways. For most of them the scene pricked their manhood to sharpness again. Others, dismissing it as naught but a trick, argued how it might be done, perhaps with a mousetrap curiously concealed; while others again mourned the loss of the gold, which should have assuaged their own manly thirst rather than Hole Mary's female famine.

Ben Pannikin muttered sleepily: 'I wish she would *lay* such eggs for us, rather than hatch them for herself.' Poverty had fallen like a blight on all their world. . . . It was John Flowerdewe who was the most determined to fight this vile enemy.

'By God, I would like to go in search of that gold piece!'

'It would cost you another to find it,' said Berryman the cook, 'and then you would lose both.'

'None the less . . .' John Flowerdewe looked round their circle. 'Come now—who has money?' he demanded.

They turned out their pockets, in accordance with the custom of shipmates in such straits, and made a mound of coins upon their table. It was a sorry collection, more fit for church at the Lenten season than for a night in the Boucan Ear. Matthew Lawe, their treasurer, made the tally.

'One guinea. One half guinea. Six silver pieces-of-eight. Some groats from God knows where. A Dutch guilder with a hole shot through it.' He counted on his fingers. 'Let us say, three pounds.'

'It would buy us a wench,' Flowerdewe said. 'And a fine strong one too.'

'But who would have her?' Oliphant the strangler asked suspiciously. He pointed. 'That is *my* guinea! Is it to buy pleasure for *you*?'

'It would buy pleasure for all of us, if properly bargained for.'

'What?—three pounds for eight men?' Coxswain Bracegirdle asked scornfully. 'We could not buy dog's meat at that price! And who would have the first bout, in any case? I'll not take your wet decks!'

'We could draw lots.'

'The girl would filch the money, and laugh at us.'

'We shall see.' Flowerdewe, who was most determined, hammered on the table with his empty ale-pot. 'Landlord! A word with you!'

John Flowerdewe had style, and a most fearsome air, but he made no progress with either. The landlord of the Boucan Ear, faced with a demand for one girl to serve them all, at the price of three pounds English, first dismissed it as a jest. Then he attempted reason and, when reason failed, gave them a warm denial. There was no such girl among his flock, he affirmed. He would not so insult their delicacy, even to ask. . . . For three pounds, *one* of their crew might have a girl, and welcome. Perhaps two. But eight of them. . . .

'We have brought you good custom,' Matthew Lawe reminded him.

'Aye, sir. And thank you. Come again!'

'But you refuse our money tonight?' Flowerdewe demanded grandly.

'No, sir.' The landlord wanted no quarrel with a man reputed to be a killing rogue. He wanted no Pirates' Benevolence to follow, either. 'You are welcome to spend it here.' He fingered his chin. 'But on wine or meat . . . For three pounds, shall we say, five bottles of the best Madeira? Or my own new rum if you choose.'

'It is a girl we choose.'

'Wine would suit your Honour better.'

'Who dares to say what would suit us better?'

Matthew intervened in this warfare. 'Eight bottles,' he said. 'One for each. Strong Madeira.'

'Do you want to ruin me?' the landlord asked.

'Yes!' Ben Pannikin shouted suddenly. He stood up, fell over, and collapsed in a sad heap under the table.

'Seven bottles,' said the landlord, adroitly capturing a moment of compromise.

'Well . . . Bring them.'

'You let him cheat us,' Flowerdewe growled, when the man was gone.

'Better a last good bottle of wine,' Matthew said, 'than a girl screaming murder before we are half done.'

'If it is not the best in the house . . .'

But the wine was good, and the row of seven shining bottles for seven men and one sleeping boy was enough to mend their humour and, when emptied, to send them on their way. Vowing friendship and a happy return, singing songs of love and murder, they came out again into the cool night air, and set course for the *Cambridge*. Being shipmates, they clung together, and when one fell he was set on his feet again, and when Ben Pannikin could go no further he was perched upon Flowerdewe's broad back, and carried onwards in triumph.

Boarding their ship at last, they had not one foolish penny left between them, and must live on small beer and harbour hard-tack until they sailed again. Their money, earned at such risk of blood and broken bones, their money which might have bought them a Devon cottage, or an honest fly-boat of their own, must now be won all over again.

It was a night in Port Royal.

III

Bracegirdle, the captain's coxswain, with a head as thick as a moon of green cheese, was waiting at the foot of the *Cambridge* gangway when Governor Thomas Modyford's emblazoned coach came into view at the further end of the quay. He turned, and called a warning to Matthew Lawe, and then they both waited as four ironshod wheels and four spanking horses set up a fine clatter on the cobbled stones of the seawall. Henry Morgan was returning to his command after a season ashore under vice-regal patronage, and did not care who knew it.

The coach came to a halt, with much shouting and whip-work from the liveried servants, much barking from spotted dogs, much leaping and jumping from half-naked children. The door was swung open, and Morgan descended, his bird-of-paradise finery setting up a blaze in the sunshine. Female cries of farewell, a glimpse of frothing silk, and the kissing of outstretched hands showed that he had been well convoyed. Then he turned to his brigantine, and became a captain again.

Matthew his mate saluted, and directed Ben Pannikin to take his baggage. Down in the cabin, the great man stretched out his long legs, and rested his head on the curved back of his chair. His olive face was haggard, and to Matthew Lawe he could admit the reason.

'Stay with me, Matthew,' he commanded, as soon as they were alone. 'I need you. But by God, I need a stoup of wine more! My throat is like a monkey's cage.' He took the flagon which Matthew had ready, and half-drained it in one great swallow. 'Well, that is one revival. I wish you could give me medicine for the rest.'

Matthew smiled. 'Time is the only cure for that.'

His captain nodded. 'It has not failed me yet. But there may come a day when I shall never overtake all the sleep I have lost—nor ever top up the well again.'

Henry Morgan made no secret of his amours, and his crew loved him for it. They knew that he could storm a ship one day, and a great lady the next, and it must be a race to see which hauled down her canvas the quicker . . . Morgan did not wench, but he loved widely, and often, with ferocious appetite. Half the noble husbands of Port Royal could never be free of gnawing doubts, and the other half, giving him best in a world of cuckoldry, could only fall into fits of coughing as they moved from one room to another in their own house.

It would not have surprised Matthew to learn that Henry Morgan had enjoyed a last brisk bout, in the swaying curtained coach between the Governor's mansion and the quayside. For him, ten o'clock in the morning was as good as any other hour, and ten minutes not too brief for a sinewy man and a pretty woman, both with the same wild intent.

Now Morgan finished his wine, at a slower rate, then waved away the offer of another flagon. Instead, he sat up abruptly.

'Well, enough of playing. And enough of wine, also. I have to talk some rare sense this morning. Is all arranged for the captains' meeting?'

'Aye, sir. I carried the messages myself, and waited for the answers. Captain Ladberry has a touch of the ague, but he swears he will come.'

'He can have the running pox, for all I care,' Morgan said harshly, 'so long as he does not give it to me—and sails his ship on time.'

Matthew asked, with hesitation: 'Is it the great plan, sir?'

'Aye.' Morgan rose suddenly, and stretched his lanky frame till his hands straddled the deck-beams above. 'You will hear soon enough. But in brief—I have done with the pirate game, and soon we go a-conquering.' He did not dwell on the foreboding word. 'Now—send Berryman with hot water, and I will make a clean face for the world. Then bring the captains down as soon as they come on board.'

'Aye, sir . . . Shall I stay for the meeting?'

'Yes.' Morgan smiled, the smile that his men welcomed and his women melted to see. 'I may need a friend, when they have swallowed all I have to tell them. But swallow it they will, every last man, if I have to ram it down with the gunner's rod!'

He was, indeed, a captain again.

They came on board in ones and twos: fierce seafarers, swaggering bullies, mincing fellows with pointed shoes: one-armed men, dirty men, men with hacked faces, men who glared like tigers or smiled like dissembling villains. Henry Morgan's captains, who had achieved their rank by what the Devil himself would be proud to call merit, were of many breeds. But one thread of quality bound them all.

There was not one of them who would not kill instantly for profit or sport, nor one who would not betray his mother's other sons, or the mother herself, for the same careless lust.

They had one other bond. They acknowledged Henry Morgan as their admiral, and trusted his proven skill, and feared his bloody spirit, and would never dare to challenge him—unless someone else's coward steel had already set him bleeding. Until that fortunate moment, they fought to keep their own position, in a trade where death by capture, treachery, wounds, or disease was a night-and-day companion.

When they heard why they had been summoned, and their admiral's plans, this companion seemed to draw very close to every listening man.

It needed only a few forceful words from Henry Morgan to turn their pirates' world upside down. Sitting at ease at the head of his table, with his captains ranged like attentive scholars on either side, he proposed an enterprise as far removed from their ambitions as a long-boat from the flag-ship of England. His plan, he told them, was to assemble an armed fleet—*their* fleet—and sail southwards: to invade the Isthmus of Darien, and to capture Panama City from the Spaniards.

From thence they would burst out into the Pacific, and ravage Spanish trade at their own sweet will.

Scarcely had Morgan given them time for their first sip of this choking brew before he was lashing himself into a fury of persuasion.

'What we do now is too easy,' he told them. 'To be sure, we have fierce fights, and take a ship here and there, and sometimes make a rich prize, and live on it till we must sail again. But that is children's play, compared with what we might do with my great plan. If we take Panama—*and I will take it!*—we strike at the very soul of Spain in these waters.' He thumped his fist on the chart which crowned the table before him. 'In the Pacific lies the rich heart of Spanish trade—the soft rich heart. There are mines of gold, mines of silver, jewelled plate from all their whoring cathedrals: cattle, hides, horses, slaves, precious stones we can scarcely guess at. That is our prize, and we can hold it forever, once we seize it. Look at the map! Darien is only a little neck of land, with Panama its single city, like a boil, like a pustule on an old man's skin. We will strangle that neck, lance that boil, and then do what we will with the corpse.' He paused for a deep breath. 'That is my will, captains all, and that is what our fleet shall do, before two months are out!'

Silence fell, while he looked round them, his eyes narrowed to fearsome slits, like wicked gun-ports which masked a flaming villainy within. It was clear to Morgan that he had astonished them, with something not to the taste of all. It was equally clear to his listeners that Captain Morgan was doing more than show them a plan. He was also giving them orders.

It would take a brave man to be contrary, and to give voice to it. But bravery, in its own fashion, sat all round their cabin table; and when Morgan said 'I will hear what you think,' he was answered without delay.

The first to speak was Captain Ladberry, the oldest of their company, whom Matthew Lawe had reported as sick with ague. It might well be so, if his yellow face and trembling veined fingers were true witnesses. But his spirit did not tremble, nor was yellow the colour of his heart, when he came to challenge Henry Morgan.

''Tis a bold plan,' he said, in his meagre voice—and contrived to make the word 'bold' sound more suited to some silly ribboned wench than to an assault on Panama. 'But whether it is the plan for *us*——' and he cocked a meaning eye round the circle of his companions, 'is another matter, hardly touched upon. We are pirates, and we do very well in these waters. We are conquerors of ships, not of lands. Myself, I make bold to say that I do not want Panama, nor Darien, nor one foot of that foul Mosquito Coast. We will all die soon enough, without insects to suck the last of our blood! Let the Spaniard keep Panama, I say. In the Brotherhood of *this* coast we have all the enemies we need, at our proper place of work.'

Against Morgan's 'bold plan' Ladberry had voiced some bold words indeed, even from a senior captain. But, like many a thin man before him, Ladberry had made more enemies than friends on his way through the world, and he was not the sort of advocate to sway their company. One of the other captains murmured: 'Of a certainty we are pirates! We do not need to be told such great truths. If there is fresh sport to be had in the Pacific, let us go there, and be damned to the Spaniard!'

Morgan, who had been frowning as Ladberry spoke, changed to a smile at the cock-a-doodle words. But his brow lowered again as another captain, Prince of the brigantine *Charles*, also voiced his doubts.

'There is more than *fresh sport* in this plan,' Captain Prince began tartly. 'We are concerned with men taking cities and oceans by storm—or should be. Sport is for children.'

'I am no child!' the captain thus sneered at, who was indeed the youngest, spat back.

'Then talk not of the Pacific as if we were to sail toy ships on a pond.'

Little love was lost between any of Morgan's fleet-captains. Old Ladberry thought that, for long service alone, he should have had Morgan's place as admiral; while Prince held Ladberry to be a miserable old fool not to be trusted with a skiff in a mill-stream. But they were both ready to round on Captain Harrison, the man who had spoken of 'fresh sport', as a drunken young braggart brave as a lion in his cups, hollow as a pot when put to the test. It was Henry Morgan's genius that he could rule such a nest of eagles, tomtits, and cuckoos.

'So the plan is to take Panama City,' Prince went on, 'then to open up another ocean, then to ravage it. What do we want with such schemes? We have our own ocean already, and still we do not command it as we should.' He glanced slyly towards Morgan. 'I know even of certain Dutchmen who have not yet heard of our rule. Why should we run our ships into fresh danger, turn our sailors into siege-men, marching-men, garrison troops a thousand miles from home? They say that a cobbler should stick to his last. I never heard that a pirate should turn foot-soldier!'

Morgan, who had not the greatest store of patience in the world, was stung by the hint of the escaped Dutch pinnace to break his silence:

'If you are afraid, then we will count the *Charles* as one ship lost to the fleet, and Captain Prince as not worth the half of his name!'

'I am not afraid.' Prince knew the value of a cool temper, when others were ready to burst into flame. 'I will sail for Panama if that is the plan we all agree. I will sail to hell, if you can prove the profit in it. What I will *not* do, is hazard my ship for some wild scheme, when we are told to say "Yes" like a row of puppet-dolls without a voice or a thought of their own.'

Young Captain Harrison answered fiercely: 'If we have any fire in our belly, we can say yes without counting the cost.'

Captain Prince had had enough of Harrison. 'Many a maid has listened to such bold pleading, and finished with more than fire in her belly!'

The rough jest set them laughing, and proved a salve for the spite which was marring their talk. Though Henry Morgan still frowned at the affront to his command, there were others who wished to have more reason and less rancour in their meeting. Any flea could bite; any man could scratch when so pricked. But they were talking, or should be talking, of the largest undertaking of their lives, and the most perilous. It was worth more than ale-house ranting.

One such temperate man who thought in this fashion was Davy Morris, a fellow-countryman of Morgan's who could not be accused either of cowardice or lack of loyalty. Captain Morris, whose given name of Davy was softened to Taffy—though not in his own hearing—did not wish to confound nor confront anyone. He wished to know.

'When we invade Panama,' he began—and he was the first to say 'when', rather than 'if'—'how do we take their city? 'Tis the Spanish capital thereabouts, and must be well fortified. They say there is a president of the province, and a president has troops at his command. How many troops? How do we find a way past them? How do we scale the walls? Do we bring up cannon from our ships? And how do we live while we march? Do you plan a mule-train of supplies?' Taffy Morris smiled suddenly. 'Captain Ladberry was quick to say that we are naught but sailors—and thank God that is true. I say, thank God for a Navy! Buccaneers can do anything! But how do we set about it?'

Captain Morgan stirred as he heard the last words, and seemed ready to answer them. Matthew Lawe, watching him closely, knew the signs, and wondered which way his captain's spirit would incline. He was balanced between a mood of the most evil intent, and the wish to control his flock by something more generous: though his curled fingers stroked the frothing ruffles at his neck

as if these might mask the throat of an enemy, they were not yet
forming an executioner's hand, any more than a lover's.

There was no one in the hot, airless cabin of the *Cambridge* who
had not seen, or heard of, this famous divide, when a baking-batch
of Morgan's prisoners might be fed to the sharks, or strangled at the
yard-arm, or spared for ransom, or even freed with laughter and
good humour—all depending on an instant's whim.

The dice fell, and Captain Morgan spoke.

'It is good to hear some spirit at last,' he said in his lilting voice,
'after all this miserable moil . . . As to the planning and the invasion and the march—less than forty miles, a little walk in the sun
—you may leave that to me. I purpose to gather more ships than
ever sailed from Port Royal—perhaps forty of them, perhaps two
thousand men . . . You are no more than the cream atop this
churn! Remember that, when you moan of dangers and hazards . . .
As to taking Panama City, we will not cannonade it, nor storm it
directly. We will use a few stratagems instead.' He looked about
him, and for a wonder his eyes settled on Matthew Lawe. 'Remember the one we used at Porto Bello, Matthew? It is not yet stale, is
it? We might use it again.'

'Aye, sir. It was a marvellous trick.'

It had been a stroke of infamy; and though Matthew Lawe laughed
with the rest at the very name of Porto Bello, he was not proud to
do so. It revived in him all the old terrors of what could happen
when a commander led his men too far, or stepped outside one
known world to venture into another; it sharpened all the new
threats now promised and opened to their gaze. He shared every
shred of fear, and none of the confidence, which the last half-hour
had brought to their attention.

Henry Morgan was a true buccaneer, a sea-animal which swam
alongside its prey and grappled it to death. Now he wished to be
a land-animal instead, a slow-moving plodding invader with
swamp-mud sucking at its thighs. But they were all buccaneers!
Why should they change, and give ten thousand extra chances to
Death?

The 'trick at Porto Bello' concerned the storming and sack of
this small town on the inner coast of Darien, not fifty miles from
where old Drake had taken his last long dive. It was more than
two years ago that Morgan had led nine ships and 400 men to

capture this stronghold of Spain. Though they only crossed a little bite of land, and their fleet was kept handy for their escape, yet a horde of dangers lay in wait for them, as thick as the snakes that hissed at their heels and the Spaniards who, faced with doom, proved themselves tigers of desperation.

As it fell out, it needed a terror so monstrous and a trick so vile that it came to be talked of as 'Morgan's Stations of the Cross'. But it also brought a victory complete, which might well have given him his present 'great plan' for Panama.

On their march from the beach to the town, they used the most rapacious cruelty to extract from any villager who lay in their path the last *centavo* of his money and the last syllable of information which might serve them for the assault. They burned men alive—Spanish cooking, they called it, as if to honour a custom of the country: they cut them in pieces limb by limb and slice by slice: they twisted cords round a man's head, and tightened it with a stick, till his eye-balls shot out—and sometimes his store of gold.

They set this brand of conquest upon a thousand such wretches, to teach them the last lesson of their lives: that Henry Morgan was not to be resisted—not by a word, not by a look, not by a mumble from a bloodied tongue.

So far, his butchery had been formal and systematic. When it came to the main assault on the town, which was well fortified behind its high walls, a horrid cunning was added.

First they captured an outlying fort, promised clemency to all who surrendered, then locked the remnants of a brave garrison within their dungeon, set charges of gun-powder, and blew the whole castle into the air. Then, faced with the inner citadel which was defended by a gallant commander, they sought a way easier than a direct attack, and solved the problem with a jest—one of the grimmest of Morgan's life.

Against the Papal whoredoms of Spain, said Henry Morgan, what better weapons to use than their own idolators? It was Matthew Lawe himself, at the head of a ferocious troop of pirates, who hunted out a great throng of priests and nuns—'Of a number,' said Morgan, 'too scandalous to be endured'—and formed them into the shambling moaning spearhead of attack.

Under pain of instant death, these doomed religious—whether high or low, vowed man or chaste maid, acolyte or Child of

Mary—were forced to carry scaling-ladders ahead of the assault, and to set them against the town walls. While the appalled garrison hesitated, and thus allowed this first breach, the shrieking pilgrimage, in an extremity of fear, were prodded heaven-wards as a screen, and—the defenders having made their cruel choice—were the first to be killed.

The pirates then followed, using this grisly jumping-board for their upward leap, and took the city for themselves.

They stayed to debauch right heartily, to squeeze from their customary vise every speck of precious metal, every gleam of jewellery, and every eatable limb of cattle: and to demand of the President of Panama a ransom of 100,000 pieces-of-eight: 'Failing which,' said Morgan the persuasive, 'we will take it in blood, at a cheaper reckoning.'

The vast tribute was paid, and the fleet then retired, for a division of the spoil in a quiet corner of Cuba. It was quiet because they took care to make it so: their plunder, in ready money alone, was 250,000 pieces-of-eight, and was best concealed from envious Spanish eyes if such eyes, by one means or another, could be removed. They then returned to Port Royal, and told a modest story, and spent their gains less modestly.

In contrast with the royal fury of the Spaniards, and some stern questioning from the government of England, Sir Thomas Modyford's account—and accounting—were models of restraint. He spoke of a skirmish scarcely to be reckoned as an assault, in which the profit to each man was scarcely more than £60. His own share —to be spent on improvements to the fishing fleet—was scarcely above £20.

'My next child,' Henry Morgan was heard to say, 'shall be christened Scarcely—and I have in mind its godfather!'

Matthew awoke from a remembrance of shame and terror—so much might have gone awry: a fighting man might have been killed by a ball intended for a nun!—to find their merry captain transported to a different mood. The to-and-fro of their discussion was still raging, without point or profit, and Henry Morgan had heard enough of it. Suddenly, his bowels turned sour, and sick of what seemed a faint-hearted crew, and now he changed his tune to suit his own fierce purposes.

'Enough!' he said, and the word was harshly barked. 'Am I to listen all day to tales of this and that, and children's prattle of who is a sailor and who a landsman? *You are all what I tell you to be*—and the first to forget that will be the first to learn the lesson, if I have to hack off his legs to make him a fish!' He waited for this shot to make its fire felt, and then continued: 'We are sailing as a fleet, when I give the command. We will sail, and we will march, and we will storm the city, and we will sail back again. Porto Bello proved that we can do it. Now there is only one watchword for this company—on to Panama!'

But he had interrupted old Captain Ladberry in the full flight of further discourse, and Ladberry took it ill. When Morgan sounded his clarion call of 'On to Panama!', the old man blinked his rheumy eyes, and frowned as if some turbulent child had disturbed a grown-up world. Then he said:

'We will not get to Panama by calling out its name, however loudly. I say there is too much risk. And what would Governor Modyford think of this? It is an act of war against Spain! Can he give it countenance? He would be mad to do so.'

'Tom Modyford thinks——' Morgan began, with a brow like thunder, and then paused. He had been about to say, 'Modyford thinks what I think' but it would not have been politic: the Governor's pretence of royal command, which was the shield of them all, must be preserved with every solemn deference. Instead he said: 'Sir Thomas Modyford knows my plan, and he approves it. We spoke of it only this morning, and his last words were, "The sooner the better".' So far Morgan had controlled his wrath, but now he broke out again, and every word was venom. 'By Christ's blood, I wish my captains had a spirit to match his! The Governor did not speak nicely of chances and dangers. He did not quake and tremble. He listened, he approved, and then he said, as I do, "On to Panama!"'

'He will not be there,' Captain Ladberry said.

Morgan had had his fill. '*You* will not be there,' he thundered, 'if all I take with me are men I can trust! By God, you make me puke with your snivelling! Let others come with me to storm Panama—and if any here present has your hesitation, let him remove his runny backside from honest men's noses, and come back when it is dry. And if you *all* have this disease in your wormy gut, I will take Panama myself!'

His foul words and murderous glances seemed to include every one of them, whether those who had supported him or the laggards who had not. In an affronted silence, Morgan poured himself a goblet of wine, and quaffed it noisily, as if to say: 'Let a good taste kill a pismire brew,' and then his captains filed out of the cabin, leaving him to a lonely field in which the sulphur of hatred and contempt could almost be smelled.

Like a host relieved of wearisome guests, Henry Morgan sat back in his chair, slack-handed, and drank the wine which Matthew poured for him, a sip at a time. But now he was smiling, as if the meeting of commanders had not been so contrary to his will after all; and not for the first time Matthew wondered at his captain's swift changes of mood. He wondered the more when Morgan said, without rancour:

'What a clutch of chicken-hearts! It was a pleasure to put the fox among them.'

Had it all been play-acting, the anger and the thunderous brow? Was it part of some trial, to discover friend from foe, firm ally from doubtful warrior? Was the whole meeting a puppet-show? Was Panama itself a jest?

He was soon to know. Made bold by Morgan's good humour, Matthew said:

'Not all were chicken-hearts. Captain Harrison spoke up manfully for you. Captain Prince only wished to know more of the plan, before he pledged his ship to it. Captain Ladberry——'

Henry Morgan, all innocence, interrupted him. 'How would you have spoken, Matthew?'

'Sir, it cannot be my place——'

'But *if* it had been your place. If you were one of the fleet-captains, instead of my mate, what voice would I have heard from you?'

Matthew Lawe saw no pitfall in this: Morgan's question had been reasonable enough, and his face continued friendly. He spoke his secret mind:

'I would have thought the plan for Panama full of risk. I would have made bold to say so. When Captain Morris asked, what are the dangers, what are the odds against us, have they been weighed? —that was my thought too. Captain Prince said that we were pirates, not foot-soldiers——'

'*So!*' Morgan roared out suddenly, his look as black as if the sea had doused the sun. 'You would have been one of those traitors?'

Matthew faltered before the appalling charge, and the voice which had taken on winter's most bitter chill. 'I——' he began, and then swiftly retreated from the world of command. 'Sir, they only wished to know, and to advise you. It was their duty to speak out.'

'It was not their duty!' Morgan shouted. His face was contorted, as if stricken by a mortal fit. 'Their duty is to obey me—to listen, and to obey. I made them what they are, and by God I will break them again, if their hearts have grown so rotten-ripe! Their *wish* is to sit in the sun of Port Royal, or to make a few easy voyages, or to turn tail when I tell them to be bolder.' He turned his murderous eyes upon his mate. 'Is that your wish too?'

Matthew could make no answer, in what had begun as friendly intercourse and was now a deadly trap. He longed to say Yes!, and have done with it. He longed to say, the plan is foolish, it will need an army, and we are not an army, we are simple pirates. He longed to say, I am happy as I am, it is for you to take the *Cambridge* and storm Panama, let me stay in Port Royal and keep a faithful eye on your affairs. He longed——

Did Morgan read his wavering mind? It might be so, since the captain now glaring at him as if, boring fiercely in, he could wrench out a handful of his thoughts like little poisonous fish from a net: as if, in a moment of ease, he had sighted a toad within his drinking-cup, and was ready to kill even as he retched . . . It was not fair to honest men . . . Humbled and afraid, Matthew said at last:

'Sir, I only spoke what was in my mind, as you asked me.'

'Aye. You *spoke*, and I found a custard-apple instead of a man.' There was nothing in his voice save viperish contempt: this was Morgan's famous divide again, and it had divided Matthew from his fellow-sailors, like a pox-mark on an innocent skin. 'Hark ye, my trusted mate . . . If these are your true thoughts, I do not want such men about me . . . If there is a sickness in you, take it out of my sight, burn it in the plague-pit, do not infect the air a moment longer. *Do not even breathe it* . . . Now quit this cabin. And send Bracegirdle to me. I need to give some orders which will not be questioned.'

Thus, from a moment of foolish confidence, of weakness, and an instant's blind fury from a man with a 'great plan'—and the Devil's own itch in his blood to perform it—disaster had been born, and flamed like the phoenix. Matthew Lawe, saluting, then running for shelter before this wild Carib hurricano, knew in his sick heart that Captain Henry Morgan might never trust him again.

IV

A new master was, for a runaway slave, a matter of good or evil fortune; the wretch might find a warm hearth and kindly usage, or he might end with worse stripes on his back than those he had fled. A new master for a runaway mate stood in the same realm of chance, though with less risk of the bondage of Egypt, the whips and scorpions of the Great Pharaoh. Yet work the mate must, and a ship he must find, and make his choice between the Devil and the blue sea.

Matthew Lawe, before a month was out, found that he had condemned himself to both.

With money stolen from Morgan's own sea-chest he had fled to Tortuga, the old pirate stronghold close to the coast of San Domingo. On board the *Cambridge* he had grown fearful, despairing of any secure future in the brigantine, where he was out of favour, and could only be carried off to Panama, to find death in that wild adventure.

So he had run away, like any tormented serf, in search of better things. At least he would be a pirate still, rather than a conqueror in the mad mould of Alexander.

His new ship was a French *patache*: small, strongly armed, and fast as the wind itself, though outlandishly rigged in the Arab style with two great latine sails. His new captain was Simon Montbarre, of whom it was enough to say that he also carried the fearsome title, The Exterminator.

Morgan the cunning Welshman had used terror for a purpose: to overawe by fear, so that his name alone was a weapon as strong as his arms. Simon Montbarre was a monster who used it for pleasure. He had become a byword even among the French Brotherhood, the refiners of cruelty.

A swarthy Gascon, said to be the younger son of a nobleman—who might have been well pleased to see his disappearing back—he was a square ape-like fellow, mustachio'd like a mountebank, incredibly strong in arm and shoulder and thigh. He boasted that there was no work to be done by any two of his crew, in lifting or pulling, climbing or cleaving, which he could not do for himself in a lesser time.

He had other boasts. He took no male prisoners—'Ransom is for clerks,' he would say, in his throaty English which was a *patois* of its own. 'I do not sell blood, I drink it.' He would never kill quickly, save in the heat of battle; but afterwards, a man who fell into his hands must die a slow death—'It gives pleasure to both of us—myself for the sport, and he for one more long day of life.'

Particularly, he allowed no man of his crew to touch any of the captured women until he had made his own choice; then, she was his plaything until he was sated, then raddled meat for all his crew, then a useless mouth fit only for the fish.

He chose, if he could, to take a woman by force. 'True love is not of the heart,' he would say. 'That is for poets, courtiers, men with tender souls. I have no soul that I know of. My love is the great sprit which stands between my legs, and when it stands between hers, it must come as a robber, not a beggar. What pleasure is it to slip through an open door? Only when she screams, and claws, and fights, and yields her doorway an inch at a time, does she have a treasure worth the taking!'

With such words as these, Matthew sometimes thought, Mont-barre showed himself a poet in his own right—a poet of terror and torment, of the dung-hill, the soilure, the mire of mankind. But at least he aimed no higher than a pirate.

He was a simple ravening vulture, without a thought of great conquest nor a shred of soul; and it was enough for him to make of his *patache* a bird of prey, a vulture like himself. This he proved, to the full foulness of its hilt, on the voyage east and south from Tortuga.

It was Catouche, the tall mulatto from Basse Terre in Guadeloupe, who was the first to spy their quarry. His eye had caught a flicker of white on the eastern horizon, a sudden blanch which, on this fair morning, could not be a breaking wave-top; and without command from Montbarre he leaped to his station.

Lithe as a tumbling-boy, agile as a monkey, Catouche raced up the towering yard of the main latine, from its anchorage on deck to its lofty peak over the stern-post: darting upwards on claw-like hands and feet, like a great yellow cat up a coconut palm.

It was something which could always make the crew marvel,

and which in turn delighted the acrobat. One moment, Catouche
had been on deck, and at the next he was fifty feet above their
heads, shading his eyes against the low morning sun. Then his
own head appeared, poking out from the very tip of the main-yard
like a pale grape on a willow-wand, and he called out in grinning
triumph:

'*Une voile!*'

A sail . . . The cry was repeated below him, as crew-men
crowded at the rail of the *patache* to follow his pointing arm. This
was the first ship they had sighted for seven long days, and they
hailed it as thirsty men might salute a cluster of green trees in a
desert.

On the voyage thus far from Tortuga, threading its way past
Porto Rico and then across the bow-string of the Leeward Isles
towards Martinique, game had proved as scarce as flying-fish in
wintertime. They had closed one treasure-galleass of Spain, off
San Domingo, but she bore too fierce a look for a *patache*—even a
patache manned by twelve determined murderers—and they had let
her by with a courteous wave.

South of Porto Rico, they had taken a coasting caravel—dreaming of seaweed and waking to find shark's teeth—and slit a few
trembling throats, and secured some wine and hog-meat, and spare
cordage which was always of use, and a meagre chest of copper
coins, and a whey-faced drab who had poxed two men and then
been thrown overboard.

But that was all. 'We would do better growing cabbage-stalks,'
growled Simon Montbarre, and set his beaked prow south-east, out
of soundings, towards the Leewards.

The *patache*, on a fair wind, could make more than two hundred
miles from one noon to another, and go where any promise
beckoned. There was hope always—and there was hope now, as
Montbarre laid off a course for the distant glimpse of topsails,
and set Matthew Lawe, who had humbly proved himself, at the
helm, while their craft clawed eastwards across a sparkling sea
towards her likely prey.

With the promise of blood, the *patache* came alive like a fire under
the bellows. The crew of twelve ran to their places without
prompting: two men to the bow-culverin, two more to the brass
cannon amidships which fired curling whips of chain-shot, enough
to shred a sail or to make two half-men out of one hale human: the

rest to their boarding-stations, armed with weapons of which a saw-toothed bill-hook was the least fearsome.

High above their heads, Catouche nestled into his lair on the latine yard, nursing a musket and a pair of pistols: down in the waist, Bonhomme the boatswain laid out his hooked grappling irons and the snaking rope which would secure them. Matthew kept the wheel as they ran on down; at his side, Simon Montbarre stood like an avenging devil, cutlass in one hand, short-handled axe honed to a carving edge in the other, waiting for the moment when he would lead the attack.

He was smiling, and the bared yellow teeth were like wolfish fangs.

Quickly the ship which might be their prize took shape and substance. First she was a distant topsail, then two; then she was a brig, northward bound, handsomely rigged, not too mighty for their comfort nor too poor for profit. Then she was an English brig, sailing serenely (as they guessed) between Martinique and Santa Cruz or the Virgins, deep-laden, not warlike, perhaps not too watchful of any other ship under her lee. Then she was a close adversary, and all came, like a flaming quarrel, to action.

With only a short mile between them, Montbarre tapped Matthew on the shoulder, and made a sweeping movement with his cutlass-arm, from right to left in a curve which came to a darting peak. The order was clear. Matthew was to bring their ship round, close under the stern of the brig, and lay her alongside as the wind hauled ahead of them.

The *patache* had done it before, and she did it now, with a sharp skill which earned a grunt of approval from Montbarre. But all did not follow their secret plan. The brig was alert, for all her torpid burgher's air; and as the *patache* sped across her wake in a cutting sweep, two quick cannon-shots boomed out. The brig had two stern-chasers, and they must have been manned all the time.

The first ball, aimed high, struck their foremast where the latine crossed its peak, and brought the foresail down in ruins; the second crashed into their hull, opening a splintered gash above the water-line.

But that was all that the brig could manage, before the *patache* fell upon her, and bit deep.

Their own bow-culverin roared out as they came up into the wind, and scored a hit on the rudder-post, throwing half a dozen bodies into the water as the helm lashed across the deck. Then, while the two ships closed, the chain-shot cannon flung its wicked scythe among the crew manning the brig's rail. The water began to boil between them as the two hulls came together. Judging the moment, the boatswain whirled the first grappling hook twice round his head and launched it at the tall bulwarks.

Its iron claws caught, first the neck of a screaming man, and then, lugging this gory cushion with it, the inner side of the brig's rail. The ships crashed together, and Simon Montbarre, with a wild shout, led his men in a bloody river which surged over the enemy side and spewed out on deck.

By the time Matthew had left his idle helm and clambered upwards, the last of the boarding-party, Catouche from his lair above had already wounded one of the gunner's crew, shot a sailor —a nesting marksman like himself—down from the main-yard of the other ship, and killed another man on the quarter-deck whose dress proclaimed him captain.

It was a prosperous start to some bloody work on the body of the victim.

The men of the *patache* customarily screamed and shouted with the utmost fury when they went into battle; Montbarre had a brazen roar, enough to topple a mountain, and Boatswain Bonhomme a piercing yell as if he were out of his mind with rage at a herd of cattle.

Others spat out vile curses, or sang the thudding chorus of a wood-cutter's song; one of them, O'Hare, a mad Irish rogue whose chosen weapon was a sort of Indian tomahawk with a handle of black bog-oak, emitted his tribal shriek as if to summon his blood-brothers from across the whole Atlantic.

This obscene choir was already in full throat as Matthew jumped down upon the deck, his bare feet bloodied to the ankles, and fell to his trade with a will.

It was an honest cutlass for him, and he went hacking and thrusting towards the brig's main-mast, where Montbarre and the rest had already cornered most of the English crew. There were some twenty of these still standing, though robbed of management by the death of their captain: broad-faced, broad-shouldered sailors in sea-rig of blue and white, a pair of officers in scarlet military tunics, whose pistols, emptied once, were now reversed and doing duty as clubs, and certain more sober gentlemen in grey, bunched fearfully together as if they wished they had never left the counting-house.

It was at this moment that Montbarre undoubtedly saved Matthew Lawe's life. Matthew had just struck down one of the brig's crew, a smaller man who had slipped on the drenched deck and thus bared his neck to a downward slash, and was standing, panting with the effort, the point of his cutlass lowered. In this unguarded pause, Montbarre's voice roared out:

'*Matthieu! Prenez garde! Tournez!*'

A hawk-eyed captain, toiling head and shoulders above the rest, had not been too busy to see a mortal danger for one of his men. When Matthew whipped his body round about, to meet whatever peril lay behind him, a thrust from a rapier which should have had his back-bone on a spit passed through his shoulder. It drew blood, but not life, and as he staggered back he could at least face his enemy.

It was a third military man, scarlet in face as in dress, whose flickering sword was already in fresh attack as he strove to press his vantage home. His weapon was longer and more crafty than any cutlass, and the soldier handled it with a dueller's wrist and a most elegant skill in probing and searching. All he sought was a living heart . . .

The glittering point advanced: Matthew slashed downwards, and missed as the rapier was withdrawn. One stroke from him could break that brittle toy; but to succeed would be like splitting a mosquito with the best-bower anchor . . . He must have length, if he was to live . . . The warm blood ran down his arm, and his head began to swim. The sword menaced him again, like a steely whip, and Matthew gave ground until his retreating back came up hard against a ship-boat lashed to the bulwarks.

He cut inside his enemy's thrust, and the raw blades hissed as they met and parted. But the rapier-point had missed his cheek by the breadth of a finger.

He must have length . . . His wounded left arm dropped within the ship-boat, and touched a loose shaft of wood. It came to hand easily, and suddenly he found that he had length, of a sort. It was a sailor's truest friend: an oar.

Taking a desperate chance which might leave him naked, he flung his cutlass full at the soldier's face, and struck him on the breast-bone. The man gasped, and gave ground; and Matthew bent to seize the oar. Then he swept it round in a vicious half-circle, near to the level of the deck, and caught his enemy just below the knee. The swordsman screamed as the solid oak blade met his shin-bone, and shattered it. The rapier dropped, and its owner with it.

Matthew leapt forward, going straight for a hated throat. Wrists of steel were no match for iron claws, and Matthew presently rose from a man throttled to death. The pain and the pouring blood from his arm were forgotten as he stood upright again. It was over—and so, he found, was all the other battle.

As swiftly as the raging tide had risen, it ebbed to nothing, leaving only limp corpses, a huddle of prisoners who might soon prefer to join them, and scuppers leaking black and scarlet as the surrendered brig rolled to the swell.

But it was more than the long fetch of the Atlantic which disturbed the brig; it proved to be the onset of three days' foul weather, a sudden summer storm which threw the Caribbean Sea into wild disorder. The *patache* was forced to leave before her plundering was half done, in limping haste and on a course chosen only by her enemy—the sea.

While the crew were stripping their prize of every moveable thing which might be of value: fine canvas and cordage, a spar to replace their broken foremast, wine and rum, tubs of salt beef and pork, fruit, smoked fish, powder and shot, and blessed golden coinage—while this happy thieving was in progress, the eastern sky blackened. Under this lowering curtain, a colder wind began to tumble the two ships together; and presently Bonhomme the boatswain was forced to bring his captain a warning.

It was not a favourable moment; though the best of the booty had been secured, there was more to be wrung from their captive—and six living prisoners still awaited the pleasure of the Exterminator. But it was clear, as the wind began to howl and the water to darken, that certain hallowed ceremonies must be abandoned.

Only one of their sails was serviceable, Bonhomme reported. The foresail must be patched and oversewn, and its mast replaced. Worse, they could only trust to the one sound half of their hull: the larboard side, holed by the brig's first shot, must be tallowed, plugged, and strengthened before it even dipped below the water-line.

With such wounds, and in this threatening weather, more than half the compass was denied them: they must sail east or south, until they found somewhere to careen, and to make, out of a crippled hen, an eagle again.

Montbarre knew this to be true, as well or better than Bonhomme, and his strong nature allowed him to make a jest of it, even in disappointment. He cursed but briefly; then he said:

''Tis a hard moment to leave. There is more booty to be found, if we can squeeze the brig a little harder. And six prisoners fit to stand! What waste of manhood!'

Bonhomme grinned. 'There will be others. And we have no room to take them on board, if we cut loose from the brig. We are crammed like a fish-hold already.'

'So—no *strappado*?'

'We have no time, *capitaine*.'

Montbarre pretended deep regret. 'It demands time . . . There is one fat man who would fall so well . . .'

The *strappado* was Montbarre's most evil trick, and thus his favourite. The victim, his hands tied behind his back, was hoisted to the yard-arm by a rope tied to his wrists, then suddenly let go, then checked at the foot of the drop. It dislocated every joint above the waist, and only a man with his tongue ripped out would fail to scream in agony. With a stalwart prisoner, it could always be repeated.

The wind began to make the rigging itself cry out in stress: the *patache* battered against the brig, while all the sails, trimmed aback, thrashed to and fro angrily. It was time to be gone.

'Very well,' said the Exterminator. 'No *strappado*, on this luckless morning. And not a woman on board to solace our aching hearts!' He looked aloft, and then at the wallowing brig, which jostled them like a fat old housewife nagging her mate on market day. 'Secure the prisoners below, then set a powder-train and blow her up. We want no evidence, beyond what the fish will tell the oysters. Matthieu!'

'Sir?' Matthew, resting in a pile of plundered sail-cloth, came to the alert.

'Is your wound dry?'

'Aye, sir.' The shoulder-gash was still throbbing like the Devil's own small hammer, and merry merciless talk of *strappado*, which he had witnessed once before, did little to calm the fever. But that

had not been the question. And the man had saved his life . . .
'Can I do aught?'

'Take the helm. When we cut loose, let her fall away to starboard. Then we steer the best course we may. It can only be east of south.'

The island came up like a golden gift from paradise—or like paradise itself, which few men found until too late for their tears of happiness. It had no name for any of them, since the *patache* was lost; after three days' stormy wandering, and two more of groping about in empty sunshine, she might have been anywhere, between Martinique and the coast of Surinam.

Montbarre had no instrument of navigation, nor any schooling in it; he could tell the sun from the moon, because the one was warm on his face and the other cold on his shoulder, but he had no friends among the stars. It was his custom to nose his way from island to island, and now he would nose his way back again to Tortuga—or capture some learned ship which would tell him, 'We are here,' before her tongue was drawn.

For him, as for all the rest, it was enough that they had come upon an island some two miles wide, protected by a coral reef which they had probed until they found a slim inlet. Within the arm of the reef was a lagoon of still blue water, and a fine sandy beach on which the *patache* could be hauled down and careened upon her sound side; and beyond the lagoon were palm trees, and no habitation, and good fresh water, and wild yams and pawpaws agrowing, and a whole world of gentle, peaceful greenery.

'What island is it, sir?' Matthew asked Simon Montbarre, when they had foraged ashore and settled in their anchorage, and those of the *patache* who could swim were swimming, and others fishing, or counting their share of the brig's spoil, or cajoling Bonhomme the boatswain into broaching the rum casks before the appointed time of sundown.

'God knows!' Montbarre growled. 'But God knows also, it is a fair haven. It will serve us well tomorrow.'

Yet fair became foul, as soon as Montbarre touched it.

When the repairs to the hull were done, and a new foremast stepped and rigged, the *patache* was set afloat once more and anchored in the lagoon. But though there was nothing to keep them there, save beauty and peace and other foreign words, Simon

Montbarre delayed. From the deck of his ship, as dusk became twilight and then deepened into night, he had noticed something which might be of profit.

It was the circlet of coconut palms which lined the shore, just above the tide-mark. Seen at this moment of uncertainty, between evening and night, they had the look of a forest of masts.

Might not another captain, in another ship, be deceived in the same way? Might he not be lured into this supposed haven, where such a goodly company of ships rode peacefully at anchor? In the aftermath of the storm, there might well be other craft lost or hurt, and looking desperately for a home. They would sail gratefully up, and crack their skulls on the reef.

Especially at night. Especially if the palms themselves wore riding-lights, which would move gently against a black sky, as the night breeze ruffled the branches . . .

'By God, I have it!' he roared out, to a crew who looked back at him, mystified. 'Bonhomme!'

Then he explained, and then he acted. The ship-boat went ashore, laden with twenty lanterns—all that they could muster on board. Catouche the great climber set them in their places among the trees, and lit them. When he and the ship-boat's crew returned on board, they left behind them, as night settled, a prosperous roadstead full of twinkling, rocking masthead lamps, where none had been before.

The baited palms did nothing for them on that night. But on the next, when the wind turned a little sour and the reef growled, and the restless palm-tree masts sent their signals to any ship within ten miles which might need shelter, Simon Montbarre was rewarded, beyond his greediest hopes.

In the small swift dawn of the Caribbean, when the world changed from an inky blackness to a bright pale yellow, within the space of half an hour, the pirates lay in wait among the trees. They had slipped ashore from the *patache*, as soon as they heard the wandering ship strike the outer reef. Now they listened, with the utmost sympathy, to the screams and cries of the castaways, the wild prayers of swimmers, the choking sobs of grateful men and women as they struggled to safety.

They listened, and forbore to laugh aloud in case it should alarm the shipwrecked mariners—and made ready their weapons.

The tattered strands of seaweed which were men and women under the flail of an ocean tide, came ashore in bedraggled twos and threes. Some swam manfully; others clung to barrels or to splintered planks of wood; others propelled ahead of them, like offerings to Neptune, chests and boxes and bundles of their goods. There were swimmers who supported their womenfolk until they reached the shore—a solicitude which the pirates found most moving.

There was also a cluster of small ship-boats, laden with all manner of treasure, and from these the rescued baggage spilled out upon the beach as it was lugged to safety by exhausted men, and began to pile up agreeably.

'Should we help them?' Boatswain Bonhomme inquired in a whisper. 'I would not wish them to lose anything in the surf.'

'No, no,' Montbarre growled back. 'There are some strong men there. Let them work as long as they can.'

Presently, among the great jumble of goods assembled above the tide-mark, there stood some ninety souls who had made their way to safety: some ninety lambs for the slaughter. Their cries and prayers, as they thanked God, or each other, for their merciful deliverance, showed that they were Spanish lambs.

'Now!' Montbarre said, in the same iron growl. 'O'Hare! In a moment, sound your alarm! You others—shoot two men each, or two women, to trim their numbers. This Spanish trash has no divine right in these waters, and it shall be our first pleasure to show them so. Now, O'Hare! Call up your damned tribe!'

O'Hare, his eyes aflame with a mad light, drew breath like the onset of a gale, and loosed his hideous Irish yell. The hapless group on the beach turned in wild dismay. A volley of musket shots rang out, and then a second. Men and women fell, silent or screaming with pain. The pirates took five steps out of the trees, and advanced towards their wailing prey.

They did not even run; they walked forward at an even pace, weapons ready, knives and cutlasses gleaming in the sunlight: laughing, shouting out to each other, calling 'Buenos días, Señora!' in the mincing speech of gallants, all with the greatest good humour in the world. Those Spaniards who tried to flee were chased and butchered: those few who had weapons and the will to fight were hampered by the throng, and easily cut down.

Those who desperately took to the water were taunted with their hopes of a last long swim—and then stoned as they swam, in a rare game of skill at which Catouche the marksman excelled. The incoming waves darkened with blood and froth; the beach, here and there, was changed from golden sand to a gory mudbank, so soft to the toes that they might have been treading the richest honey in the hive.

Within an hour, some fifty prisoners, men and women, were tethered to the palm-trees, or haltered together and anchored with a coral rock; and their captors turned, with an appetite to match their humour, to feasting and drinking in the green shade of their domain. They used what fancy viands these obliging Dons had brought ashore, and swilled their own stolen rum, and spoke merrily of the pleasures still in store.

They were roaring drunk by the time they came to the torture, but sober again at its end: as if an afternoon of blood-flow and torment unspeakable could work some holy magic of its own, purging out wicked liquor with a healing brew of agony.

Simon Montbarre, on these occasions, did not act on random caprice. His prisoners were mustered like cattle at a fair; the women—who would only suffer violation and outrage—were separated from the men, as ewes from rams, tabbies from tom-cats, and then the men were sized off, by weight and agility.

The strongest were chosen for *strappado*—'a game of honour not for weaklings', as Montbarre had once phrased it—or for another favoured sport which he called 'swimming on dry land': the victim was bound, by his feet to two palm trees, by his outstretched arms to a windlass: a stone of four or five hundred pounds was laid upon his belly, and then the windlass slowly tightened.

The body of a very strong man might even lift the stone for a while; a lesser mortal was ripped in two without managing his 'swim'.

Fat oily men were destined for the roasting spit; they saved the cook the trouble of basting. Men with big backsides were handsomely suited to Montbarre's favourite toy: a monstrous bull's pizzle some four feet long, hardened to the texture of iron—'Oh fortunate cows!' Montbarre had exclaimed, when he had found this prize in a Martinique *bordel*—and now used for the flogging of less fortunate rumps.

A priest—and there was one of these rare treasures among their present captives—would be given an instrument of torture to play with: a spiked iron collar, to be tightened with a screw, which Montbarre had long ago taken from a dead brother's neck, and always carried with him. The holy man of God was allowed to ponder this awhile. Then he must wear it, and tighten it himself until it bit him to his death.

Small men were pared down even smaller: 'We need fish-bait,' said Montbarre, 'and who would waste a man when there is a manikin to hand?' Lastly, the smallest prisoner of all would be flayed by the butcher, stuffed and sewn up by the sail-maker, and hoisted to the great yard of the *patache*, to serve as a dancing dolly-bird in harbour and a weather-cock at sea.

With such a bill of fare did the long afternoon pass: to its music, the foreshore of a delectable isle resounded without a pause. The demented screams of living men, the groans of the dying, the loud entreaties of fresh victims as they 'walked Spanish'—half lifted, half pushed on tip-toe by lusty executioners towards their doom— were like the rise and fall of an infernal melody.

Above it, beneath it, all around it was a faint tune more subtle and more lasting than any: the wailing of the womenfolk, watching from the trees as their last loves on earth, their last hopes of salvation, were gutted one by one.

Matthew Lawe, still fevered by his wound, had early lost his stomach for this sport. He had seen too much of it; there was nothing new in this mindless butchery, nor any trace of honour or good acquittal. Though he watched the priest die, he thought him brave—braver than any who laughed and capered about him as the neck-iron pierced the blood-stream of his throat. Though he took his turn at the windlass-haul, he could have vomited upon the half-human which was drawn out as its severed link.

He could not hate any man to this pitch of frenzy. Nor any woman.

The shadows lengthened across a field of infamy. Dusk fell, as the pirates quaffed afresh their forgotten liquor. Then it was the turn of the women.

They were lying in an anguished heap under the palm trees, the muted music of their sobbing now faded like the rustle of mice —mice in terror of the Giant Despair which was the agile cat. If they had any tears left to shed, they must be hoarded for their own travail.

There was a boy among the women, who seemed to cherish and

protect him even in their own extremity: a prince of the blood, perhaps, or the last survivor of a noble line. He was a pretty sunburnt boy, lithe and slim, in a bright blue doublet with a sailor's stocking cap atop his curls.

'There is much running and jumping in that one,' Montbarre had said when he first saw the lad; and by custom he might have had him chased up a tall tree, with knives thrown at him until he fell: or would have sliced off one foot and set him to playing hop-and-skip upon the other. But the Exterminator had spared him as a sweetmeat for Catouche, whose private taste—and his privates with it—tended ardently in this direction.

At his ease in the warm dusk, with the moon rising like a great glowing ball above the lagoon, Montbarre chose the two comeliest women, tied them back-to-back so that one might serve as a yielding mattress while the other performed her nobler duty: drove them deep into the crab-grass beyond the palms, and there—if noise might serve as a true measure—raped them both right heartily.

Catouche the mulatto secured his pretty boy, cuffed him until he was compliant, sent him running towards the fringe of palms as if he were to be given a chance of liberty, and then—with two minutes' start for the glory of the chase—began to stalk him.

Others embraced their less exotic lusts, their earthy copulation. There was no need for quarrelling; if a woman was consigned to one man for tonight, then she would be another's on the morrow—and hardly the worse for wear. Live and let live was the rule, in this wild pigs' paradise. Before long, the coast of their delectable isle grew loud with lamentation again; but it was only the lamentation of women, the smallest currency in the world, not to be counted by man the overlord, man the hot invader, man the great blade of execution.

Happy hours passed. When the moon was up, and hung like a kindly lantern above the palms, the pirates shambled back one by one from their endeavours, ready for more liquor, more carouse. Leaping fires of driftwood and coconut husks and oily palm-branches flamed skywards along the length of the foreshore. The corpses which were an earlier tribute to manhood had been piled at its further end, ready for burning or burial next day—another task fit for house-proud women.

But they would need to be prompt. Already there was a steady rustling, mysterious and intent, as a thousand flesh-eating soldier-crabs formed ranks and marched towards this, the most blessed banquet of their lives.

Catouche the great marksman, Catouche the licensed pederast, was the last to return from his own banquet. He came suddenly into the firelight, looking angry and out of countenance, with his dishevelled boy in tow, to be greeted by men who felt themselves, for all the varied reasons of mankind, to be superior to this laggard tar-brush sodomite.

'What took you so long, Catouche?'

'Could you not find the mark?'

'Use a woman next time. She will guide you.'

'With them, it lies due south of Mount Belly-Button.'

'Aye. No need to circle the great globe.'

Catouche stood silent before this raillery, while the boy, his head dropped on his chest, his weary body spiritless, waited in captive shame. Finally the mulatto said sulkily:

'She led me the devil's own dance. And all for nothing.'

It was Montbarre, huge and shaggy as a mountain bull, enthroned before the firelight like a god of sacrifice, who heard him first and best.

'*She?* What do you mean, Catouche?'

Catouche, who now had little choice but to turn all this to a jest, answered: 'I bought a hen for a cock, *capitaine. Ce n'est pas un gamin. C'est une fille!*'

'*Sans blague!*'

Now every eye turned upon the slim figure at Catouche's side. By God's blood, it could be true! The pretty boy now seemed too pretty altogether; the flanks were a thought too bold for the shoulders, while the bosom under the doublet was seen to have certain gentle swellings which could hardly be the muscle of manhood. Though this was not the moment for jaded men to grow lickerish again, there was always prying curiosity—and always tomorrow.

'Take off his cap, Catouche!' one of the men cried, and in a moment it fell to the ground. The dark hair tumbled down, framing a face as fair as that of any maid they had ever seen, with great brown eyes, a skin like a darkening peach, and lips which, in the flickering firelight, blossomed as red as the cherries of early

summer. No wonder Catouche had fallen back from this treasure!
He would not even know its use!

Montbarre's greedy eyes were alight, and he took command on
the instant, drunk and drained as he was. His sweating face
became a monstrous leering mask as he said:

'I need more proof than this. Even pretty boys can have long
curling hair, in these idle days . . . Let her show her shapes.'

The girl, already wild with rage and shame, clawed at Catouche's
face as he made to strip her. A brutal slap subdued her strength,
and the bright blue doublet was ripped away, and a most unmanly
bodice with it. Now she stood half naked before them all, her hands
pinioned behind her so that she had not even use of these two small
curtains to hide her glory.

Her beauty stunned them all. She was still young, perhaps no
more than sixteen, but her body, just emerging to shy and marvellous womanhood, glowed like the moon which revealed it. Her
slim waist matched the stem of a flower, and the flower itself, with
its delicate cups tipped with pink, was fit only for the altar of love.
Beneath this paragon, her gently-rounded flanks promised a lively
seed-bed for the very root of man.

Aware of the hot caress of all their eyes, the girl continued to
wrench and twist her body from side to side, seeking refuge from
their violation. But this only served to inflame the watchers. It was
rapture to imagine such tormented movements carrying their own
manhood to the heights.

They howled their lust. Amid the babel of voices, there was only
one critic of the sublime.

'Her breasts are too small!'

But he was answered swiftly: 'Montbarre will make them
sprout!'

'If not he, let it be me!'

Simon Montbarre, who was not too drunk to drink her in, now
asserted his command. It was to be himself, and he alone.

'By God's own holy organ!' he blasphemed, 'why did I waste my
precious fluid on those drabs? This one could have sucked it all in,
till she was full to the neck!' He leaned forward, his huge body like
a threatening cliff. 'Come here, little one.' The girl did not obey
either his voice or his beckoning finger, and he said: 'Bring her to
me, Catouche. And keep those damned claws out of my face!'

Catouche pushed the girl forward, until she was standing within a hand's clasp of Montbarre. Then their captain surprised them all.

'It may be too soon,' he said, in a voice as easy as if he spoke of the chance of rain. 'Let us discover . . .' He stood, and unbuckled his belt, and dropped his breeches. Then he leaned against the naked girl.

She recoiled in horror and disgust, but Catouche, at a sign, held them close together. Montbarre waited, until the gross assay had been given its chance. Then he stood back, and looked downwards.

There had been no stirring. His member still hung slack, as idle as a rope's end.

Montbarre was laughing, even as he secured his belt again and took his chair. He was man enough and captain enough to concede before them all, without shame, and he did so now with wry humour.

'Too soon,' he reported—and the simulated disgust in his own face had a mirror, fit for merriment, in the girl's terrified revulsion. 'My sprit is sprung, and I am too drunk anyway . . . Let our rendezvous be later. She will ripen, and so will I! Matthieu!'

Matthew Lawe came forward, as the girl was allowed to fall back. He did not look at her: the vile scene, and the more vile future which it promised, had sickened him.

'I set you on guard,' Montbarre told him. 'A wounded man is all I can trust, with this little gazelle.' It was not true: when Montbarre fixed his eyes upon a woman, all others backed down. Though they might prove their manhood many times, they could— upon the point of a cutlass—only lose it once. 'I set you on guard,' Montbarre said again. 'Take her into the trees and tie her up. We will see how strong I am at dawn.'

Long before that threatened dawn, Matthew and the girl were crouched together in the long-boat of the *patache*, and creeping out, stroke by stroke on tender oars, from the lagoon to the open sea.

He did not know why he had thus betrayed his guard and be-friended her, save that compassion for this tormented waif had overcome all fear. He did not even know how he had won her trust. They were barely able to talk to each other: she had as much of the English tongue as he had of the Spanish—perhaps a hundred words between them. When he had first taken a grip of her arm, to lead her from the beach into more private captivity, the fear and loathing in her face were absolute.

But then she yielded, as he had done. Perhaps it was because, as soon as they reached the shelter of the trees, away from the firelight and the glittering eyes, he had taken off his shirt and given it to her, to cover her shame.

This sorry garment smelt of all things, from fish to sweat, from honest brine to infamous blood; it bore fearsome stains upon it, the fruit of the afternoon's murder. But she took it from him as if it had been the Holy Shroud, and wrapped it round about her like armour. From that moment, her violent trembling gradually ceased.

Then, when he said 'Come with me,' and beckoned her with signs, she had followed him deeper into the trees; and then to the far end of the beach, where the long-boat lay unguarded. The moon was down, and the distant fires glowed feebly: the stooked corpses —perhaps one of them was her own father's, even her mother's— were hidden in the darkness.

Their oars took them out over the lagoon to the reef, and then, as they crouched like thieves, across a shallow patch of coral until they had scraped free. Now they were in the open sea in an open boat—and open to pursuit as soon as their flight was discovered. It was a time to stop and consider, and decide their best plan.

The girl lay still in the bows, as Matthew surveyed the distant shoreline and their fearful dilemma. How to choose the next course?—for if he chose wrong, Montbarre's boundless rage would ensure that Matthew, flayed alive, became a sawdust dolly at the mast-head, while the girl was turned to a sick and exhausted animal—but slowly, through many hands.

He had not sworn to save her, but now they were linked for ever, as fugitives in an extreme of peril, as twin souls fleeing from the same lash. He *must* contrive their safety.

Simon Montbarre—how would he plan? He would find Matthew gone, and then the girl, and then the long-boat. Since this could only mean a sea-escape, he would give chase upon the instant.

He would know that the speedy *patache* could always out-run the boat, whose small working sail was a limping aid. A runaway in terror would stream-down-wind; a resourceful sailor would claw to windward—and still be caught in the end.

Montbarre would try both courses.

So Matthew must go where Montbarre would think him not to be. The long-boat must disappear. Time must run on, as Montbarre searched in vain. *So they must stay where they were*, until they were believed lost at sea, and the pursuit slackened and died.

The long-boat, weighted with coral rock and holed at its stem, was gently set adrift from the point where they had landed, on the far side of the island away from the accursed beach. It would sail a league or so, and then vanish. He had taken from it all its meagre store: two fishing lines, an empty water-keg, an axe, a spy-glass, and a pair of oars. To add to this, he had his own knife, and a pistol with powder and twenty balls.

With these, they must set up their housekeeping, and somehow save their lives.

In the perilous darkness, no word was spoken. The girl, who seemed to have placed a desperate trust in him, followed all he did without question. Laden with their small burdens, they drew back from the beach, and then began to creep inland through a tangle of rocks and low bushes, moving in stealth and fear like hunted night-animals.

When it seemed that they had climbed and striven far enough, they stopped, and lay down at the foot of a tall palm tree, and there fell into exhausted sleep. If she had been Venus, he could not have touched her; and if he Zeus, his most awful thunderbolt would not have pierced her maiden armour.

At dawn Matthew was wakened by roaring voices close at hand. It was the most dreadful sound of his life, and he sprang up, as ready to be killed as to kill. But the voices did not come from the sheltered side of their paradise; they were sounding beyond the crest of the hill. He crawled fearfully forward, a body's-length at a time, and presently found himself looking down on the beach, and the *patache* at her anchor.

All was in uproar. But it was the uproar of departure, not of search. Treasure, stores, and a few women were being carried down to the tide-mark, and loaded into the *patache*'s second boat. Montbarre had taken his bait, and was off in pursuit.

There was a rustle at his side as the girl joined him. She began to whimper in terror as she saw the men below running to and fro on the strand of death, and the women who screeched at a new captivity. But when he put out his hand to comfort her, she struck at it, and shrank away. He was one of these same monsters!

In sick silence they watched the *patache* heave her anchor short, and break it out from the sandy bottom, and thread her way through the gap in the reef. She worked eastwards and up-wind all that morning, casting this way and that like a questing hound. Then, when she was almost out of sight, she turned and ran back towards the island. Matthew's throat was dry, and his heart thudding, as he followed her relentless progress. Would she, after all, make a second landing and then a proper search?

The precious answer was No. Though she cruised slowly by the outer reef, and Matthew and the girl buried their heads for fear of sharp eyes, the *patache* did not stay. Montbarre turned her again, off the wind, and began to run steadily south and west to the open Carib sea.

At sunset the last flicker of the latine sails had vanished, and they were alone.

V

At first, and for a long time, she went in fear of him. In one brief encounter, she had seen enough of such foul men to poison thought and feeling for all her days. She would never trust again. Even if this man had saved her life, it would only be for vile purposes, hideous ends.

Often she hid from him between one sunset and the next, until hunger drove her back, and she would come trembling into their camp among the rocks, and snatch whatever food he offered, and retreat once more, the wildest of animals—and never to be snared again.

Then, when no ill befell her, she grew confident, and after that haughty, and then imperious; as wilful a maid as ever might drive a man mad. She had divined that, though fair-looking and strong in body as in purpose, Matthew Lawe was humbly bred; as the daughter of great lineal splendour—or so he guessed—she could only be scornful of such gutterlings.

Thus she would never tell him her name, so marking the distance between them, and imposing it for ever. Instead, she used him as a slave. The fact that she needed this sturdy and resourceful man to support both their lives was of no account—save in her secret soul.

Matthew made no quarrel over this. He had been granted as blissful a toil as any in his life; if he must share it with a proud, disdainful mistress, it was nothing in the great reckoning of happiness. Perhaps there might come a day when she would melt, when

delicious young flesh and blood would demand its season. Or perhaps she would remain a Snow Queen for ever.

In the meanwhile, they had a life to make. He set himself to
make it with ardour of another sort—the resolve to fashion paradise
out of their small, welcoming wilderness.

He had set up two tents—far enough apart, God knows!—of
sail-cloth which drifted ashore from the Spanish wreck, and covered
them with palm-fronds to hide them from enemy eyes. Between
them there came to be a garden, where yams flourished, and pawpaws whose gleaming pink flesh could set a man's mind on other
things, and wild figs, and a yellow root which he could not name,
save that it tasted like a West Country turnip and made an honest
soup.

He lit one fire a day, for their evening meal, using the eye of the
spy-glass and some dry tinder from their coconut husks to kindle
it.

For meat they had nothing, and the lust for it was a long time
a-dying. If he had ever chanced to find wild pigs, he would first
have worshipped them, then set them to breeding, then stocked a
whole larder with their fine fat shanks and glistening shoulders.
Instead, their larder must be the teeming sea, and from it they
drew baby dolphin, and succulent flying-fish, and red snapper, and
albacore, and sometimes a wandering turtle, fit for any banquet of
the Lord Mayor of London—and for them.

He built a fish-trap, a basin for the tide to flood in and there
leave stranded tiny crabs, and pale shrimps, and the small fry of
bonito. He robbed the nests of gulls for their blue-speckled eggs.
He set up a wooden hut, thatched with palm branches against the
winter wind, and put in rough shelving, and in it stored dried fish,
and the seeds for the harvest of next year, and firewood, and milky
coconuts.

He changed the course of the little stream coming from the
spring above them, so that it flowed through the garden. He made
the girl a bed, wide enough to house any three of such slender
maids, and got nothing but scorn for his labours.

For the comfort of her soul, he cleansed the other beach of its
remaining dead.

Much of this evil work had already been done for him, by the
regiments of soldier-crabs, and ravenous sea-birds, and the kindly
bleaching of the sun. But they could not live with this grove of

tormented skeletons still showing above ground . . . He put together a serviceable raft of driftwood bound with strips of sail-cloth, and took the oars, and ferried these poor remains one by one, like old Charon across the Styx, out beyond the coral reef. There he set them loose upon their last ebb-tide.

But she would never visit the beach, even when it was made as fair and innocent as their own. It stood for the darkest side of hell; and it was still marred by the wreck of the Spanish ship, though this had slipped back into deep water and yielded nothing but the memory of fathomless sorrow and pain. For her, only their own quarter of paradise could be stainless, and blessed by peace.

So the uncounted months passed. The seasons turned, and turned again. Using a roughly made sundial, no more than the stump of a tree with a flat rock beyond it, he could judge mid-summer, and then the winter solstice when the sun stood still. It was enough reckoning for a twelve-month or more. Beyond such sparse accounting, who would wish to know, and who to care?

Their solitary state was sometimes broken, sometimes threatened. Once or twice a year, fishing canoes manned by Carib Indians wandered their way, and put ashore; they stayed to feast and to dance, to collect water and paw-paw, and then to voyage onwards. Matthew was sure that, just as he knew of their presence, they knew of his. By their nature and skill, they were perhaps the wisest men who ever sailed these mysterious seas.

But they always kept their distance. They were poor, innocent, and peaceful. Perhaps they divined that any castaways must be of the same quality. Thus there were none to rob, none to kill, and none to disturb. Of such gentle forbearance, the whole world should be made!

Three ships in all, big ships more menacing than any other visitors, also closed the island, and thrust through into the lagoon, and stayed there long enough to suit their mariners' needs. At such times, Matthew and the girl lay hid like mice who knew that silence and stillness can outrun the cunning of a cat.

Not a shadow moved, not a leaf stirred, as these robbers foraged past their camp with blind eyes, and never saw it, and went on their way. At such times, Matthew took command, and justified it, and made his wisdom the rule of the day and the night.

The most feared enemy, the *patache*, never fouled their paradise. If Montbarre had not met his death—for which all Christians must devoutly pray—then his nose for navigation was out of joint, and failed to lead him back again.

Their need for each other, though unspoken, ruled every moment of their lives, whether in time of danger or in the long seasons of solitude, when they must still wrest a living from their small kingdom. Innocence also ruled. They remained apart, the sturdy man and the glowing girl, like novices vowed to forget the flesh.

Then, for no other reason beyond those mysteries which Henry Morgan had sometimes cited from Holy Writ, in a mood between piety and lechery—'The way of an eagle in the air: the way of a serpent upon a rock: the way of a ship in the midst of the sea: and the way of a man with a maid'—for no reason to be fathomed by the lead-line of chop-logic, all that they lacked was added.

Perhaps it was born of their tattered clothing, which began to fall from them like sun-dried, weathered leaves. There came a day when she could no longer piece together the shirt which Matthew had long ago given her. She took the meagre best of it, and fashioned a sort of bodice to contain those cherished breasts. This too, before long, went the way of dissolution; and after that she must needs go bare-breasted—and dare him to stare at the disclosure.

But he was making his own disclosures, and was vexed by them, yet in the end indifferent. His stout canvas breeches became the most revealing of small-clothes, and then resigned their duty altogether. But what matter? He was a man, and could prove it; she was a girl, and he knew it. Let a blind bishop deny what the first shaft of sunlight revealed . . .

Presently all that they had, like Adam and Eve before them, were little aprons to cover their shame—if shame were still a thread in their lives, which now seemed scarcely the case.

There came a morning when they were naked, and suddenly they laughed together, and it was the spring of the year.

After that, as on a hundred days before, they spent the time apart: she to the garden, the replenishment of the water-keg, the plaiting of tendrils to mend the roof of her tent: he to the beach, the fish-traps, the gathering of driftwood, the eternal watch to seaward in case they were taken by surprise.

When he came back, she was thoughtful—but not too thoughtful to exclaim at the sight of a huge log he had salvaged from the sea, and at the beauty of a golden-beaked parrot-fish which might or might not be worth the cooking.

It was praise, of a sort: the first praise he had heard since Montbarre commended his rope-work on the last *strappado*. Then she said:

'It is so hot. I go to swim.'

For the first time, it seemed that he should follow her. He began to wonder, and then to cease to wonder.

They swam, and stood, and swam again, thigh-deep, waist-deep, breast-deep in the warm still water of their small lagoon, their coral fortress. When her slim, sun-burnt, naked body took the first caress of the sea, it glowed like fine yellow sand which had thirsted for a rising tide, and now need thirst no more. His broad barrel chest and sinewy loins took on the same colour—and the same pride.

It was a moment, towards dusk, when a man could be glad that he was fashioned in a different mould from a girl; and she knew this, and shared it with a shy acknowledgment, and so they both knew it.

She shook the gleaming water-drops from her face and hair, and stood still in front of him at the shallow edge of the reef, and said, in her careful English which was no more absurd than his sparse Castilian:

'It is time to say I thank you.'

He could not fathom this mood. He could only accept it, and stare at her grave, glowing face, and then at all the rest, and then at her face again. He answered:

'No call for thanks. We had to live, and we have done so.'

'But you have worked so much. And I have nothing to give back to you.'

They were together, and alone, in a world of colour and warmth. The reef was alive with the last fire of sunset. There were purple sea-fans waving their gentle invitation, and wide-mouthed sea-anemones, the hue of tender flesh, opening and closing at the urging of the tide. The brain-coral shone pink, and the delicate finger-coral was a pale yellow, like a maid's slim hands.

Idly she picked up a piece of this, and turned it over, and then held it out to him in sudden merriment. It was, beyond denial, of the lewdest shape imaginable: two round pinkish globes, and between them a sturdy prong of brown. When she put it into his hand, the prong hung down; at the last moment she turned it upwards, to complete a picture of erect manhood.

Then she seemed to stumble, and fell against him, and his thirsting hands came up to enclose her breasts. She gasped, as if these were at the core of her passion. Then they were entwined in kissing, and true manhood made any finger-coral a puny miniature.

By the time they had walked ashore, and across the beach, and into the first shelter of the palm trees, they were moving very swiftly. When they lay down on a bed of sun-dried grass, he would have tried, despite his wild urgency, to spend some time caressing into readiness her superb, willing body. But she had urgency of another sort.

She was trembling: perhaps with fear, perhaps with shyness, perhaps with fearful memory, and perhaps with a passion which could kill all other thought. She said, in a fierce whisper:

'Do it quickly! If I cry out, do not heed. Now! Now! Take me now!'

She did cry out once, calling '*Madre! Madre!*' in a choking voice —but whether she called upon her own lost mother, or the Eternal Mother of God, only the sky could tell. Her pain was brief, and so was his rapture, which crested under the long denial of his bursting loins, as if molten fire had cracked a towering mountain peak.

He shuddered, and poured, and was finger-coral again, between one cry of joy and the next.

But it was the happiest way . . . Uncovered by his broad body, the girl lay back, a sensuous delight with all the heavenly promise of tomorrow, and proclaimed to the first pricking stars of the evening, and to him:

'Now we are free!'

Love had come slowly; now it never left them. She had spoken the greatest truth of their lives when she said that they were set free; by her compliance she had turned their prison into the paradise which God and nature had planned, from immemorial time. Nothing was labour any more; all was giving and receiving, the ebb and flow of that great tide which was man and woman in sure confluence.

It was a mingling of elements as different as fire and water, sea and sky, the facts of earth and the capers of fairyland. Soft black hair twined happily with coarse yellow on the same couch; slim hands were entrusted to a sailor's serviceable claws; warm Spanish blood met English sinew, and tamed it with the tenderness of a maid at her first milking.

In blessed sunshine and the warm curtain of night, they lived their idyll. No man saw them; they only saw each other, naked children on their coral strand. It seemed that love and good fortune would never desert them.

On the mortal night which condemned them both to hell, Matthew Lawe was adrift from reason. Earlier, he had taken a trifle of poison from a manchineel tree, whose treacherous fruit had dripped some of its venom upon a fish drying in the sun, and thence into his mouth. Cured of a griping fever, he had drunk some fermented coconut milk, of which they made a cordial to sweeten a life already rapturous.

It had made him full of talk, full of wild and cloudy dreams, and, at the last, full of love's imperious appetite. For the thousandth time in a thousand such moments, when he must enter hers, the smallest kingdom which could be called divine, or else perish from starvation, he lay down with her. But he delayed long enough to turn aside from her supple body, and to exclaim in a kind of drunkard's happiness:

'If you could only know, what it is to be cleansed by you! Even once in a hundred years!'

Her body had been listening to his, rather than her mind; her secret parts had become the ready ear of love. But he spoke with such tempestuous fervour that she was also prepared to delay, and to question.

'What do you mean? What talk is this? A hundred years?'

There was still light enough to see, and suddenly he could not bear to keep his dread secret from this loving bedfellow. He turned back to her, and mastered her eyes, and said:

'Look, love, look! Tell me—what do you see in me?'

There was no answer, for a long and awful time. She did look back into his eyes, and thus into his soul, and then into his story— or as much of it as she could encompass without losing her belief, and after that her reason. She did divine a fragment of it, or the misty whole of it, or its tormented shape: the ghastly wraith of his imprisonment, his century of cunning and survival and doom.

Frightful images came to her, and no resolving of them, and no trespass upon love, and thus—for a girl of this quality, this proud enslavement—nothing to fear. She was still practical, and avid in the way of ripened womankind.

Their time was *now*: not a hundred years past, nor a hundred years to come; and no play of words, no unearthly vision, could mar another sort of vision—of him on her, him in her, him giving all and she drinking it up, like the parched earth its storm of rain.

She said: 'I know nothing. Or I know all. Come into me now. As deep as you have ever been.'

'Thank you, for ever.'

Presently she said, from a willing throat: 'By Christ, I do not know what to believe! But I know what I feel!' It was the first time, and the last, that he had ever heard her speak their Saviour's name, except in prayer. But at the same moment he released into her such a spring-torrent of life, whether old as God or new-born as His Son, that for a single chime of eternity it could not die.

In the morning, after the deep sleep of the gods and the drowsy stirring of mortals, she sat up on their shared couch in all her naked glory, and said:

'Perhaps it is time you called me by my name.'

Matthew opened his reluctant eyes, saw her delicious breasts standing proud above him, and buried his face between them. From this hallowed vantage point, he murmured:

'But I do not know it.'

He felt her body tremble a little, but it was the trembling of laughter.

'You will not mock me?'

'Why should I mock? What is this name?'

Her bosom moved again delightedly as she answered: 'Concepción!'

They spoke no more of that night, nor of any part of it; his secret, once shared, seemed dissolved by love, and as to their great love-making, this slipped into the calendar of such joys and was there enshrined. Its true and tender price was confirmed within the month, when Concepción found that she was with child; and within a season of bearing, when she died in her travail.

Matthew knew nothing and he could do nothing; the babe seemed locked within her, and her tormented body, sweating in its agony, unable to expel it. She was in labour for two terrible days, while her strength was shredded and her spirit brought to nothing by these hellish terrors.

He bathed her face tenderly, and held her hand as she twisted and wrenched and toiled. When finally it was done, with blood and shrieking and hideous torture, the last price fell due.

The child cried, and began to live; Concepción sighed for her release, touched with fearful fingers her ripped body, and began to die.

He bathed her face and held her hand, and gave her water, and begged her distractedly: 'Love, do not leave me!'

But it seemed that she must. Within a forenoon the pinched mouth refused the last sip, and her exhausted head fell sideways, and the lonely journey was accomplished.

He wept, and as her body grew cold he cursed heaven and all other enemies, and wept again. Then he turned grave-digger, and buried both the dead and the living in the same cruel cot, among the palm trees, in the little garden they had fashioned together. He slew the tiny boy for revenge, and for pity; why should he nurse into the world another cursed wanderer? The girl he laid to rest in bitter mourning and despair. How could he now live on?

It had been so ordered. . . . On the morrow Matthew climbed the hill above his graveyard, and saw what he wished to see. Giving no backward glance he descended slowly, without concealment, to the old beach of shame, and walked towards the Carib Indian canoes drawn up at the margin of the sea. As he advanced, he showed the staring savages, in peace and anguish, his empty hands.

VI

In Barbados, the bright jewel of the Windwards, Matthew met the world once more: Barbados the lush and languid isle, laughing in the sun, Barbados the fount of sugar-cane and fiery rum: Barbados where these same gentle savages had put him ashore in the dead of night, and fled away before the dawn might make them slaves. In Barbados he became afraid.

Round and about the inner harbour, called the Careenage, where ships were hauled down for cleaning and caulking, Matthew Lawe wandered furtively. He went in terror of the throng; there were too many pressing bodies, too many faces unknown, too many eyes upon him. To have lived in blessed solitude so long, and then to lose it so cruelly, was burden enough. This jostling exchange could only be a matter of punishment.

But he must join this world none the less. His small store of guineas, carried for years in a scrap of sail-cloth from the Spanish wreck, was dwindling to nothing. He must embrace the horrid press of his fellow-men. He must sail on.

Then, one day when he was watching an island trader making ready for the sea, and wondering if he could find a berth on board, he met a friend, an astonished friend. It was John Flowerdewe from the old Morgan crew.

'Matthew!' Flowerdewe hailed him incredulously. 'Is it truly you? I thought you dead these five years or more!' He searched his friend's face anew, then clapped him on the back with a well-remembered fearsome fist. 'By God's blood, it is! And not changed a whit, either! Where have you been?'

'Here and hereabout.' Matthew could only be glad to see him, though he feared entanglement with the wicked past. 'I never lacked for ships.'

'On the account?'

'Nay, I have done with the pirates.'

'Have they done with you?' Flowerdewe seemed about to add to this, then changed his tack. 'Have you the price of a pot of ale?'

'Aye.'

The villainous scarred face broke into a grin. 'Then why do we delay?'

In the tavern, their talk was long and, on Matthew's side, a great rope-walk of guarded lies. But when it came to their joined past, Flowerdewe had astounding news. Matthew had only to ask, 'What became of Captain Morgan?' for his companion to come flooding out with the strangest story ever heard.

'Do you not know?' He affected to look closely at Matthew's head. 'You still have two ears! Are they stuffed with fish-gut? Morgan is the greatest man in Jamaica! In all the Carib sea!'

'What—that rogue?'

'He is no rogue now. Watch your tongue, Matt! You speak of Sir Henry Morgan.'

'*What?* It cannot be!'

'It can be, and it is. He fell into disgrace at the end, him and old Tom Modyford; their pirate tricks and lies were too much for the King. They were sent to England under guard in a frigate, and put in prison. But after two years, you can guess who stayed in the Tower of London and who came out!'

Matthew had to stop this flow, which confused him mightily. 'But what disgrace is this? I thought Morgan was above the law.'

'So did he, but he went too far. Did you not know? *Morgan took Panama!* Then he burst out into the Pacific, and harried the Spaniards till they were brought to tears. Then he marooned most of his crew, to make a smaller division of the spoil. He left them on an island, cursing and staring, and came home with half a million gold pieces!'

'But the Tower of London? How did he——'

Flowerdewe took a swig of his ale. 'You know how Morgan could talk. The birds in the trees would come down to listen, not to speak of the women. Somehow he got the ear of the King, and was tried and set free. Then he was forgiven all. Then he was dubbed Sir Henry Morgan, and made Lieutenant-Governor of Jamaica. And what did the King's commission instruct him to do? Hunt down the pirates!'

Matthew laughed, in spite of his doubts and miseries. 'He would know enough.'

'He did. They say there is not a naughty word spoken in Port Royal.' Then Flowerdewe looked at his companion more closely. It was something like a pirate's look, crafty and sharkish. 'He would not forget you, Matt. I have never seen a man in such a rage . . . He does not forgive deserters. If he knew you were alive . . . There is talk of Montbarre, too. They say the old Frenchman is still searching. What enemies you make! Morgan and Montbarre. You might as well say, the Devil and his bastard son!'

'They will not find me.'

'That is fool's talk. If I were not such a friend, I would sell you myself!'

'Are you a friend?'

'Aye.' The pirate grinned again, easily enough. 'For a day and a night. And another pot of ale.'

'But what shall I do?'

Flowerdewe shrugged as he snapped his fingers for the pot-boy. 'Ship out. Put half the world between you, for a start. But leave me enough to pay the score. A lonely man grows thirsty.'

FOUR

ADMIRALTY CLERK

1682

'By his precept and example he was to transform an inchoate and ill-directed service into the most enduring, exact, and potent instrument of force seen on this disorderly planet since the days of Imperial Rome.'

> *Samuel Pepys: The Years of Peril,*
> by Arthur Bryant, on Pepys'
> appointment as Secretary of the Navy in
> June 1673, at the age of 40.

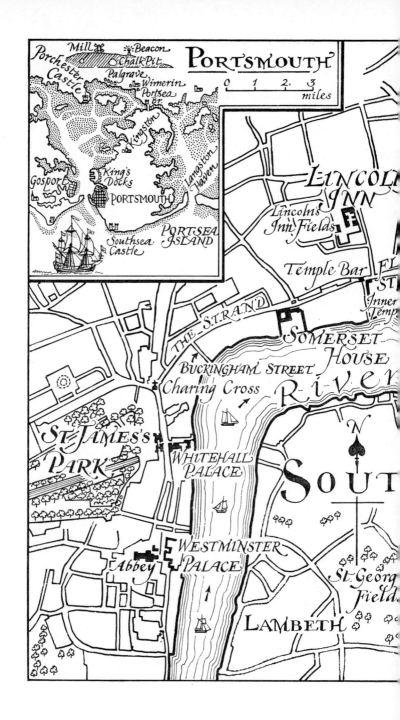

IT WAS A FAIR BRAVE English day, and all the world of Portsmouth knew it. The sun shone as if by royal command, the flags and pennants fluttered like clapping hands. The King was here, and his comely mistress the Duchess of Portsmouth, and his more sober brother James, Duke of York, and a throng of nobility to line the dock-side.

There was a royal yacht to be launched, and free ale to be quaffed; and nothing for the common sort to do but to gape and enjoy, and to swallow, with the freedom of men who need not pay, their regal bounty. It was a Portsmouth holiday, with the honest smell of brine to sweeten it, and the goodly feeling that all men here, and women too, and the children who would be the next breed of sailors, were celebrants at the very shrine of the sea, set within their own church—the finest sea-port in all England.

Yet since this was free England, and free Portsmouth, and since sailors—even sailors swilling free beer—were not to have their mouths stopped by such a paltry bribe, there were certain coarser comments to be heard, loosing a puff of voided wind into the holy incense of adulation. Matthew Lawe, who had landed not six hours before from a skin-flint *barca-longa* plying the Guinea trade, and would have gone thirsty if a running pot-boy had not thrust a mug of ale into his startled hands—Matthew Lawe listened with avid ears.

He knew nothing save that it seemed to be a holiday, and there was a pretty ship enthroned on a launching cradle, and lords and ladies in the best vantage points, holding aloof from the riff-raff of the port. But with good ale for the asking, it was more than enough. At first he only listened; men who spoke least, heard most; and, in the press round him, there was plenty to be heard. After so long in his African wilderness, he had need of such scholarship.

His new friends had no names, only rough labels, even after an hour of close fellowship; they were Portsmouth citizens, of high and low degree, and Matthew was admitted to their company by a chance mishap. The running pot-boy had jostled the elbow of a tall man with a seasoned drunkard's face, who in turn had given Matthew a liberal christening of ale as his foaming pot capsized. In the swabbing and mopping thereafter, Matthew found himself an adopted son of the city.

His friends had to be called Toss Pot, Snarly, Bishop, Big Bum, and Merrylegs. It was Toss Pot, whose flaming nose proclaimed a life-long devotion, who gave him his ticket-of-entry.

'Sorry, lad!' he apologized, for the third or fourth time, as he rubbed at Matthew's jerkin with a big red kerchief. 'I would not do that to my dearest enemy. And what a waste of good ale!'

''Twas the pot-boy's fault,' said Snarly, a mean wharf-rat glad to find a meaner target for his spite. 'I would have fetched him a buffet if he had not run away! To teach him his place!'

'Now, now!' the Bishop reproved. He was a pursy fellow, and a townsman of some consequence, if his fine grey cloak and lofty nose could be judged aright. 'An accident can happen to the best of us.' It sounded a most handsome admission. He eyed Matthew's face, and then his outlandish clothes, as if to judge if such a clown could be admitted to his congregation. His gaze fell last on the soaked breeches, which were from the ship's store and baggy as a pair of udders before the evening's milking. 'Such galligaskins!' he intoned. 'We do not often see such a style hereabouts . . . Art 'ee from Portsmouth?'

'Nay. Barnstaple.'

'And come all the way for this?'

'Nay,' Matthew said again. He was vexed at being put in his place, and more vexed still at his ruined clothes—all he had in the world, and now like a beggar's hand-aways, fit only to scare the crows on a rainy day. 'I came ashore this morning. From the Western Sea. The Guinea trade. Our last touch was at Lisbon.'

The ocean words were his passport, and established his honourable rate.

'A sailor!' cried Merrylegs, the joker in this pack, a little twisted fellow with a face as creased and split as a wind-fall walnut. 'Welcome ashore! But did you think to find a dry berth here? You should have stayed snug at sea!'

'I have been drier in a Bay of Biscay storm,' Matthew admitted with a grin.

'Myself, I am dry again already,' Toss Pot declared, having finished his mopping labours and wrung out his kerchief. 'Why should we die of thirst, on a day such as this?'

'Here comes another pot-boy,' said Big Bum, a larded citizen who sat fat to the world like a tun of Rhenish wine. 'Let us snare him, before he wastes his trade on others.' Though he wheezed like a grampus, and his monstrous bulk would have made a service-able ship's anchor, Big Bum could be agile enough when the need inspired him. His dropsical right arm shot out, and imprisoned the running lad who was trying to pass them by. 'How now, young fellow!' he roared. 'Can you not tell a thirsty man when you see one?'

'What other sort is there?' the pot-boy answered cheekily. 'You have had your turn. There are others——'

'Finer men than us?' Snarly interrupted menacingly.

The pot-boy looked at Snarly, who would not have been diffi-cult to top, opened his mouth to answer, and then thought better of it. He was not among friends. Big Bum's fat arm was strong; Toss Pot was known to every tavern in the town as a most determined fellow; the Bishop looked as if he had money in his purse, and with it could speak loud enough to lose a poor scullion his place. With-out a word he thrust his platter of six ale-mugs into Snarly's hands, and turned again, and ran back for another cargo.

The villains laughed, and drank, and were at ease once more.

Matthew, in whom strong ale upon an empty stomach had pro-duced a muddled head and a marvellous contentment with it, looked about him happily. For the moment, a sailor's luck had put him in a snug berth; if it lasted, well enough, and if not, tomorrow's chance might mend it. Certainly he would live in this merry mariner's delight for as long as daylight held.

The focus of that delight was the yacht on the wheeled launching cradle, a most lovesome thing, all curves and tapered spars and paint-work as bright as a Spaniard's cuirass: ready rigged, her sails hoisted and drawing gently, her royal banners blessing the vulgar air—and the whole dainty conceit thirsting for the embrace of the sea. What it must be, to command such a ship! What it must be, to command the gold to build her! Matthew gestured towards the treasure with his ale-pot, and asked:

'What is she to be called, the little yacht?'

It was Merrylegs who answered, slyly: 'Why, the *Fubbs*, of course! Have you not heard? What else would the King's favour-ite vessel be called?'

'*Fubbs?*' Matthew repeated the strange word, which the smirking round about him declared to be part of an unknown jest. 'There's a name! What does it mean?'

Now it was Snarly's turn. 'Where have you been, for the love of God? It is the King's pet name for his trollop. Do you not know the jest? They say even the little scholars in school make a new conjugation of it. "I fubb, Thou fubbest, He fubbs." But there is one thing certain. We fubb *not*!'

The others laughed, though the Bishop looked disapproving. Matthew was still lost, and did not mind confessing it. He knew that Snarly would not have been content if he had been King Charles himself, and that the bitter cry of 'We fubb not!' was the very curdled froth of the world's envy. This apart, Matthew was as ignorant as Adam before the Fall. 'What trollop is this? I have been away these five years. A man could lose count.'

'The Duchess of Portsmouth,' said Toss Pot. 'And I drink to her! I would drink to anyone who keeps us in such good ale.' He raised his mug, and proclaimed with a tapster's drunken gravity: 'Good fellows all! I give you the Duchess of Portsmouth, and the King's majesty, and the *Fubbs*, and all who sail in her, and all who do not, and God bless Sam Pepys who gives us the ships, and the landlord who gives us this ale, and——'

He then lost his balance, and fell down, and, sitting squat like a broody hen with a flaming beak, finally drank.

They could only laugh, as the throng all round them laughed, and then drink, and then raise up the fallen warrior. But the Bishop, who took the smallest possible sip, like a maiden lady entrapped in her first carouse, looked coldly down a very cold nose.

'You may jest,' he told them, as if they should all thank him kindly for his dispensation. 'But open sin is open sin, if the King himself gives it licence. Indeed, it is worse, when the Lord's anointed teaches his subjects such lechery, and makes a brazen scandal of it! He may call her Fubbs, if he wishes so. He may call her Cock-a-doodle, for all I care. I would call her Bathsheba!'

There was less of a stir at this denunciation than he might have hoped, and Matthew felt that he should come to the aid of the prophet.

'Bathsheba? Did the King then send her husband into battle, to be killed like Uriah the Hittite?'

The Bishop looked at him, for the first time, with approval.

'You know your Bible, lad, if you do not know your tailor! No, there was no need for murder to aid adultery. Perhaps the King might have gone so far, if there was a Duke of Portsmouth who wanted his goods back again. But there is no such valiant cuckold. Mistress Fubbs rose to the nobility by her own efforts.'

'The King rose,' said Merrylegs with a grin. 'She fell back.'

'The jade does not matter,' Snarly said spitefully, breaking into laughter which might, by softening envy, destroy a profitable grudge, 'If it was not her, it would be some other strumpet. What sticks in my gullet is the yacht itself. Another royal yacht! What does the King want with another? He has fifteen of them already, and all to play at racing-games with his brother the Duke of York! Is this what we pay our taxes for?' Snarly the wharf-rat, who had not the air of a large tax-payer, was looking villainously betrayed. 'Royal yachts! Such wicked waste! Royal toys, I call them!'

But the King had an unlikely ally, Big Bum, who, it seemed, had been a loyal sailor before he became the monstrous bag of guts now stacked in their midst. He took up the challenge readily.

'Royal yacht she may be,' he wheezed, 'but that is not all. She is navy-planned and navy-built. They would not carve toys, not even for the King. It is their chance to try new plans, new ships. This one——' like Matthew, he gestured towards the *Fubbs* with his pot of ale, 'is Phineas Pett's new dream. Ketch-rigged—the first ever to be so. Will she sail better? That is what we shall find out, and perhaps copy it later. She will serve with the Fleet if need be, like the *Henrietta* and the *Katherine* in the last fight with the Dutch. Or she will voyage round the coast, and plot the tides and the shallows. So talk not of royal toys!' he finally growled at Snarly. 'No ship of this quality is ever wasted.'

Matthew, made careless by the ale and the good sea-talk, plucked a familiar name from the air. 'Phineas Pett? Him that built the old *Bear* for Sir Walter?'

'The *Bear*?' Big Bum repeated, puzzled. 'What *Bear* is that? And which Sir Walter?'

'Why, Raleigh, of course,' Matthew answered—and stopped appalled, before a gaping chasm. Then he managed a foolish smile. 'Nay, that cannot be! My wits were wandering—'

Among the circle that looked at him in wonder, the Bishop's eyes were the most searching. He also had the most authority.

'What know you of Sir Phineas Pett that built the *Bear*?'

'I must have heard his name,' Matthew said lamely.

'Then you have long ears. This one is his grandson.'

It was a moment of great danger, but Matthew was rescued by chance and time. At the further end of the quay, there was a sudden lively stir, and the crowd swayed as they pressed back and forth. There came the glint of steel on soldiers' breasts, and the glitter of halberds prancing tall above the throng. Behind this, a burst of new colour caught the sunlight: the gleam of silks, frothing plumed hats, lace ruffles of a quality to top all others.

Then there was a roar from a thousand liberated throats, and loyal huzzas, and female mating cries, as the tall handsome figure strode forward through the parting crowd. The Bishop, forgetting all past damnations, altering course as swiftly as a vagabond actor might exchange his clothes, murmured devoutly: 'Here comes His Majesty! God bless him!' as the royal party mounted the draped and bannered dais in front of the yacht.

The great moment was at hand, and the greatest man in the realm was its spear-point.

If the King who had ruled all their lives for two-and-twenty years was, at fifty-two, past his lusty prime, as some whispered, he was still every inch a man and a monarch, with the keenest mind that ever topped a throne. As he stood smiling before them, his huge curled wig and foppish clothes might, on any other man, have earned him disrespect, and the outrageous petticoat breeches—tight to the knee-bone, then flouncing out into a full twelve inches of foamy lace—would have seemed fit only for ribaldry. But for Charles their undoubted King, all such outlandish frippery, being his own choice, was also theirs.

They loved him as he faced them, and the full-throated roar came from full hearts as well. For a near quarter-century, he had shepherded them through countless hazards: the scourge of fire, the desolation of plague, war and the threat of war, plot and counter-plot, broil and ferment enough to topple a whole dynasty of kings, and their subjects with them.

But King Charles had lived, and endured, and prospered, and—almost beyond belief—had fashioned Merry England out of the dull grey tyranny of Cromwell the Lord Protector. A King who could do that, and rebuild a great city from its ashes, and turn back the Dutch from its very water-gates, might bed ten mistresses and sire twenty royal bastards, and don all the petticoats he chose, if it kept him so supple and strong, so loving and so loved.

Portsmouth loved him especially as a friend of sailors. They knew from the common talk that the King and his 'little advocate' of the Navy, Mr Secretary Pepys, had toiled hand-in-hand, for wearisome years, to give England a fleet, and its seamen their bread. They loved above all the fact that, on this great occasion, the fat and fussy little man in his fine clothes now stood side by side with his tall monarch.

He had earned his day of glory in the sunshine. King and subject had both earned it, and Portsmouth would give it to them— and be damned to all who thought otherwise!

Perhaps only Matthew Lawe, for whose instruction the Bishop had been pointing out the dignitaries, was not yet to be numbered among the worshipful throng. He was in some turmoil of spirit, and it was the first sight of the King which had given him such a qualm of well-remembered fear. Without a doubt, Charles in his finery was, from top to toe, the image of Henry Morgan—or, to put it more loyally, Henry Morgan was the image of his King.

If Matthew were not so full of ale, the sight might have unmanned him altogether. Had he travelled so far, to meet his most deadly foe face-to-face at the first breath of ease? While he knew in his heart that this was a false and foolish vision, yet that heart could still beat to bursting when such vile memories were conjured up.

The Bishop continued his catalogue of the great, as one by one they took their place of privilege on the dais.

'There is your Mistress Fubbs,' he said disdainfully, when a small figure in the gayest clothes of all made its way to the forefront. Though she was too far off for true appraisal, even by the most hawkish of males, yet Matthew found himself stirred at this glimpse of a lively shrine of sin. Did a man have to be King, to command such sprightly mating? . . . 'A bed-fellow for twelve years, and now a Duchess! And she came to court as a maid-of-honour!'

'Both maid and honour went quickly enough,' Merrylegs said lewdly. 'At one royal entrance.'

'The gentleman in the black hat and cloak,' the Bishop announced, ignoring such unseemly jesting, 'is the Master-Shipwright, Sir Phineas Pett.' He looked at Matthew like a schoolmaster, which perhaps he was. 'Your *old* friend.'

'My very oldest, I dare swear.'

'Now you have a better sight of Samuel Pepys,' the Bishop went on, balked of his advantage. It was true that the diminutive figure in shining pale blue came into sudden prominence, as it moved forward to the front of the dais. But then the 'better sight' suddenly melted to nothing. Pepys stumbled, and would have fallen head-long, if a royal arm had not come to his rescue. The moment of dis-order stirred the crowd, and the ripples of unease reached their own circle.

'By God, the fellow is drunk!' Snarly said, with that particular satisfaction which could turn an envious man into a happy hypo-crite. 'What a day to choose!'

'Have some pity, for pity's sake!' said Toss Pot, an excellent judge in such matters, with a companionable heart to soften his verdict. 'He is near blind, as we all know. He did no more than miss his step.'

But Snarly was not prepared to pity anyone, least of all a man of quality set in superior company.

'Mr Pepys is not so blind, when he *peeps* into the public purse! Look at those fine clothes! First he is Secretary of the Navy. Then suddenly he is a traitor, walled up in the Tower of London. Then he is near back in office again. In and out like a weather-cock! But does he starve? Never! Now he drives in a gilded coach with fine black horses. He has treasures in his house——' Snarly almost choked, with righteous indignation. 'Whence came such a moun-tain of money? We all know the stink of corruption that fouls every Navy ship that was ever launched. There is a finger in every pie, and in every purse of guineas too. Sam Pepys may be blind, but he has a nose as long as an elephant! He can smell a profit, when-ever he sees a hank of cordage or a stack of timber. Do not talk to me——'

'We do not talk to you,' Toss Pot broke in forthrightly, 'because you do not leave us room. For the love of God, forbear! We are here to enjoy! If a hank of cordage would secure your tongue, I would pay for it myself, with a fat profit for Mr Pepys as well. I'll wager his fingers are not half so crook't as some others.' He eyed Snarly with a friendly relish. 'Did we not see you snug in the stocks last Michaelmas? A small matter of some turnips gone astray?'

'The matter was cooked,' Snarly said venomously.

'Aye—like the turnips!'

In laughter the question dissolved, and the Bishop, to whom talk

of stocks and stolen turnips brought almost as much pain as a royal
fornicator, took up his more virtuous tale.

'The Mayor of Portsmouth,' he announced, 'at the further end.
You can see his golden chain of office . . . I have his acquaintance,'
he was constrained to add. 'His wife's elder brother married a cousin
of my mother's half-sister.' By his tone, he defied them even to hint
that there could be a closer or more impressive connection. 'And
there comes the last of the royal party. The Duke of York, brother
to the King!'

'Papist dog!' Merrylegs exclaimed with genial contempt, as
Matthew watched another tall man, with a grey sombre face and
clothing to match it, advance and take his place at Mr Pepys'
side. 'Do you mark that swelling above his paunch? There he
carries the Pope's own crucifix! He earned it by a kiss.'

'A kiss?' Matthew asked in wonderment.

'Aye. He took the Holy Mass and kissed the Holy Arse!'

The coarse jest did not offend even the Bishop. With England set
against Rome as fiercely as it had ever been, the merest breath of
Popery wafted as foul a stench as came from any common sewer.
Only Big Bum, the old sailor grown too gross for the sea, said a
word to the contrary.

'Papist he may be. Papist he is, and makes no shame nor secret
of it. But dog he is not! He has been a good friend to the Navy, as
good as the King or Sam Pepys himself. You should not call the
Lord High Admiral of England a dog.'

Merrylegs' humour was not to be marred. 'Prince of Cats, then,
for all I care. I'll wager he can scratch a nun in the dark.'

'To be exact,' the Bishop intervened, before the talk took another
lamentable turn, 'the Duke is *not* the Lord High Admiral of England. He *was*, and then he was put out of place, for that same
Popery. Now he directs our naval matters. But the office is gone.'

'And the rogue is still there!' This was Snarly once more,
smarting from defeat upon one field and turning doggedly to
another. 'Tell me, why is that?—save that he is the King's brother.
He did not win us so many wars! Think of the old battle names, and
you think of defeat. The Texel, the Four Days' Fight, the Solebay—
the Dutch gave us a thrashing every time. So the great Lord High
Admiral—'

'We won the Four Days' Fight,' Big Bum broke in. 'I was there,
and you were not, and I can tell you. We lost twenty ships, but the

Dutchmen quit the Downs before we did. If Prince Rupert's squadron had not been separated from us, to keep an eye on the French——'

'Well enough, well enough!' Snarly said irritably. 'I will give you the Four Days' Fight. But within a year the Dutch fleet came up the Medway, did they not? They sailed up to Gravesend itself, with a broom at every mast-head. They even broke the Great Chain at Chatham! Where was our Lord High Admiral then? I tell you, Papists cannot fight. Or they do not choose to!'

'You mean, treason?'

'I mean——'

But whatever was Snarly's meaning, on this perilous ground which could still cost a man his head, it was lost to the waiting world. On a sudden, a cannon-shot boomed out from the bastions above, as if to silence all petty carpers and traitors alike, and the chosen apprentice-lad, beribboned like a little May-pole, advanced towards the launching cradle, with a wooden sledge as big as himself poised for the stroke.

All else seemed, at that moment, foolish vanity, as empty as words, or the fleeting wind itself.

A breathless waiting; then King Charles raised his hand. The sledge-hammer fell, with one fair stroke, and the chock beneath the hind-wheel vanished. The well-greased chariot began to roll and rumble down the slipway, then took the water with a merry splash. First the forefoot dipped, then the whole shapely length of the hull—and the *Fubbs*, fluttering free of her cradle, swam at last!

She swam as handsomely as anything afloat, and there was a rousing cheer, and a marvelling of admiration when men concealed below, as within the Great Horse of Troy, ran out on deck and began to trim the yards and bear down on the helm. Within a moment, with all sails drawing, the *Fubbs* turned shorewards to face the dais, as if to curtsy to her King, and then began to cruise very prettily down the length of the harbour. It was all done in a right seamanlike manner, and the crowds loved every moment of it.

'Now they christen her,' said the Bishop, who had been staring open-mouthed, as happy as any common mortal. 'Look—there is the goblet now!'

There was a gleaming flash of gold from the dais, as a shapely chalice was passed from hand to hand, from the King to the

Duchess, from the Duchess to the Duke of York, from him to Sir
Phineas Pett, and so on down the line of the royal party.

'By God's blood!' Toss Pot said suddenly as he watched, 'I wish I
was younger, by a year or two. I would dive for it myself.'

'Dive?' Matthew questioned him. 'What dive is this?'

'Have you not seen the christening splash? When that golden
goblet returns to the King, he will drain it, and then throw it into
the water. All the lads who take pride in their swimming, and some
fools who are ripe for drowning, will dive in and try to fish it out.'

'What then? Can they keep it?'

'If they choose so. But what use is a golden goblet to common
folk? It is too tall for under the bed! Nay, they sell it back to the
Master-Shipwright.'

'At what price?' Matthew asked.

'Whatever the old shark will pay. Perhaps twenty guineas.'

It was riches beyond avarice. 'By the Wounds, I will try it myself!' Matthew cried. He had a pressing need to relieve himself,
apart from all else, and if he earned twenty guineas at such business
he would pass water till Portsmouth Harbour burst its bank. 'How
do I set about it?'

'Canst swim well?' Big Bum asked doubtfully. 'There are rogues
down there who have been waiting for this chance since the keel was
laid. For twenty guineas, they would be ready to rip off your parts,
if they see you come near the gold.'

'I can swim,' Matthew answered. 'As to the rest, I have played
water-games before this.'

'Good lad,' Big Bum pointed. 'Make for the steps there. Mark
where the goblet falls, and jump straight-way, before they stir up
the mud. With luck, you may dive and follow it down.'

'Would you be buried here, or in Barnstaple?' Merrylegs asked.
But Matthew was gone already.

When he had set his elbows to work, and forced a way through
the throng, he joined a band of suspicious men, all glowering
ruffians save for a scatter of merry-makers, gathered at the head of
the steps. Asked what was his business, he said nothing, and
avoided every eye. Instead, he watched the dais and the progress of
the goblet. Already it had made its laughing circle of the great, and
was back in the King's hands.

His Majesty, enjoying a royal moment, was in no hurry to
conclude it, and condescended to a little by-play as he first peered

into the goblet, then shook his head in seeming dismay at what remained of his share, then set to work to drain the last of the wine.
He had the crowd laughing, as any great man who was great
enough to act the clown might do, without loss of reputation, then
waiting in warm silence as the part was played out.

It was not delayed a moment longer than public patience would
allow. The empty goblet dropped, then rose in the air, then was
launched in a generous arc which bespoke a manly arm behind it.
At this, Matthew Lawe stripped off his jerkin and dived head-first
into the harbour. As he came to the surface, the golden target made
its own dive, five yards beyond his head.

There followed a time of some violence. Angry cries, peals of
drunken laughter, shrieks of merriment, were succeeded by
splashes, sullen spouts of water, a thrashing of arms and legs,
bullet-heads butting, flailing limbs, bubbles bursting out, bodies
sinking like stones and rising like volcanoes, as the host of ambitious, greedy, desperate, or careless men launched themselves upon
their goal.

Though Matthew, with lungs well filled, was already sub-
merged beneath this watery avalanche, he felt its weight, and some
of its malice. A hand reached out to grasp his hair, and he clawed it
loose with nails which did not scruple to draw blood. He had the
goblet still in view as it wavered downwards through murky water,
and it was within his grasp when another pair of hands fastened on
his throat.

This time, a backward kick found its mark in some defenceless
bowels, and he was free of enemies.

He rose with his prize, thrust it down his breeches, and leaving
behind the milling shoal of swimmers as if he had surrendered his
aim, dog-paddled gently ashore.

His hour of triumph began when he topped the steps once more,
a bedraggled figure naked to the waist, below which the absurd
galligaskins which the Bishop had condemned bulged monstrously.
He retrieved his jerkin from a sympathetic onlooker, and, since he
was the first warrior to return from battle, his movements were
watched, even from the royal dais.

When he reached down, and gripped the golden cup, and
brandished it on high, no vagabond actor suddenly touched by the
divine spark could have won greater acclaim.

'By Holy Cross, he has it!' his coat-holder shouted, and the cry was taken up on every side as the sunlight put fire to the goblet. The crowd decided that they loved this wily winner, and cheered him to the very echo. Not far from him, his late circle of friends could be seen jumping for joy. Even the Bishop was waving a black hat like a sepulchral banner; even foolish fellows still wrestling for shadows in the water stopped their frolic, and gaped at their own defeat, and gave him best.

There had been no other moment in Matthew's life when his friends outnumbered his foes, by such a regiment.

Yet the peerless peak of that day was still to come. Mugs of ale, new-found friends, murderous swimmers, even golden goblets, might by now seem to be priced at two-a-penny, in his present happy progress; but Master-Shipwrights and moguls of the Navy were not, duchesses were not, royal dukes were not, and as for Kings . . . In the next hour, Matthew Lawe made a giant step forward. It could not last, but while it endured, it lit his private sky as the Northern Lights of old could set a whole icy world on fire.

Even as he mounted the steps of the dais with his prize, aided by allies unknown, spilling a trail of water like a child caught on the hop of mischance, he thought: 'They cannot kill me now, whate'er I do.' Within an hour, his thought was: 'By God, they could not love me more, if I were one of King Charles' own spaniels!'

The first to meet him was the guardian of civic pomp and circumstance, the Mayor of Portsmouth himself. His Worship, close to, loomed somewhat less than his chain of office: a small swill-tub of a man, humble as a dwarf marooned on the Giant's Causeway, red and sweating under the pressures of this day, which must have brought him challenges and terrors undreamt-of on the morn of his election. But to Matthew Lawe, at least, he could show the face of authority.

'Well done, young man,' was the most he would allow himself, as his gaze rested with distaste on this intruding scarecrow. 'But stay where you are on the stairway. Give me the goblet—' he held out his pudgy hands, 'and I will see that you are properly rewarded.'

Matthew might have said Yes to all, on any other day. But he had much ale, and an equal stoup of triumph, to spur him on. A way had opened among those on the dais as they turned to survey the victor, and a way had opened in his mind. He was within sight, a brief glimpse told him, of greatness, and a real King crowned it.

He would make his own mark, and be damned to all lackeys standing in his path.

'I would deliver it myself,' he answered firmly.

'What!' the Mayor said, appalled, in an urgent whisper. 'The Master-Shipwright——'

But he had spoken too loud for secrecy. 'Let him through, if you please, Mr Mayor,' the voice of true majesty broke in. 'He has earned his entrance.' And then, in a jovial tone: 'Pett! Button up that benevolent purse and prepare for boarders!'

Yet it was not towards Sir Phineas Pett that Matthew steered his course. A certain naughtiness had taken hold; a wilful determination that on this day of modest good fortune, the best was yet to come, and only waited to be won. Once again, they could not slay him for trying. . . . Vaguely he saw the tall figure of the monarch; clearly he spied the delicious armful which adorned his side.

Without hesitation, Matthew marched forward, bowed low as he had seen Henry Morgan make his humble duty before a like treasure, and offered the golden cup of *Fubbs* to Fubbs herself.

Louise de Kéroualle, Duchess of Portsmouth, was not one of the world's holy innocents; no woman who had made the progress from lady-in-waiting to King's mistress, from French spy to English royal pet, from swooning girl to seasoned mount, could be found wanting in any circumstance short of rape. On an arduous path, she had known every approach, from grey Jesuit slime to the froth of court intrigue, from the shy simper to the prong erect, and she had made an exact measure of Matthew Lawe before he had straightened his back.

Strong, well-formed young men, in a humour beyond their station, were minted every day and melted every hour. A gentle flame was enough.

Her answer to this newcomer was French, English, feminine, and wise, all at the same stroke. She assumed a radiant face, took the cup, and then—wonder of wonders—curtsied to her benefactor. Royal smiles, encouraging laughter, a burst of cheering, and lustful growls of approval from the commonality, showed that she had acquitted herself to perfection.

The strong young man in a high humour lost his sauciness and became a pliant slave again, from that moment on.

Yet he had enough of his wits about him to note once more a lesson of life—that people close by, whether high or low, scarcely matched their public portrait. The radiant Fubbs, though arch and elegant, was now a haggard beauty; if she still reigned as mistress, it must be by the wiles of a ready tongue rather than the engulfing limbs of a whore. In early middle age, she was hardly a meal for a lusty man.

The lusty man himself, who now bent upon Matthew a pair of piercing eyes, was no more a king in this realm; at fifty-two, he had grown worn by care, so that the dashing monarch, long a by-word for every whim of excess, was a tall shrunken shell whose only prodigality was to dress for a part which he had outplayed.

He might be loved for his own consuming love of his country; but as animal man he must be limping towards his grave, with the best of life a matter of fond memory. No peacock display of finery could mask the gaunt crow within.

Yet he could still play the king to all his loving subjects, and he did so now. The keen eye had taken in all of Matthew Lawe, as had that of the duchess, and the monarch made his generous best of it.

'You swim like an otter,' he said smiling. 'Where did you learn such skills?'

'As a sailor, sir,' Matthew answered, abashed.

There was a hissing from behind him, which was the Mayor of Portsmouth on his civic guard-duty. 'Say *sire*!'

'Sailors are the blood of this realm.' The King smiled again, but only in gentle mockery of himself. 'Could I say less, in Portsmouth? . . . Your name?'

'Matthew Lawe, sire.'

'Well, Matthew Lawe, you have conquered more than the sea today.' He gave an intimate glance towards his duchess. 'I have never witnessed such gallantry, on either side.'

'He does but follow his king, sire,' the duchess answered readily.

'Indeed, I hope not. That would be the worst of *lèse majesté*, in all this kingdom.'

The by-play was mysterious, and far above ordinary mortals; it even seemed to include the Duke of York, a grey unsmiling figure who stood apart from everyone and everything—from his brother the King, from the royal mistress, certainly from this shabby half-drowned sailor who had invaded an occasion of dignity. His aloofness made its mark. There was a pause, the slight unease of the great world when it knows that it has melted enough, and would close its ranks again, to signal a return to all proper stations.

Once again, Fubbs took feminine command. She turned and whispered to the King, whose brow cleared on the instant.

'Why, yes!' he exclaimed. 'You have the goblet, while the sailor has nothing, for all his gallant pains. He has earned more than a curtsy . . .' Suddenly, smiles sprouted again, as King Charles bound them all together in accustomed hands. 'Mr Mayor!'

The Mayor of Portsmouth, hovering between the heaven of a royal occasion and the hell of public mishap, came to trembling attention.

'Your Majesty?'

'Pray remove your hat-of-office, and put it to profitable use.' King Charles, whether jesting or not, could charge his lightest word with all the weight of command, and it was a command which now ensued. 'We are in honour bound to ransom this cup, if it turns us all pale. Walk the length of the dais, Mr Mayor, and say——' somehow he managed, even in majesty, to mime a suppliant Mayor of Portsmouth soliciting the favours of the great, 'say to all, "Spare a guinea, kind sir—a guinea *or so*—for the swimmer." I would direct your diligence towards my royal brother, to the Master-Shipwright who has so far escaped notice, to Mr Pepys, and to all such great benefactors. . . . I will lead.'

He led indeed, to some purpose. A gentle lift of a finger under the nose of a courtier close to him produced a silken purse, and the purse shed five golden guineas, dropped into the Mayor's beaver hat like canary-birds alighting on a parched brown lawn. Others followed steadily—what choice had they, in all the royal world?— as the collector made his rounds.

At the end, the Mayor's high-crowned hat bore more weight of gold than it had ever held brains; and when it was all spilled out into Matthew's cupped hands, he was accorded the gracious smiles due to a victor. For a space, he became the centre of attention again: a royal favourite, in a world where happy chance could make or break a man, and do many better or worse things for a woman.

He even won a cold word of approval from the Duke of York, and affable condescension from Sir Phineas Pett, who had escaped a ransom of twenty guineas by the lordly advancement of two. The hero of a brief hour, his head in a whirl, Matthew Lawe had never been so warmed and cherished, in all the hungry years so far.

Presently the *Fubbs* yacht was seen to be returning down the length of Portsmouth Harbour, and all eyes and all attention abandoned one new toy to centre upon another. Matthew, with his wealth safely tucked into his breeches, was uncertain how to take his leave. His hour was out, and he with it. Did a man just walk away?

The question was answered by another stroke of chance. As the shapely little yacht made her return, even the noble assembly on the dais forgot its courtly manners, and pressed forward this way and that, like common folk, to gain a vantage point. Before long, the surging spilled out one small awkward figure, tumbled backwards by stronger bodies.

It was Mr Samuel Pepys, who would have fallen if Matthew had not stepped up to support him under the arms. Though, with such timely aid, he managed to keep both balance and dignity, he was not pleased.

'Throngs, throngs!' Mr Pepys said testily. 'One might as well be set loose in Bedlam!' He turned to face his Samaritan, not yet encountered, and peered upwards with clouded eyes. 'Thank 'ee for a strong arm . . . Who are you, lad?'

'Matthew Lawe, sir. I brought back the christening cup.'

'Ah, the sailor who can swim. That is a rare fish indeed!'

Matthew had made no judgement, as yet, of what such a great man, Secretary of the Navy Board, should resemble, save that he *must* be a lordly tyrant, with his power of life or death over common sailors erupting like volcano fire under threatening brows. Mr Samuel Pepys was no such thing. He was tiny, barely up to Matthew's shoulder, and fat, and fussy, and decked out like a paunchy marionette whose master could afford silken strings.

Though the face was wise, it was not commanding. Strip him, Matthew thought, and subtract his rank, and he might be an ageing lawyer's clerk, robbed of his eye-sight by musty parchment and cramped penmanship.

But whatever his looks, or lack of them, Mr Pepys had authority beyond all question. He showed it now, with an easy humour.

'You did well, Matthew Lawe. I heard the King call you sailor.'

'Aye, sir.'

'Well, you sailed upon us like a true man-o'-war's man. The boarding of Mistress Fubbs was most valiant . . . I passed a twelve-month in the shadow of the Tower for less . . . I hope you did as well in profit.'

'Sir, I have not yet counted.'

'You must indeed be a sailor! But when you come to reckon it, you will find a gold piece from myself—and gold, in these times, is scarce as feathers on an egg.'

Matthew could not forbear to say, 'Thank you kindly, sir,' though the gift seemed somewhat less princely than Mr Pepys' manner of telling it.

'It was my duty and my pleasure,' Mr Pepys went on grandly. 'The old customs are best to be observed, even in these days of licence . . .' He caught a brief sight of the yacht passing the end of the dais, curvetting to a fresher breeze like a frisking colt, and he exclaimed: 'Now *there* is the prettiest thing that ever swam! And when the royal pleasure is slaked, we can use her as guardship at Chatham or Sheerness. . . . Did you ever sail in one of the King's ships?'

'I—no, sir.'

Pepys looked at him with sharper eyes. 'You hesitated. What was it—did you run? You need not fear the Press-Masters, in these times of peace.'

'No, sir. I only intended, that I have seen no special service. All the ships in the realm are the King's.'

'Amen to that! Whether war or peace, they serve a common cause. We need to bind all together, so that one single weapon is forged. . . .' He patted his pale blue ruffled sleeves until they were to his liking, then rubbed his plump hands one against the other. 'Well—I to my desk, and you to your mariner's toil. Have you a berth waiting?'

'No, sir. We take what we can, or else go hungry.'

Mr Pepys gave another of his sharp glances, belying his fabled short sight. 'How would you mend that, as a sailor?'

Matthew, emboldened by this condescension, answered: 'Sir, treat us as men of worth, not to be cast off and cast on again, as the mood takes.'

'But when there are no ships, and no places?'

'If that is true today, 'twill be a lie tomorrow.'

'Well said! You are older than your years . . . But believe me, for your own comfort—there are men at work on such questions. And I am one.'

When he was alone, Matthew counted his reward. There were forty-four golden guineas, and some meaner coinage besides—

what fugitives from a church-offering were these?—to swell a noble prize. Now, for a space, he was rich in worldly goods—and, he decided, richest of all in human kind. But by this he was not remembering the King, nor his wayward duchess, nor even the Bishop and his motley choir of friends.

Of all the fine quality and the merry dross, he remembered best of all a small, lonely, funny old fellow with, it seemed, great new thoughts of the sea: Mr Secretary Pepys, of His Majesty's service—and of Matthew Lawe's also.

II

When, under the assault of that legion of lifelong friends to be met at every tavern in the town, his store of guineas had dwindled to ten, Matthew Lawe called a halt to foolishness, and shipped out of Portsmouth. His berth was the *Grace & Favour*, a humble collier plying from London to the far West Country and back again: voyaging down-Channel till she had sold all her black gold, to the last slimy hod-full of slack, and then wandering homewards.

On that return journey, the *Grace & Favour* carried anything that would not float of itself: Cornish cheese, live bullocks, ripe or rotten fruit, goose-manure or the geese themselves, soused mullet, salt pilchards, poor passengers, or hogsheads of Plymouth ale.

Thus she was not only humble, with little grace and no favour, but dirty as well—a very ditch-vagabond of the sea. It was in this small stinkard of a ship, better than a year later, that he rounded the Kentish North Foreland and so made the Pool of London.

He had not sighted it since the old days of Henry Hudson, and now, in the spring of 1684, the prospect was astonishing.

This was a great new London which had outdone the phoenix in rising from the ashes of the old. After the most malevolent of plagues, which in a single week had condemned seven thousand citizens to noisome death, and the searing of the greatest fire ever seen in England, hope might have died, and the city with it. But Charles their King, and Christopher Wren their noble architect, had other dreams, and a greater spirit to fulfil them.

The brave reality now blessed the eye, with whole terraces of lordly buildings, finer than any he had ever seen; and all crowned at the eastern end by the rising peak of the new St Paul's, to lift the heart of Everyman closer to heaven.

At the other margin of the river's great sweep, like a stern anchor to which the city might ride secure, the towered Palace of Whitehall set a kingly stamp upon a dream come true.

But in this vast metropolis paved with such golden promises, Matthew was an orphan again, a mere river-bank wanderer. Good sea-berths were scarce, and poor ones no better than starveling servitude. It was as he had dared to say to Mr Secretary Pepys, speaking from an ancient experience which no man, great or small, could ever share: once cast off, a sailor might rot like a hulk on a mud-bank, for all that God or man would care.

Then chance, the most precious clove-spice in the whole pomander of life, came to his aid, and set him on the high road, perhaps for ever.

On an April morning, warm enough for a man's bones to credit, at last, the advent of an English spring, he was wandering the quays which flanked the water-gate of Somerset House when he came upon a scene of despair and desperation. Its central actor was a liveried waterman, his choleric face as purple as his coat, who seemed near to bursting with rage as he looked this way and that along the river front.

He darted a furious glance towards Matthew when he came in sight, then turned away with a curse. It needed no magician to divine that he was waiting for someone, and could not wait much longer. Below him, his boat lay idle on the lapping tideway: a handsome little barge, gilded like a coach, with two benches for the oarsmen, and stern-sheets cushioned for half a dozen passengers.

In it were two untended oars, an empty tiller, and one other man in the same purple livery. This was a mournful staring man, as slack as the figure on the quay was taut. He also was waiting, but time did not press against him. Nothing pressed against him save the fatigue of idleness. His hunched shoulders and empty eyes proclaimed: If I must row, I will row. If I must sit, I will do that also, just as willingly.

Perhaps he was hired by the twelve-month. Perhaps he was in bondage forever. In either case, the world which passed him by was not his burden.

The master-waterman on the quay stamped and swore again. Then, as Matthew drew near, he came to a resolve.

'You!' he barked out. 'Can you pull an oar?'

Matthew, in an easy mood, was not above a taunt for a thwarted man. 'Aye, master,' he said, as simple as any rustic. 'Where shall I pull it?'

The waterman turned from red to scarlet, and raised clenched hands to heaven. To save his reason and his bursting veins, Matthew relented.

'Aye, I can row. What is it you need? A bow-man? A man to keep stroke? And what is the great turmoil?'

The waterman looked at him, and suddenly the steam went off the kettle. 'Well, you have the sound of a sailor, thanks be to Christ! . . . I am called to the Privy Stairs.' He pointed towards the noble outline of Whitehall Palace. 'I am called *now*! And no one to pull the second oar. My whoreson mate is sick or drunk! So some other dog will get the fare, and I lose a silver piece—and my credit!'

'How much for me?' Matthew asked.

The waterman turned sly. 'I'll see that you do well.'

'*I* will see,' said Matthew, who knew a man caught on a lee shore when he met one. 'You may have me by the journey, or the hour, or the day. But not for a promise. I must eat more than promises tonight.'

'You damned dog!' the waterman shouted. But somewhere near at hand a church bell struck the hour, and he jumped as if the bell rope had twitched his vitals. 'Well enough—the day's rate for the time you serve.' And as Matthew nodded agreement, the waterman jumped down into his little barge, and beckoned him to follow. 'Hurry!' he shouted. 'And put on that boat-cloak, or you will scare his Honour as well as the crows.'

'His Honour', when they had stemmed the tide and made fast at Whitehall Privy Stairs, was discovered to be Mr Samuel Pepys.

The little man did not stand small this morning; his moment of leave-taking from the Palace water-gate was attended by bowing courtiers, servants who gave him an arm across the planked pontoon, secretaries who preceded him with a chest of papers and saw it safely stowed in the barge, and even a glimpse of the Duke of York bidding him farewell from an inner courtyard. It was clear that Mr Pepys had grown in the world's esteem, perhaps within the last half hour.

He was also in the highest good humour, laughing as he stepped on board, waving to the company with all the condescension in the world. As soon as they were unloosed, and turning out into the tideway, he gave more evidence of this to the waterman.

'Jem Belcher,' he charged, with mock sternness, 'you have delayed me!'

'Ask your pardon, sir. I have a man sick.'

'I care not if he be hanged this morning! If you delay me, then I delay the grooms of the ante-chamber who attend me, and they delay the King, and so the realm rots . . .' But he did not seem to give a fig for any of this. Whatever had transformed him had transformed all the globe, and nothing could mar it. Sitting on his cushioned throne, some three feet away from Matthew Lawe whom he had not noticed, he was his own king, and could order everything in view.

Jem Belcher eased the tiller as the barge began to stem the ebb-tide. 'To the Westminster gate, your Honour?'

'No. I have two hours of holiday this morning. Take me down-river to the Tower stairs. I will find a coach for the return. And do not break your backs. I would enjoy this journey.'

In the bright sunshine, under gentle strokes which drew their main strength from the swift tideway, they began their progress down the great blood-artery of the Thames. There was all to admire in the fine buildings on their left hand, and the green meadows and rush-beds of Lambeth Marsh on the South Bank. There was room to breathe also, and freedom from the narrow alleys and pressing crowds of the city. The best view of London was to be won thus—to sit apart under the river's magic wand, and let it flow past.

'Better than a hackney!' Mr Pepys exclaimed presently. 'Better than my own coach. There is only one way to travel through this town.' And then, to prove that he could think of small things as well as great, he asked the waterman: 'Did you say you had a man sick?'

'Aye, sir. Doggett failed me. But I found another.'

Mr Pepys turned, and looked directly at Matthew for the first time. It was uncertain how much the pale eyes could see, but it must have been enough. He frowned, in puzzlement. 'I know you,' he said.

Matthew, rowing light with no great effort, looked up from his oar and nodded. 'It was at Portsmouth, sir,' he said. 'The launching of the *Fubbs* yacht. I am——'

'Matthew Lawe the swimmer!' Mr Pepys exclaimed. 'The sailor also . . . Do you follow me, Matthew?'

Matthew grinned. 'No, sir. But I would do so.'

As a scandalized and hideous frown from Jem Belcher showed, it was a bold say. It had been conjured up from nowhere, and it might have disappeared into the same pit, with no trace left save the rebuke of silence. But on this fortunate day, it found its mark.

'Would you now! . . .' Mr Pepys, no fool on any other morning, seemed to have nothing in his heart but the generous brotherhood of man. 'So—the first of a new breed of place-seekers!' But there was no malice in the words, as his next sentence proved. 'You may be in luck, Matthew Lawe—I am like to say Yes to all the world this forenoon.'

Then, for a long time, there was silence in the boat, save for some hard breathing from Jem Belcher, and an occasional groan from the man at Matthew's back—the bow-man with the iron of resignation in his soul, for whom this day was no more fortunate than yesterday, and would so continue for a century of tomorrows. The river and the city unfolded: the barge rocked and faltered in a chance eddy of current: the oars creaked, the sweet birds sang, the sun shone on a scene of blissful content.

Two miles down stream, they shot the swirling tide of London Bridge, with its fifteen narrow arches and its double rank of houses and shops, like ramparts balanced on a spine of stone; and so to the Tower stairs, and the end of their journey.

My Pepys awoke from God-knows-what musing of glory or hazard, and stood up as the boat was berthed. 'I have a box,' he declared, looking down at Matthew. 'Full of State papers, which means that if you let it fall in the river I shall lose my head, and if you surrender it on the way, you will lose yours . . . Can you carry such a box for me, and find a coach? Are you free?'

'Aye, sir.'

The bow-man held them close to the quay: Mr Pepys stepped ashore, while Matthew shouldered the box and made to follow him. Only Jem Belcher was out of humour with this scene. He touched Matthew on the arm, and his face was thunderous.

'What are you at, you crow?' he growled. 'Would you lick his arse?—the first great man you meet! Would you jump like a lap-dog when he calls "Up"? You are bound to me. I'll not pay——'

Matthew swung and faced him. A furious man was nothing new in his life, and nothing of consequence either. 'There is no charge,' he said grandly. Then he added, for good measure: 'Thank you, waterman. You did well,' and left Jem Belcher, as he had first found him, out of his wits with rage.

On the long rumbling coach journey through the new heart of the city, in which Mr Pepys seemed to take continual delight—whether at the sight of a fine building rising high, a fish-man crying his wares in the voice of Stentor, or a pretty maid arching her grateful breasts towards the spring sunlight—on all this homeward way, the great man who had sat so silent in the barge now chattered like a magpie fresh from the egg.

Matthew Lawe, a cautious companion, was not to know the cause of this holiday mood until he had picked up all the threads which had woven it—a task of many days. But in brief, Mr Samuel Pepys had returned, between one sunrise and the next, not only to full favour but to the greatest glory of his life; he was newly arrived from Windsor Castle, where he had kissed hands on a truly royal appointment.

From this morning onwards, after favour and disfavour, re-nown and venomous attack, he could sign his name as Secretary of Affairs of the Admiralty—a true Sea Lord, with power to re-fashion his beloved Navy, from captain to powder-monkey, virgin keel to tapering top-mast, according to his own dream. His journey from Whitehall Palace to the Tower had been a gulp of air, a sigh of purest pleasure, before he buckled on his armour and turned this dream into fact.

Being mortal, being Mr Pepys—old in sin as well as new in power—he could also spare a side-glance towards his princely salary: some £2000 a year, with house-rent, travelling funds, and all the other perquisites of office, to match a peak of authority never yet bestowed on any servant of the Crown.

But today his talk was of the past. He was fifty-one, of great repute, and tomorrow was bright with hope. Yet he had taken that river ride to touch his roots again, sometimes with regretful hands. When Matthew had found their coach by the west wall of the Tower, they drove north up Seething Lane; and here, it seemed, the memories were vivid and, for Samuel Pepys, crossed with scars scarcely healed.

'Seething Lane,' Pepys announced, peering out of the rattling window. 'I came to live here in 1660, when I was first made Clerk of the Acts. The old Navy Office was under the same roof. I was twenty-seven, and married five years. My wife was only twenty still . . . In that house—' he pointed, 'I lived for fourteen years. We were happy there—we had nine rooms to play at housekeeping— and then she died, and I was happy no more. They were cruel times, in any case.'

Matthew could only ask: 'How cruel, sir?'

'You must then have been a lad,' Mr Pepys told him, 'and not a Londoner either. We saw the Great Plague of '65 from that house, and it passed us by, though it was the most evil scourge that ever struck. The Bills of Mortality rose every week, by tens, and hundreds, and thousands . . . But the worst evil of the Plague— forgetting the danger, and the death-pits where people dragged themselves to die, and the stench of it, and the fearful sights and sounds—was what it did to living men . . . It made us more cruel to one another than if we were dogs . . . When it struck another house, we thanked God it was not our own, and barred our windows, and prayed to be spared, while pitiful wretches knocked at our doors pleading for help . . . The Court removed from London, and my wife too, and I stayed working alone in Seething Lane, with the streets as silent as if the wheel had never been invented. Then, when all was done and men crept back to town, I put on the airs of a brave archangel! But in truth, though I worked all hours, I never lived so merrily, nor got so much, as during that Plague time.'

'Got so much?' Matthew repeated, puzzled. 'In gold?'

'Women. From honest wives to street drabs, they thought they heard the Last Trump, and it was the most desperate call to arms I ever knew.'

Matthew, the country boy turned humble sailor, was conscious of a horrid shock. So this was Mr Samuel Pepys, the great officer of state . . . He was no better than some low-born randy rascal with gaping breeches . . . To go a-whoring in the midst of a plague, with his wife sent away for safety . . . He sat silent in the tumbling coach, as it rolled along Cheapside and into the holy shadow of St Paul's, and wondered what respect should be paid to rank, if all were equal hog-wash.

Then Mr Pepys, whether nudged by an Almighty Hand or divining the disapproval of this simple mind, sought to justify the occasion, as a prudent ambassador might inscribe 'Paragraph the Second: Some Points in Favour', after 'Paragraph the First: The Disadvantages.'

'Well, whatever the blame, I was punished—and by more than my wife's tongue when she returned and heard the talk of the town. Or I *began* to be punished, for it was a long business, and no man is quit until he dies . . . My reading eyes were starting to fail, from too much close labour; though I could still see a country view, and a street of people, and men in a room—and women anywhere!— print and writing were becoming mere torture. To discover a meaning, I had to peer down at a page, letter by letter, through a leather tube. By the day's end I would grope out of the Navy Office in agony, with a skull sick to vomiting . . . My own precious diary could scarce be kept up, for fear of going blind. At work, all long communications must be read out to me, while my best clerks had leave to counterfeit my signature when I told them what to write in reply.'

Matthew was once more in thrall to a valiant man. 'But sir, why did you not resign to it? So much of the work was done.'

'It is never done!' Mr Pepys' eyes, the subject of such turmoil, came round on him sharply. 'I had Navy business to finish, or to begin, and I still have. And I thought that my eyes might mend, if I plotted to spare them.' Then he smiled, the man overtaking the paragon of duty. 'And who gives up office when he can crawl to work?'

It was difficult to catch such a mood, or to answer it worthily. Did every man of stature melt into laughter at his own smallness? Were all such men ready to mock themselves, and win this race before the mob began to howl? Matthew did not know, and perhaps never would. He could only listen, and hope to learn.

'But before that, we were all punished, the year that followed the Plague,' Mr Pepys went on. Their coach was now descending Ludgate Hill, a poor cobbled pit of bumps and holes, and he steadied his plump body against the leather strap at his side. 'Where were you when the fire took London?'

'In the West, sir.'

'But at sea?'

'Aye. By turns.'

'*O Fortunatus!* . . . It is only afterwards that one declares, "I would not have missed this, for the whole world!" At the time all that one can say, or think, is: "I would give the whole world to be a thousand miles away!" I had hoped never to see worse than the Plague. But the fire of London topped all else, for evil and misfortune.'

'Did it touch here?' Matthew asked, looking out at the rise of Fleet Street and the brave houses on either side.

'*Touch?* We drive through the very graveyard! Remember where we took up the coach?'

'At the Tower, sir.'

'Aye. My fine former lodging! But now we are coming to Temple Bar, where the fire was at last contained. That leaves more than a mile of ruin behind us, and this was the heart of it. There was not a house left standing, in all this mile. From the Tower to the Temple, from High Holborn to the river, all was burned to nothing.'

'How long did it burn?'

'Six days. In that time, God made the great world, and in that time He destroyed this little one.'

It had started, Mr Pepys said as their coach voyaged onwards, on a fine September night, when he had been called from his bed, in his nightgown, to watch 'a little fire in the city'. It was near enough to them, he judged: somewhere in Mark Lane, a neighbouring street. But there was a good easterly wind. The little fire would blow away, or blow out. He had gone back to his bed—and that was the last time he slept, for four nights.

By mid-morning the flames had taken hold among numberless little wooden houses roofed and coated with pitch: many of them store-houses which held lamp-oil and tallow and spirits, as ready for the torch as a quick-match. At noon the northern buildings on London Bridge were all in flames, and the citizens of the river-side were already flinging their goods into barges and lighters, and making their escape.

The first to fail in this escape were the pigeons of London, which could not believe this disaster, and stayed too long on roof-tops and window ledges, and fried their wings, and thus fell down.

They were not alone in their disbelief. No one was doing a hand's turn to oppose the peril. Mr Pepys, a sober citizen with greater spirit than most, and greater access to authority, took boat to Whitehall, interrupted the King in conference, and gave him the news. He was sent back with a message to the Lord Mayor. 'Pull down the houses,' was the royal command. 'Make a fire-break.'

'I found the Lord Mayor, and told him,' said Pepys. 'I might have been speaking to the fire itself. He was already distracted, wringing his hands like a girl in a fit. "What can I do?" was his answer. "No one will obey me!"'

It was true. Those who had been scorched were in terrified flight. Their laden carts were running over each other. At the water's edge, boats and barges crashed together as desperate men tried to claw away from the burning land. The rest sat and waited for their doom, while the whole air about them grew hot with flakes of fire, and the noise and the crackling roar consumed all sense.

Pepys and his household, to escape the heat, had watched it from the south side of the river. At dusk, it was already enough to put terror into any heart of less than stone.

'You know the sea,' he said to Matthew broodingly, as if he still saw the old horror in the green fields of Lincoln's Inn. 'This was a sea turned to a furnace. But not an honest blaze—a horrid malicious bloody flame, in an arch of fire near a mile across . . . It could only gain . . . Next day, as it crept near the house, I sent our main goods away, by coach to the country or barge from Tower Dock. The whole world was on this same run—we fought for our space like animals.'

Then he went back to Seething Lane and the Navy Office, to do his duty.

The fourth day was the worst. Though the King took direct command, with the Duke of York at his side and a whole regiment of troops to do their bidding, the fire spread ever more fierce, and ever westwards. Old St Paul's went, sending a river of molten lead down from its roof as it fell. Flames leapt across the gap, nearly trapping the Duke and his soldiers within a ring of fire as they were blowing up the houses on Ludgate Hill.

But their concern was for the western side of the city. To the east, it was Mr Pepys' Navy Office which was in desperate need, and wild confusion had left it bare.

On his own authority, Pepys sent a call for sailors and dockhands from the yards at Woolwich and Deptford, to pull down the houses near Seething Lane. He might be running his neck into a noose, even at this time of chaos. By ancient statute, a man who destroyed houses in the City of London must himself pay for them to be rebuilt.

'But I sent a memorandum of it to the King,' the prudent civil servant told Matthew, 'and had back a scrawl of initials to cover me . . . The sailors came running with ropes and axes, God bless them, to save the heart of their Navy, even though it owed them wages long overdue. But it seemed that I left it too late. At two o'clock of another sleepless night, the fire reached the bottom of Seething Lane, and all was gone. Above the noise of the flames came monstrous explosions from the soldiers blowing up the whole of Tower Street, to save the powder magazine within. There were no voices left . . . So I gathered the last of the gold—mine and the Navy's—and locked the doors behind me, and took boat to Woolwich with my wife and Will Hewer, my clerk. We were near dead with tiredness . . . By morning, there could be nothing else to save.'

But by morning, when he returned at first light to survey the ruin of his house, he returned to a miracle. The fire had stopped at the lower end of Seething Lane. Both house and Navy Office stood intact, thanks to those loyal sailors. Indeed, all the fire in the city had now been contained. With the help of King and Duke, rope and powder, soldiers and seamen, the monster was tamed.

Walking the streets later, eagerly noting—and mourning—the profit and the loss, he trod a grisly blackened skeleton, still hot enough to scorch his feet through the soles of his shoes. And when the count was made, as the last fires smouldered underground, some 13,000 houses had been burnt, and 100,000 people rendered homeless.

'But I was proud of England that day,' Mr Pepys said, as their coach rolled downhill towards Charing Cross and the Palace of Whitehall. 'Where else in the world would a King and his brother, using their own hands, standing on their own feet, leading their own men, save so much of our capital from ruin?' There was silence for a space; then, as the Palace itself came into view, Pepys added: 'Not all was a curse, in '66. The flame, at the least, cleansed the city from its infection. Perhaps it cleansed *us* . . . And it gave us this reborn city . . . . My friend Mr Evelyn used to say, the happiest sight after the fire was to see the King, sitting up there in his embrasure with a great broadsheet in his lap, sketching out a new London.'

Mr Pepys stopped their coach at the margin of St James' Park, first to return the greeting of a fine gentleman who was watching, with great curiosity, the carriages flowing past him, and then to enjoy the view, which with its meadows and duck-ponds was as pastoral as the Garden of Eden.

The gentleman, dressed in the height of fashion with a towering periwig like a very pyramid of the Pashas, began with the sneer direct and ended with the most suppliant backside since Moses humbled the Egyptians.

'A hackney coach, Mr Pepys!' he minced. 'Are things so shrunken at the Navy Office?'

Samuel Pepys became, on the instant, the turkey-cock, and made no bones of it.

'Why, as to that,' he snapped, 'I do not know. Perhaps the *ghost* of the Navy Office is shrunk a little spare. The new Admiralty, of which I am, from today, the Secretary of Affairs, is tolerably prosperous—or will be, as soon as I come to grips with it. In the mean time, I take a hackney coach because it suits me, and I will ride a pig if it suits me, and—' he beamed, with all the merciless good will in the world, 'what a beautiful day it is, sir!'

Matthew had never seen a man so chap-fallen, nor so swift to repair it. His fawning politeness was a delight to witness. His parting words, after a string of inquiries and compliments as fulsome as a lover's with a rich widow under siege, were: 'May I have the pleasure of waiting on you tomorrow?' Then he bowed again, and backed away as if from a sunrise too glorious for mortal eyes.

'What a toad it is!' said Mr Pepys, as soon as they were alone. 'But by God, he can sniff a wind! "May I wait on you tomorrow?" *He will wait!* . . . When I was down, that man's little fist was ready to squeeze my privates. Now hear how he sucks at the teat!' The bitter contempt bespoke a whole census of such time-servers. 'Well, he may not know of it, but today I would not give a turd for him, whether friend or enemy!'

'Sir, I think he knows.'

'I am vastly relieved. To think that he might be in doubt!' But Pepys the man of capacity had no need to prolong a spite for more than it was worth. He looked across the spacious grace of St James' Park, shading his weak eyes against the sun and the rippling gleam of water. 'In the old days I used to walk here with his Majesty, or his brother the Duke, every morning of every day, and by God I will do so again! We will talk, as we used to, of the Navy—there is naught else of account . . .' He turned towards Matthew once more. 'Never forget—at the time I was speaking of, the plague year and the fire year, the Dutch were forever tearing at our throats, without mercy. Marvellous seamen—and ruthless in attack, just as they should be, if they have any fire in those good round bellies! And all we had to oppose was an idle shadow of a Navy . . .' Mr Pepys drew on his remarkable memory. 'You spoke something of it when we met at Portsmouth, and it was like hearing an old story from a young mouth. We starve our men, and let our ships rot, and then we are surprised when we run into danger, and have to build again at speed, till our backs are broken, slaving like Titans with less brains than a mouse. But what you cannot know of, is the enemy within.'

As he stopped speaking, and fell to brooding once more, Matthew made bold to prompt him. 'What enemy is this, sir?'

'Thieves!' Mr Pepys answered promptly. 'I would not call them less, nor flatter them more. The Navy has been a target of corruption since I was a young clerk, twenty-five years ago. I found it so then, and I will find it so again tomorrow morning. By God, if I had taken every bribe offered to grease my palm, I could have built my own ship-of-the-line. I could build my own fleet! We were fair game for every cheat who ever bought a contract, every hog that smelled the public trough, *because we had to have the ships*. We were sold green wood, rusted iron, rotten meat, bread not fit for the swill-tub. Officers stole the very cordage and canvas off the King's ships, and sold them back to the dock-yard as new. Pursers had a finger in every poor sailor's pocket. My own men, trying to get back stolen goods, were beaten, or threatened with a knife on the next dark night. What does a man do, when he is beaten for honesty? Either he swims with the foul tide, or he starves, or he dies . . . The matter that stuck in my maw was this—that there was contract-money a-plenty to be made by fair means. But no—such pigs must root for an extra snout-full, and if it cost a sailor his life, then sailors were simpletons, and deserved it.'

This was very strange hearing for Matthew Lawe, who thought he must know more than most about following the sea, and now saw that, of all the troubles which could beset a mariner, he had glimpsed no more than a speck on that same ocean. There were greater villainies than using chain-shot, a worse scourge than a boatswain with a ready fist, more cunning enemies than sharks. There were pirates ashore in England . . . In simple amazement, he said:

'It is a wonder the ships were built and manned.'

'They are always built and manned, though the delay can break the heart. . . . Oh, I put down plenty of villains, and so I knit it all together in time. The King was an ally beyond price, and the Duke has the sea hidden somewhere behind that grey mask. . . . But I made too many enemies on the road, so I was put down in my turn. By a man whom the King himself called the wickedest dog in England.'

'Who, sir?'

'Only my Lord Shaftesbury, the Lord Chancellor of the realm! He needed a stick to beat the King, and I had the honour of election. . . . Well, that dog is dead a year, and I live again. But while I was down, the Navy was down.' Mr Pepys added, without undue modesty: 'What else could be expected? Though Parliament voted money by the million, it has all been swallowed up. There is no fleet worth the name today. They have dismissed all the old tarpaulins, rough captains who could sail and fight, and put in gentlemen instead. . . . So I am recalled, and told to build again.' But his face had grown melancholy, as though confidence had ebbed to a tide-mark below his brave words. 'It is the greatest chance of my life. But I am not young any more.'

In all sympathy, Matthew could only declare that it happened to the best of men.

'What?' Mr Pepys replied instantly. 'Do you think the King would give employment to some old dotard with his senses gone to the grave?' He rapped smartly with his stick on the coach roof, and called out: 'To my house!' And as the coach wheeled around, he added tartly: 'If you have any such thoughts, pray keep them private, or we may fall out. . . . Now—answer me these questions.'

They came thick and fast, sometimes in friendly guise, sometimes like little pincers: of a sort to bring a clerk to heel, or put a prying committee out of countenance. Could Matthew read and write with proper fluency? Was he a runaway? Where was his gear—if he had any gear? ('At the *Cock* by Wapping Stairs' had to be the humble answer to this.) Though bred to the sea, had he worked as captain's clerk?—as store-keeper?—as anything better than tarry mariner? Was he married? Then did he enjoy women? ('I would not trust a man who said "No" to that, whether he were a liar or not!')

Could he judge honest timber? Was he ever diseased? Was he Papist? ('I have had enough fouling in that nest.') What would be a fair price for one hundred barrels of bloat herring, caught off

Yarmouth and delivered to Chatham? (When Matthew answered sorrowfully: 'Sir, I do not know,' Mr Pepys capped this with: 'No more do I. It always differs, and it is always too much.')

If some poor fellow's widow came to the office with a ticket for his last voyage, would Matthew honour it, or buy it at a fat discount, or tell her to go to the Jews? Or perhaps tear up the ticket, and take the woman? If a certain ship was never ready for service, should the Captain be dismissed, or his officers privately examined for the cause, or his pay stopped until he sailed?

'Well, you have the sea at your heart,' Samuel Pepys said finally, as their coach slowed to walking-pace for a turn into Buckingham Street, where they were bound. 'So have I. Until we learn to flap our wings and fly, sailors are the world-makers . . . What I need now is another pair of young eyes. And two strong arms. And most of all, I need faithful hearts about me. Like all others in my house. I have had enough of traitors, great and small.' When the coach stopped before the fine front door of Number Twelve, Mr Pepys delivered an even yet friendly verdict: 'We shall see . . . Now hand me down.'

Matthew could only feel honoured to obey. It seemed, from all the happy omens, that he had been twice adopted, between the Tower of London and Whitehall Palace: once by sea and once by land. Cap in hand under the London sunshine, he swore every oath in the world to earn such a trust, and honestly to prosper with it.

III

A confused morning was followed by an evening more strange than Matthew Lawe had known for many years.

The house at No. 12, Buckingham Street, which Samuel Pepys shared with Will Hewer, his faithful clerk and valiant friend of twenty-four years, was all things never met before. It was rich, elegant, busy, and of good repute. The evidence of wealth was everywhere to be seen: in the furnishings, the silver-ware, the polished wood of closet and linen-press, cupboard and chest-of-drawers; the gleaming candle-brackets, the wine which stood ready for serving, the very smell of linseed oil patiently rubbed upon every cherished surface.

Even the mirrors which reflected this magnificence were no shy advertisers of their own renown.

Samuel Pepys might well have seen hard days. But he had

prospered also, and this full flowering proved that he had never lost all nor over-reached himself, only waited in humble prudence for the best of times to come again. Matthew, surveying in wonder this noble home, remembered its proprietor's words in the coach: 'If I had taken every bribe offered to grease my palm . . .' It would be hard to dispute that a little such palm-grease must, upon occasion, have found a compliant hand.

But it had not dishonoured him, nor tarnished his merit. Smiling servants bore witness to another claim of Mr Pepys: that he had faithful hearts about him.

Will Hewer was the anchor of this harbourage. A personable man of forty, strong, fair-haired, and resolute in service to his master and himself, it was he to whom all turned when Mr Pepys was occupied, or idle, or absent. He was not surprised nor hampered by a new inmate of the household, entering without notice under the wing of the Secretary. Perhaps there had been many such; perhaps he had been one himself, a quarter of a century earlier.

It was all in a long day's work to Will Hewer. A private discourse with Mr Pepys, a close examination of Matthew himself, and the disposal was made, with every sign of good humour and the most concise direction.

'You will lodge in one of the garrets,' Will Hewer told Matthew, 'and work by day at Derby House, as we all do. There, at the start, you should undertake *nothing* save what is ordered. But in this house, you will hold ready to do anything that comes, without prompting, whether bringing chairs for the company, or serving wine if the servants fall behind, or finding a coach, or arming a lady to the door, or reading a paper which Mr Pepys cannot decypher, or turning music pages. We are all one family here, and we all serve the same great man—greater today, by God's will!—and if there is need to carry his bass viol, or usher a King's messenger to his room at midnight, or expel a stray cat from his garden, then the one nearest at hand will do it.'

It sounded like a right royal service . . . But talk of music pages and King's messengers had done nothing to raise Matthew's confidence, already taxed by surroundings so far from his acquaintance. He said:

'I will do what I can. A cat, I can manage . . . But I fear to make mistakes.'

Will Hewer surveyed him straitly, and, for a wonder, did not seem to find him lacking. 'Keep that fear, and you will prosper . . . You have a chance. Take it! If you fail, you will be told, and so you will learn. If you fail again, I shall peer into your ears to discover what makes a block-head . . . For the rest, watch me, follow what I do, and never run too fast.' He became more practical. 'Your chest, or sea-bag, or whatever it is, will be sent for. Meantime, we are of a size, and I have a good grey suit which will serve you for tonight, when there will be great company. The suit is nothing to set the world a-blaze, but Mr Pepys does not like his clerks to shine. There are men in Parliament ready to put a question, if a Crown clerk keeps a canary-bird in place of a sparrow!'

It happened, at that moment, that Mr Pepys himself appeared at the end of the long gallery where they stood, splendidly arrayed in some laced finery, newly purchased and now donned for the expected gathering. The little man, for all his awkward figure, looked resplendent, and as much at home in his elegant house as if house and man had been designed together.

Matthew sighed despondently. There were no dun sparrows in this golden cage. 'It is all so rich,' he murmured. 'What must a man do, to earn such paradise?'

Will Hewer, a stalwart defender, chose only one aspect of this question to answer. 'He must work,' he said briskly. 'Like the man we watch, who has worked more and taken less than any Crown servant in England. What you see, all around you, comes not from accident, or theft, or plotting in some dark corner. It is a reward for ceaseless service, and tomorrow he will earn it again . . . Now he needs his foot-stool. Will you carry it?'

'Aye.'

'There in the corner. Then put on your grey suit, and prepare for duty.'

Supper, at which Matthew Lawe, almost a ghostly figure in candle-shadow, sat far down the table—supper was sumptuous. Though ordained as a modest meal at the end of the day, Mr Pepys had extended himself to honour his well-wishers, and his household had laboured mightily in his support. Matthew had never seen such abundance, nor enjoyed it more.

At one end of the table stood a neck and shoulder of mutton festooned with carrots, and at the other a leg of the same, with cauliflower steeped in the best Cheddar sauce. In between were more victims destined for destruction: a pair of rabbits, a breast of larded veal with wine, and six little chickens which nested among artichokes and peas. Fruit and cheese were at hand for those who could manage more.

Bottles of claret for the quality, and beer for the humble, completed a repast which not an admiral feasting in his own stern cabin, let alone a common seaman with a mind no higher than brine-pork and biscuit, could have dreamed of in a sailor's heaven.

Matthew sat safely among the nobodies. There were two other clerks who eyed him warily but ventured no curiosity; some cousins of Will Hewer's, as shy, silent, and voracious as poor cousins everywhere; and a solemn man who said: 'I am Sam Stallybrass from Deptford Dock,' and then, as if he had staked his claim on immortality, uttered not a word more.

Matthew himself was placed between an elderly shrivelled lady who, not knowing that such a fate had befallen her, was as arch as if her lightest glance could set the whole male world aflame; and a little lad, solemn as Mr Stallybrass of Deptford, whose reach for food might have gathered in a mainsail before the boatswain could finish his roar of 'Clew up!'

But Matthew was greatly content. There was food in plenty, and plentifully more to watch than the champing jaws of his neighbourhood.

At the head of the table, under the brightest of candles, Mr Pepys held his court among a brilliant company. At least one of them was a lord, because he was addressed as 'My Lord' with every drawing of breath, while his lady was flattered some decades beyond her true deserts. But she had the jewels to make the purchase. . . . There were others better favoured, pretty young creatures to whom Mr Pepys paid great attention.

When Matthew noted that this attention was returned, with the warmest of fluttering glances, he wondered enviously how such a man—accomplished, yet with no looks beyond plump home-spun —could conquer these paragons. But tonight it was so, and even husbands supplanted seemed to find nothing amiss with it.

There was present an admiral of great authority, before whose growling pronouncement all others fell silent—save for Mr Pepys, who seemed to have as persuasive a way with admirals as with their wives, or even their daughters.

In a general pause, their host was heard to ask: 'What think you of my scheme to put a lighthouse on the Eddystone Rock? Should Trinity House be given the care of these sea-marks?' To which the admiral replied, with all deference: 'My dear sir, I think whatever you think. There is no man better qualified to judge—as his Majesty has today endorsed.' Wonder of wonders...

There was a rotund gentleman who could only have been a Member of the Parliament, so roundly did he condemn—'As an appointed watchdog of the public purse, Mr Pepys!'—all expenditure beyond bare necessity, the while he quaffed the choicest wine as if it poured *gratis* from the Fountains of Rome. But always there were women, to dispel every hint of argument, to add the harmony of desire to the cutting edge of rivalry: to soften disputation while firming all else.

The company grew merry. Mr Pepys' good fortune was pledged once and again. The candle-light fell gently and warmly on silk and satin, ruffle and lace, craggy male brows and the more tender valleys of shoulder and breast beside them. Matthew sat entranced, forgetting his own plain companions, watching the great at play. This was the life a man was born to—if only he could lean forward by twenty feet, and embrace it.

He could even, for a space, forget the other life beneath the stairs, to which he more properly belonged. On an earlier errand to the kitchen in search of Guinea pepper for the punch-bowl—all new words, the first of many—he had found below decks a dark hive of industry.

There was a cook in a foul temper—and who would not be, condemned to conjure up this feast? There were two men-servants, who knew nothing of Matthew save that they had not been warned, and had much to do without his presence. There were three serving-girls, marshalled like little timorous soldiers on the brink of the firing-line: treated as bond-slaves by the cook, and as tempting half-wits by the men.

One of them was a sprightly child, young as spring-time, pretty as a leaping lamb, sweetly fleshed, innocent as snow on the breast of Vesuvius. They had exchanged glances, in the manner of the confederate young, and he had her name, which was Lucy. He would not forget her!

He would not forget, either, the darkest presence of all: fat Trott the Necessary Woman, who stalked the house at all hours, as slatternly and sour as the task which ruled her life—to empty and then to clean the several privies of No. 12, Buckingham Street.

Hers was the true throne on which this magnificence took its ease. Perhaps it was best to forget all save the shining stars at the table-head, and to enjoy.

After supper, and a walk upon the roof-leads to drink the evening air and the lights of London, Music reigned. Mr Pepys had earlier asked Matthew if he could play or sing.

'I could blow the Irish bag-pipe when I was a lad,' Matthew volunteered.

Mr Pepys pursed his lips. 'I spoke of music,' he said primly, 'not of some heathen howling among the potato patches. . . . Let us forget your answer, and my question . . . When we come to play, pray assist with the chairs and the music-stands. We shall need candles close at hand also.'

Thus Matthew, set a task within the compass of a sailor more used to bawling drunken doggerel on the homeward journey from an ale-house, remained in the shadows and enjoyed an evening of magic. Samuel Pepys, it seemed, was a musician of renown, and not reluctant to show it to the world. He played with equal skill the flageolet, the lute, and the bass viol: he sang a very fair tune in the barytone. His friends would have been ashamed if they could not have married their skill with his, and added to it their own harmony.

In the gentle candle-light, soft voices died as music took command. One could be near to tears in witnessing this delight, while remembering its occasion—that today a man of honour had his honour restored, that he was safe among his friends, and that his only wish, by way of celebration, was to make music in this loving company.

Every note, every lilting melody, every falling cadence of phrase was a kissing act of homage—to the man, to the music, to the fair and the brave, the rich and the poor, and to the night.

'Well, were we not merry?' Mr Pepys exclaimed, when all at last were gone. He asked in the manner of many a host, as if he alone had devised all merriment—which might be true in its essence but was a vain conceit for a sensible man. Now the long farewells were past, with Mr Pepys courteous to the gentlemen and fondly gallant to the ladies. Will Hewer had locked and barred all doors, and bid them a smiling goodnight, leaving Matthew and his new master alone.

'Were we not merry?' Pepys repeated. He had cast himself into a chair, loosened his shoes and neck-cloth, and now signed to Matthew to pour him a glass of wine. Already mellow in liquor, he was wholly pleased with the world. 'I have not seen a finer supper-table this year! And the music was a delight.' He looked up as Matthew set a goblet in his hand. 'Or did you yearn for your Irish pig's bladder, or whatever it is that breaks wind in your bag-pipe? . . . Will you take wine, Matthew?'

Matthew still had a serviceable thirst. 'Why sir, I——'

'You are wise,' said Mr Pepys. 'It is best to know when one has had enough.' He drank heartily, and sat back at ease. 'Music, my love! . . . I have thanked God for it all my life. There were times when only song could take away the foul taste of men. When my wife was alive, we would——'

A shadow came over his flushed face, and he did not finish his words. Matthew divined that for all the company and merriment, all the pleasures of a noble house, this might be a lonely man.

'It must have been a great sorrow,' he ventured.

'She drove me mad!' said Mr Pepys, instantly restored. 'Though I loved her always, and I love her now, she could bring me to rage beyond measurement. Money for this and that and the other. Hats and ribbons, silken stockings, buckled shoes! . . . Yet when I reproached her for forcing us to the debtor's prison, she would say: "What of your new coat? Did that cost a sixpence?" and I would say: "But I have to attend at Court. I cannot dress like a beggar," and she would answer: "Must *I* dress like a beggar, so that you may shine among those royal harlots?" and so it would continue until dawn. If she saw me even give a glance to a woman, she would roar until my ears burst! She was French,' Mr Pepys concluded. 'A jealous race.'

'Were you not jealous, sir?' Matthew asked.

'Aye! And pray why not? She was very fair, and there are those who are more like beasts than men . . .' He smiled at some memory, glanced at Matthew as if to measure his discretion, and said: 'Do you know what I sometimes did, when she went visiting?'

'No, sir.'

'I would creep into her closet and make an inventory of all her clothes, to see if she was wearing her drawers. I knew she had fourteen pairs . . . There's jealousy for you! But it is the wine of marriage, none the less. A man not jealous is a man out of love—and with her I was never out of love, even in the fiercest of storms.'

Silence fell again, while all around them, and beneath, the rafters and the timbers creaked, as if they too were settling for the night. Presently Mr Pepys drew from his waist-band his newest toy, which was a handsome watch of Switzerland, enamelled in blue and gold. He fumbled with its key, but his eyes would not serve him, and he handed both watch and key to Matthew.

'Wind it,' he commanded. 'But gently, till it grows stiff. The spring is tender as a curl of hair.' And when this was done, and the watch restored to its fob, he said: 'Ten o'clock. Time to make an end . . . We have played tonight, and we work tomorrow. I foresee a paper mountain already, and so must you.'

Matthew said: 'I hope I may serve you, sir. I will need direction.'

'You will receive it, never fear! . . . You said in the barge, "I will follow you", and I liked it, as I like all willing men. Following is service, and service is life. I follow the King . . . Whom have you followed, in your time?'

'None of note, sir. Some good captains, and some less good.'

Mr Pepys eyed him with a smile. 'What makes a good captain?'

It was late, but Matthew did his best. 'A strong will and a kind heart. And courage all the time.'

'There are a few such men. I can warrant that Will Hewer is one. Follow him at the start—copy him—you could not do better. He was seventeen, and a young rogue, when I first took him in. But he lost his bad company, and went to work and prospered, and near twenty years later it was *he* that took *me* in—into this very house— when I was put down. He was the first to offer, and never forgotten! You cannot buy such a man for gold, but you may win him with a smile.'

Matthew Lawe, faced with such peaks of achievement, felt bound to admit his doubts. 'But how can I take this road? I wish to, above all else. But how is it done? What is the first step?'

'You took it this morning . . .' Mr Pepys, who had been ready for his bed, decided otherwise—which was the measure of an infinite sympathy. He pointed towards the side table on which sat the wine. 'Pour yourself a last glass of claret—' Matthew had thus far enjoyed none of this nectar, 'and listen to me.' He waited until Matthew was settled again, and went on: 'You must never doubt that a man can rise in the world. Otherwise, we are no better than little ants that creep and crawl—and if that is God's will, then God is the littlest ant of all.'

The blasphemy was terrible, and comforting at the same time. Matthew, sipping his wine, waited.

'You can rise,' said Samuel Pepys, 'just as I have risen. I rose from a tailor's son, and fell from power, and rose again . . . We passed by the Tower of London today. Sometimes I walk that way, to do no more than smile at it. I was held there, after the Popish Plot, and I would have hanged, if some had their way. I was reviled by men—some of the greatest villains in the realm. Liars all! Where are they now? Dead as crows. Where am I? Secretary of the Admiralty. *And I did not plot to win it.* I worked! With head up, head down, and head up again.'

It was difficult not to take fire from such flights of oratory. It was also the end of a lengthy day. In duty bound, Matthew said:

'I would do the very same, if I could.'

'Then set a course, and hold to it.' Mr Pepys began a summary, as he must have done with pen or voice a thousand times before. 'Do not steal. Do not cringe. Do not vaunt yourself. Do not neglect the great. But do not sell them your soul. And never forget, there is a larger scene than our cramped little lives. For some fortunates, it is the whole world. For us it is the sea, and in this cause we serve. We are Crown servants, no more, no less, and tomorrow we will prove it.' He sided his goblet, and rose to stretch and to retrieve his shoes. 'Sleep well, Matthew. But before you do so, go down, and send Lucy to help me undress.'

There was one more thread to the weaving of that strange day and night, and Matthew, drifting into sleep, could not decide if it contradicted all, or confirmed it, or was some mad embroidery of its own manufacture. It had happened after his errand to the kitchens, when he loitered below in search of a privy. Afterwards he had stolen upstairs again, but softly, so as not to wake the household. It was this secrecy which gave him his last view of the Secretary of the Admiralty preparing for slumber.

A shaft of candle-light from a door half-open attracted his eye, and in all innocence he approached. On a chair within, Mr Pepys sat at ease, with his coat shed and his legs comfortably spread, while pretty Lucy combed his hair.

Presently he reached up, with a smile almost fatherly, and began to unbutton the throat of her bodice until it was wide and free, and the tender flesh peeped out in marvellous display. Gently he started to fondle her breasts, which were small, and pointed, and as pretty as her yielding face.

At which moment Lucy, like a well-schooled child, laid aside her comb, and having loosed the belt of his small-clothes, bent and slipped her coaxing hands within.

It seemed that Mr Secretary Pepys, at fifty-one, was unreformed—as his beloved wife might have been the first to tell him.

IV

Theirs was close work: often dull, sometimes perilous, ever rewarding. In the five years before Mr Pepys was 'put down' again, this time for ever, he achieved the most brilliant toil of his life. Always in the interest of the Navy, he fought against corruption, battled with the Parliament for money, talked boldly, wrote endless minutes on a thousand matters, cajoled, threatened, wooed two kings, juggled with impossible debts, and got his way—or enough of it to give him joy, and England a fleet to be proud of.

He and his faithful staff, with Matthew Lawe at the end of the tail, worked through the best of times, and the worst. Outside, storms raged as fierce as any that whipped the ocean. Charles their King died: James his brother ruled and was overthrown. There was turmoil in the city streets, riots against Popery, murder by stealth, rumour of Irish cut-throat bands who would kill anyone of the wrong persuasion, and fear always.

There was a rebellion, with bloody reprisals in the West Country, and a revolution which brought, astonishingly, a Dutch king to an English throne. But until the very end, all bent their heads and gave their honest best.

The Navy was all, and their only motto was the one by which Samuel Pepys constantly rallied them: 'We shall have the Dutch up the river again, unless we succeed!'

When Matthew reported for his duty at Derby House, and was given his desk by Will Hewer, the chief clerk said most cheerfully: 'Welcome to a dusty berth! We deal in ships, but we live in a dry world.' Though it was an ordered universe, certainly, it was never dry.

From that first hour, as Matthew gradually found his way about the office, and examined books and papers, and explored a whole library of bound volumes—letters, minutes, accounts, plans of the past and schemes for the future—he was amazed at this life-time of industry, and its variety.

There was nothing, it seemed, that Mr Pepys neglected, nothing he had not tried, nothing that he did not hope for.

Much of the past was astonishing, and his vision of the future was magical. Years before, he had instituted an examination for naval lieutenants, to promote knowledge of navigation and to ensure that ships were not entrusted to the hands of unskilled sprigs of the nobility. He had assisted at the founding of the Hudson's Bay Company, whereby roving adventurers, still searching for the North-West Passage, were encouraged to remain and to trade on a continent of fabulous riches.

The Royal Observatory at Greenwich—again to promote navigation—owed much to his endeavours. As an Elder Brother of Trinity House, and soon its Master, he had contrived to supersede the efforts of Christian charity in the matter of coastal lighting—often no more than a lantern on a church tower—buoyed channels, and the control of pilotage, and to set it upon a more secure footing.

He was President of the Royal Society, that dovecote of 'Physico-Mathematical Experimental Learning', in the same year as he became Secretary of the Admiralty. He took a special interest in shipbuilding and ship design. Old and new maps were his delight. He had laboured for years to improve the lot of the common seaman, so that a faithful man who had served his country was not paid off at the end of a voyage with a 'ticket' almost worthless, save for what it might squeeze from a cold-hearted dock-side swindler.

When Mr Pepys returned to power, on the morning of his meeting with Matthew, it was to find that all the old puzzles persisted: how to build ships without money or credit, how to recruit unwilling, cheated men to serve in them: how to save rotting vessels from disaster, how to find contractors who would not promise timber and deliver sawdust: how to persuade a Parliament of comfortable time-servers, safe from the thunder of the guns, that funds voted to the Navy were not a foolish luxury.

But there was a worse enemy than this sleep of fools. In his absence, decay and corruption had once more taken command; it was not too much to say that the whole conduct of the Royal Navy, from the laying of a keel to the dismissal of a ship's company, was now a giant swindle in which all must join, or lose their place.

But it was not too much for Samuel Pepys, either. Within a few weeks he held it all in his hand again; and within six months, when he had removed the Admiralty itself to No. 12, Buckingham Street, he held it in his own house also.

Fortunate was the country, and happy its king, which could promote this wayward man to such supreme power, and be rewarded so richly.

Through the open door of his private office, Mr Pepys kept an eye upon the Long Room where sat the copying clerks perched high on their stools. Even as he dictated to Matthew from the great machine within his head, he was watchful of all that went on, and all men knew it, and took care accordingly.

Meanwhile Matthew, a closer captive, sat at the small table opposite the Secretary, and with weary brain and cramped fingers pursued the third hour of his morning toil. The shorthand for speedy writing which he had now mastered could scarcely keep pace with the nimble tongue and searching intellect of a man who never rested, never flagged, and never forgot the tasks he had set for himself and for all others under his command.

'Paragraph the thirty-fourth, and last,' Mr Pepys proclaimed in his slow, steady, relentless voice. 'In conclusion, it must be said that the life of a virtuous officer in the Navy is a continual war defensive, namely against the Ministers of State, and in particular the Lord Treasurers, in time of peace, and all prejudiced inquisitors and malcontents with the Navy management in time of war; the former grudging every penny of money almost that is spent, and so keeping it short or postponing it *sine die*, whether it concerns the repairing of old ships, the building of new, or the laying-up of stores. On the other hand, in time of war these officers are not only left to shift for themselves, but are expected to find ways of imposing upon the world the belief that all things are well with the Navy, the supplies of money plentiful, and the work done.'

Mr Pepys sat back, watched for a moment a distant clerk who had dared to pause from his labours in order to look out of the window, and then turned to Matthew again.

'Leave a space below in the copy-book,' he said. 'I may add a thought or two when you have read it to me. God knows we have heard this song before, and this new verse may never see the light, unless one day I write my great book. But it should be preserved . . . Now there is one more thing in my mind, and then we may pause.'

Matthew stifled a yawn, which he knew Mr Pepys did not care to see. He took fresh paper, and waited.

'New regulation of the guard-ships,' Mr Pepys announced. 'These are no more than the heads, you understand. Set them down in order, one after the other, and I will make sense of them to-morrow.' He gathered his thoughts, and was off again. '*One*, Rigging, whether standing or running, to be in good repair always. *Two*, Six months' stores to be kept handy at all times. *Three*, Ships to be docked and burnt off once a year. *Four*, The Captain is to sleep on board without fail. *Five*, Not less than one quarter of the ship's company to be on watch at night. *Six*, All open flame, including candles, to be doused by eight o'clock at night. *Seven*, For Chatham and Portsmouth, guard-ships are to send off guard-boats—call them scout-boats—to row a night-patrol, fully armed, round every ship in harbour. *Eight*, These should challenge all passing ships. *Nine*, They should board any ship that does not give a wide-awake hail. *Ten*, In open harbour, they must search every creek and backwater. *Eleven*, The guard-ship must keep a daily journal of these patrols, and submit reports to the Comptroller four times a year.'

Mr Pepys considered for a moment. Then, as he sometimes did, he asked Matthew the salt sailor:

'Can you add to it?'

It was always a surprise, but Matthew was ready.

'The boats rowing patrol in harbour should muffle their oars, with no lights and no talking.'

'Why so?'

'Sir, it is the way that an enemy might come. Otherwise, the scout-boat's crew will make noise enough to signal their arrival, so as not to catch their comrades napping.'

'Would honest sailors do that?'

'Aye, sir. To save a flogging.'

'Estimable reason! We must thwart it! . . . Well enough. Add it as you have said, under number eleven, and alter the last to twelve.'

Matthew made this correction, and took pleasure in it. There was little enough that the small could do to help the great; when such help was set down in black ink on white paper, a man could warm himself at its modest glow. . . . Then silence fell, save for the distant scratching of pens, and he lapsed into a daydream in which life at sea, duty ashore, pulling an oar and pushing a pen were all tumbled together. He awoke, guiltily, to find the Secretary's eyes fixed upon him.

But the gaze was not fierce. 'You look downcast, Matthew.'

'No, sir.'

'Very well—you look absent.' With rare understanding, Mr Pepys put his finger upon it. 'After two years, do you hanker for the sea again?'

'Aye, sir. On some days. I was thinking of rowing patrol at night, or of duty in the guard-ship, or of setting sail again . . . Things I can master myself.'

Mr Pepys waved his hand, indicating all he could see from his desk. 'You have mastered this. And very well.'

Matthew took heart at the praise. 'Sir, I have been proud to do so. But—' he recalled Will Hewer's words when first he came to the Admiralty, 'it is a dry berth. That is all.'

'You would rather live wet?'

'On some days,' Matthew said again.

'Who knows?—those days may come again. But do not think that a dry berth, as you name it, is a shameful one. Land-sailors are necessary, and I hope you have come to know it. Behind every man at sea sits a man on land. Or even two men, but never more, or we waste our gain. We must feed that man at sea, and pay him, and care for him, and nurse him when he is sick. We must do the same for his ship—and first we must build that ship, and make it honest, and then keep it in trim, fit to voyage, fit to fight, fit to endure.'

It was an old sermon—not less true on that account, but old and familiar, like bread and cheese, which sustained life yet did not glorify it. Matthew was indeed proud of his Admiralty service, and of the Secretary's trust. There remained these days—spring days, days of beckoning sunshine and briny promise—when a pen could not match a coil of rope, nor a staid office floor a heaving quarter-deck.

Ships moved and jumped and lived. Desks only mouldered into dust.

Mr Pepys was speaking again. 'So take heart—and take these

notes! I will need my credit and debit accounts by next Monday morning, for the money bill in Parliament . . . The Purser of the *Swallow* has invoiced for seven new anchors. No frigate can use seven anchors, unless she feeds them to the fish. *Quaere!* . . . And send an express to Captain Banning of the yacht *Pluto*. He is to wait on me here at eight o'clock on Tuesday morning.'

Matthew's pen came to a halt at the word 'eight'. 'Sir, he must come from Chatham.'

'Then he must rise early, like a good captain—with a good explanation of his being asleep in his cabin when his ship touched the mud at Sheerness. . . . I do not sleep on duty! Why then should Captain Banning, son of a baronet and a grimy blanket and God-knows-who on its wrong side? . . . Then tomorrow you and I will take a holiday.'

'A holiday?' Matthew repeated, surprised.

'Well, a hackney coachman's holiday. We will inspect a dock-yard, and I shall dismiss a rogue. Then I will redeem my promise, and show you Bedlam.'

Neither the dockyard, which was Deptford, some seven miles down the river, nor the dismissible rogue, who was Sam Stallybrass, last met at Mr Pepys' own supper-party—neither was ready for the onset of the Secretary of the Admiralty.

'We will board them while they sleep,' Mr Pepys had said grimly as their coach rumbled eastwards. 'It is the best way with knaves . . . I judged Sam Stallybrass honest, and I judged him wrong. Does he think because he put his feet under my table that he is licensed to steal from the King's own purse? Enough is enough!'

It was surely enough for Sam Stallybrass, who first smiled like a villain, then blustered, then turned surly, then cringed and begged for mercy. It was chiefly a matter of a thousand pounds—'A great sum,' said Mr Pepys. 'Even thieves should have some delicacy'—a thousand pounds spent on timber, which was spirited away on a night conveniently dark, then sold back to the Crown for another thousand, then found to be green, unseasoned stuff which any honest dockyard officer would have condemned at his first glance.

'I made a mistake,' Sam Stallybrass said finally. They had found him in an idle dockyard where not a man seemed to be giving a fair hour's work for his wage. Stallybrass, idlest of all—feet on desk, flagon in hand—was caught: first by surprise, then by an array of

facts and figures, bills of receipt and bills of payment, which proclaimed a glaring guilt.

It was a joy to see Mr Pepys, the best land-sailor of all, engage
him like an admiral afloat and rout him bare-arsed, with his shirt
hanging out of his tattered breeches.

'I made a mistake,' Stallybrass repeated. 'Any man can be forgiven for that.' His fat sweaty face took on a leering good-fellowship.
'Give me another chance, Sam. I'll not fail again.'

'Do not call me Sam,' said Mr Samuel Pepys, bristling and contemptuous at the same time. 'You befoul an honest name. . . . It was
one mistake too many. Do you think you are not watched? Do you
think I cannot read a balance, and discover the lies in it? There is a
year's perjury and theft in this one paper alone! There are twenty
like it! Do you think we are all fools?'

'Some of you are more than fools,' Stallybrass snarled back. He
was cornered at last. His appeal for clemency had fallen on stony
ground, and he had only insult left. 'You can read a balance because you have cooked a thousand of them yourself. Would you
have me speak of——'

'I would not have you speak at all,' Mr Pepys suddenly thundered. 'I have warned you twice in the past. Too gently, perhaps,
but I took you for a friend. Now you are dismissed, from this hour.'

'Who tells me this?'

'I tell you. By His Majesty's command.'

Matthew, hovering in the background, but not less avid of hearing on that account, recognized words as familiar as his own name.
The Secretary always spoke, or wrote, 'by His Majesty's command'; sometimes it was true, sometimes it outstripped truth by a
day or a week. But it was never questioned, nor altered. The little
man with the heart and stomach of a king had no need to usurp
either. They were his by right.

'Then who takes my place?' Sam Stallybrass asked. He fired his
last shot at random, in aimless spite. 'You have some relation in
need of preferment?'

'None takes your place, for a month or more.' Mr Pepys was truly
in splendid trim. 'I would not so hazard an honest man. We must
fumigate before we elevate . . . But you will leave the dockyard
tonight—with empty hands, if you please—and fresh orders will
follow when His Majesty makes known his will. In the meantime,
Captain Lysart of the guard-ship will rule.'

'He will not rule me!'

'True enough,' said Mr Pepys. 'You will not be here.'

Matthew Lawe was not the only man in Deptford Dockyard whose hearing had been at full stretch. As he and Mr Pepys walked out, first to the guard-ship where Captain Lysart, receiving his orders, merely said 'Aye, sir,' and sent his Lieutenant to mount a guard on the Office of Works, and then to the main gate where their coach waited, industry blossomed like a coy flower in spring.

Men who had been playing at cards were suddenly dealing hammer-blows upon innocent iron. Other men tripped in their haste to carry coils of rope from one neglected corner to another. Apprentice boys ran hither and thither on fat stumbling legs. A man caught bare-handed doffed his cap five times. Axes were sharpened on grass, nails driven into wormy wood, sailcloth folded and unfolded like a hopeful bride's best linen. Movement was all.

'Oh, they will act the angel for a space,' said Mr Pepys as he and Matthew quit the portals of Deptford. Having made short shift of one superior rascal, he was in high good humour to note how many smaller men had drawn a lesson from it. 'Then the poison will creep in again, and I will pay another doctor's visit and bleed them a little, and so it will go, till we are all dust. There are tides of endeavour, tides of honesty, as well as salt water . . . Now let us forget all, and repair to the other Bedlam—the true madhouse!'

But first they must pay a visit to Lloyd's Coffee House, where the Secretary had some private transaction to complete, and a modest dinner of cold fowl, cheese, and Rhenish wine to enjoy. Newly opened in Lombard Street by Edward Lloyd, a printer with a taste for company, it had quickly become the resort of City men whose business was the sea, and any others who might have a share in it. They were venturers all, whether they smelled of the counting-house or the captain's cuddy.

There were ship-owners whose purpose was profit from ten thousand miles of hazard, ship-builders whose dream of stalwart hulls must first be floated on a tide of gold, and ship's captains with tales to tell or fresh voyages to promise. There were solid merchants discussing the sale of such things, or their cargoes, or planning new schemes which might thrive on good fortune or founder on the first rock in the English Channel.

There were ship-brokers—the men in the middle—who took their chance on the price of a cargo, whether it was wine or winkles, cloth of gold or Calicut, and on the skill of the sailors who carried it.

There were other men of substance, with a bond to prove it, who were ready to insure the whole worth of a ship, for its lifetime, its successive voyages, its value as a staunch trader or its loss as a wreck. They would make out a policy for the sum agreed, and write their names one under the other, till the cost was warranted in full.

These 'underwriters' would prosper in good times, and might face ruin in bad. But there was no lack of such venturers either. It was a wager on the quality of ships and men. Compared with the speed of a horse or the virtue of a woman, this was a gamester's paradise.

In this guarded back-water of the ocean, Mr Pepys was received like a prince—which, in the sea-world, he was. Men waited their turn to approach his table, to pay their respects, to express the hope that if he needed assistance in any matter, great or small, their names would be remembered, and then to withdraw and give place to other courteous suppliants.

The merchant banker with whom he had particular business was the envy of all as Mr Pepys was seen to peruse an offered paper, set his name to it, and shake hands upon whatever bargain it was. Matthew knew what it was: a loan from Mr Pepys of £800, at one-half *per centum* a month, to be repaid on sight whenever the lender chose.

Mr Pepys, the man of property, no longer borrowed from a banker. It was his private gold which kept the fellow in credit.

A tall fellow of genial good humour but with a sharp eye for business of any sort, Mr Edward Lloyd waited on them himself; and when the traffic at their table eased, stayed to gossip. He was full of schemes, and glad of an illustrious ear in which to spill them.

'A club is what I plan here,' he told them. 'For like-minded men whose interest is the commerce of the sea. I would collect all shipping news, all the daily intelligence of arriving and departing, the latest prices for the latest cargoes, and display them where they can be read in passing. Even a newspaper, Mr Pepys, even a newspaper! What would you say to "Lloyd's News of Shipping"?'

It was difficult to withstand such spirit, even though sparked by self-interest.

'I would subscribe,' said Mr Pepys. 'And read it too! But you should be on your guard to preserve integrity here. Otherwise your "News of Shipping" might be no more than another lying broadside, and your club a nest of sharpers.'

'It will never be that,' Mr Lloyd answered forthrightly. 'I will be on the watch, I warrant you—I can tell an honest man from a rogue by the way he hangs up his hat . . .' He gave Mr Pepys a sharpish glance. 'May I say that I have your patronage, sir?'

Mr Pepys returned a glance equally sharp, and then smiled. 'I expect that you *will* say it, Lloyd. But I would think it churlish to deny.'

With that he nodded, rose, bowed to certain acquaintances, and took his leave. Preparing to step into his coach, he confided to Matthew:

'Lloyd is on the right course . . . An honest place with honest company . . . A man's word is his bond there, or he would not prosper beyond the sniff of a coffee-pot. . . . Now I will match it with the foulest spot in London.'

The foulest spot in London proclaimed its repute to the very street outside, which was Moorfields; it puffed out the smell and sound of misery for all to catch, and as Matthew and Mr Pepys entered the gates of Bedlam and paid their two-pence to pass through, both noise and stench grew horrendous.

From corridors far and near, shrieks of laughter, groans of pain, bellows of rage resounded like some infernal symphony. Words could not be caught, any more than one sweet note could be heard from a mob of mad musicians, or a child's prayer-full whisper from the depths of hell. It was no better than the sound of an animal forest, the speechless uproar of the damned.

But near at hand there was speech enough, of a particular sort: the shouts of the Bedlam hucksters crying their wares.

'Poking sticks! Poking sticks! Stir them up!'

'Foul water bottles! Splash them! Give them a wash!'

'Tin trumpets! Little tin trumpets! Make them start!'

'Pinching irons! Nip them as they reach out!'

'Holy wafers! Feed them God's mercy!'

Though the holy wafers were crosses made of straw, and drew laughter in plenty, poking sticks were the brisker trade. . . . But Matthew was already confused.

'Sir, what is this selling?' he asked Mr Pepys. 'What is its purpose?'

'To bait the lunatics,' Pepys replied, as if to a simple child. 'What else? Here is the greatest show in London, but it can be dull without a little sport . . . You would not think that this place was named for Bethlehem.'

They advanced into the place named for their Saviour's cradle.

The mad lay in their iron cages, ten and twenty together, on straw fouled as if by cattle; or they stood plucking and wrestling at the bars; or they fought bloodily among themselves, condemned to prey upon each other even in this hell. Between the cages the strollers and the drunkards and the family parties moved to and fro, laughing, taunting, tormenting, calling out to each other at some fresh delight.

'I wonder,' said Mr Pepys, 'on which side of the bars are the madmen.' He was watching a fat slattern of a woman, armed with a poking stick, prodding at a gaunt young man whose vacant eyes were ringed by the most tortured face to be seen on this side of purgatory. The miserable youth would not stir. The woman would not relent. Her friends encouraged her with cackles of laughter.

Suddenly the young man flung out an arm, at lightning speed, and seized the stick from his tormentor. Then he thrust the point into his breast, and drove it home, as if to say: 'There—have your way.'

As the blood gushed from both breast and mouth, the transfixed youth bowed over his gross wound and slowly fell insensible. The crowd, robbed of activity, moved on.

'God help us all,' said Mr Pepys. There was compassion in his tone. But Matthew, sickened at the sight, could not forget one fact—that he and his master were there as willing onlookers, and that this mortal baiting was the 'little sport' of which the Secretary had spoken. On which side, indeed, were the mad?

There were many more souls in torment who caught his eye, as their Bedlam voyage continued. One was an ancient grey-beard, dressed in filthy rags, whose burning gaze bespoke a tempest of affliction. He clung shrieking to the bars, shaking them as if he wrestled with the whole folly of the world, and calling out: 'I am John Baptist! Hear me before you perish!' A frothing spittle fell from his lips with each frenzied mouthing.

'*Prophecy! Prophecy!*' the crowd roared in unison, seeming well accustomed to this mad display. When the old man was silent, they poked at his wasted limbs, and threw water full in his face, and made trumpet noises as if to wake the walls of Jericho. But all he could do, in his agony of spirit, was to repeat: 'I am John Baptist! Repent before too late!'

The crowd did not want this. They wanted visions, they wanted mad imprecations, scandalous words. They wanted a sort of blood.

The cry of 'I am John Baptist!' echoed wildly as Pepys and Matthew passed out of sight. That one was truly mad. But what if his tale were true?

There was another captive, who looked no more mad than the holy mother of God: a piteous girl, gentle in her desolation, with three whey-faced children clinging to her skirts. She was watching the faces of the passers-by in a desperate search for help: it was to kindly Mr Pepys and troubled Matthew Lawe that she addressed her last appeal.

'Good sirs,' she called, advancing to the bars, dragging her children with her, 'can you not help me? *I am not mad!*'

Mr Pepys, to his honour, stopped to listen. The children whimpered and stared, the young mother spoke in a torrent.

She was not mad, she repeated—and one could believe it, even from that trembling mouth. Her idle husband had left her, penniless with three children to feed. When all were hungry she had stolen some bread. She was caught, and had struggled, and been taken to the back room of the baker's shop. There the baker had offered to let her go free, if—if—— He had advanced upon her there and then. She had struggled again, and screamed, and tried to beat him about the head, to kick his rampant parts, anything to defend herself.

'You must be mad,' said the baker venomously, nursing his wounds—and from that moment she was declared mad, a danger to herself and to all lawful citizens.

Stern justice, faced with such a fearsome assault on the very fabric of society, had decreed Bedlam for this miscreant, adding that if, as it appeared, there were children, it would be unnatural to part them from their mother. Thus did the hapless family come to this pass. They lacked bread, and after bread, freedom. Could not the kind gentlemen help?

Her tale was done. Was it true? Was it the child of a mind diseased? Or was it whining lies?

'I will make inquiry,' said Mr Pepys—and Matthew knew that he would. A coin changed hands, though too many envious eyes must have caught the gleam of gold for its later safety. Meanwhile they were free to pass on, and the woman was not, and the children were not, and all the sour malice of mankind could be smelled in that distinction.

They passed on to lecherous chaos. The last two souls of their visit were twinned: a man and a woman joined together in a desperate coupling—the only solace of their despair. Eager eyes watched them, raucous laughter rose, as the two unholy lovers lay, and thrashed, and groaned, and bit—yet there was more love here than was to be found in the cruel world outside.

Beyond the laughter, the foul jests, the shouts of encouragement, there was the most pitiless spite—that of a drunken man who was trying to set fire to their bed. Though the stinking straw would not catch full flame, the man persisted, poking burning tinder through the bars until his stock was exhausted.

Thus, in their copulation, a farmyard stench of smoke was their holy incense, and their marriage hymn never ceased throughout Bedlam—such a howling and shrieking, cursing and snuffling, laughing and weeping, screech of trumpet and rattle of iron bars, as might be heard in hell on the most evil night of the year.

All for two-pence.

Yet there were advantages to be had for their two-pence, beyond the lunatics chained in love and hate, despair and exaltation. Matthew grew aware that there were other loiterers round about them who gave no more than a careless look to the cages, but were intent on something else besides. This was sexual commerce, at its most shameless.

'Oh, there is more to Bedlam than its proper purpose,' Mr Pepys confirmed, when Matthew asked his question. 'It is a good dry walk, to escape the rain. But there are, as you see, ladies of the town going about their business, and pimps putting their wares on parade, and customers in plenty. This is a sure market for lechers.'

At that moment he caught the eye of one such lady-in-waiting, a handsome harlot less haggard than the rest, with a bosom so well displayed and so opulent as to promise a most movable feast. It was possible—all things were possible with Mr Pepys—that he made to her a covert signal.

Whatever the truth, Mr Pepys's next action was to consult his pocket watch, and to ask Matthew: 'Would you wait to see the lunatics fed? I am told it is a rare battle-piece.' When Matthew answered, No, he had seen enough of Bedlam, Mr Pepys declared: 'Very well, soft-heart—so home.' They passed out of the gates, into the gathering twilight, where their coach was waiting, and within it the girl.

Her bouncing promise seemed to fill half the coach, and Mr Pepys and his mounting desires the other half. Matthew, from long experience, knew the signs. Something had pricked his master: whether the tumbling lovers in their cage, or the forlorn mother who still bore traces of a worn-out beauty, or the gross parade of whores, or this girl herself, who looked as ready and willing as a young mare skipping at pasture, no man could say for certain.

But whatever the spur, Mr Pepys was not to be denied. He lusted *now*, and he would be discharged *now*, and there was the end of it, as neat, pleasing, and final as a tally of numbers which balanced to the breadth of a pubic hair.

About to step into the coach, Matthew had whispered: 'Sir, shall I leave you, and find a hackney?' and Mr Pepys, already heavy-breathing, had replied; 'I think there is room here for my *confidential* clerk. But if you choose, you may admire the view outside.'

Matthew did so, gazing determinedly into the dusk of London as, close by his shoulder, the amorous play began. It was no more than play, though it had its appointed climax: to fumble and be fumbled was Mr Pepys' desire, and the girl seemed to know this, or to have been told in a whisper, and was conformable.

She sat on Mr Pepys' knee, and—Matthew could follow their progress without the use of his eyes—her magnificent breasts were unlaced, and enjoyed with hands and lips, and other rubbings and strokings and clenchings ensued, until the end.

The coach rocked, the girl nestled and squealed, the lover gasped, and that was all. After a long silence, the next sound to be heard was Mr Pepys knocking on the coach-roof with his ebony stick.

They were stopped at a street corner. 'You're a good girl,' said Mr Pepys—which was manifestly untrue. 'Here's for your trouble.'

The girl said: 'Is that all the work? What about your friend?' and Mr Pepys answered: 'My friend is a man of the cloth,' and she replied, with the lewdest giggle in the world: 'It is a cloth he may need!'

But she did not persist. The coach door opened: she stepped down, and disappeared into twilight; and Samuel Pepys and his confidential clerk were alone.

'Can you see to write?' Mr Pepys asked, with scarcely a pause.

'Aye, sir.' Matthew knew this sign also. After such a bout, the Secretary was released immediately into thought and action: intellect took command, with his brain as quick, clear, and innocent of lust as a single stroke on a church bell. Matthew had ceased to marvel at such agility, such instant diligence, except to say to himself: If we judge by results, then we need more sinners in the service, and fewer saints.

'Some notes for the morrow.' Mr Pepys was dictating already, his tongue as apt as his brain. '*Item:* Convey to Mond eight hundred pounds in gold. *Item:* Bring the documents to the notary public. *Item:* The King to approve notice of dismissal of Sam Stallybrass. *Item:* Confirm Captain Lysart's commission to oversee the Deptford Dockyard for thirty days. *Quaere:* Sir John Haydon, Governor of Bedlam Hospital: the favour of a report upon that woman and her children . . . I hope I was not fooled,' added Mr Pepys. 'It was a ready story . . . And now for future business. There were still some sick and dying seamen unpaid, outside the dockyard gates. *We must honour our promises!* What I would propose to the Admiralty Board is . . .'

Their coach rolled and rocked and rumbled, and made steady progress through the streets of London, while the business of the Navy advanced as steadily. Samuel Pepys had not loosed the helm, for one single careless moment.

That day remained a holiday until its end—and a most strange one it proved. At Buckingham Street Mr Pepys, tired with his many exertions, went early to bed; he did not even call on little Lucy to assist him. The girl, coming up the stairway, met Matthew coming down; their meeting owed something to his design, since he also had been pricked by the sights and sounds of the day to a restless desire.

'Does the master need me?' she whispered.

He shook his head, smiling in the dim light, staring at her. They were near enough for touching. 'No,' he said. 'He sleeps already. But I am still wakeful.'

She gave him a close glance, which melted into surprising softness. 'That I can see . . . Will you never have done with this?'

'Not this side of your bed.'

He was bold, because the gentleness which surprised him was the first hint of her readiness. Little Lucy was a coltish child no longer, but a grown woman who had blossomed to full flower under this roof. Yet for years she had held him off: sometimes merrily, sometimes angrily, sometimes with a shyness which made nonsense of those free caresses, given and taken, which he had spied on during his first night in the house.

Often she protested: 'Be off with you! You are as bad as the master!' and when he answered that he was very ready to match Mr Pepys in anything that he might do or that she might wish, she always said: 'I do not *wish* anything! What do you take me for? A sailor's wench?'

Now, on this night, she was different. She was looking up at him from the step below his on the stairway, with eyes glowing like stars at dusk; and his vantage point gave him a most quickening view of her bosom, now sprouted from shy maidenhood to a creamy fullness. He imagined his hands busy with these beauties, and her own not reluctant to grasp anything that touched her fancy. She seemed to be taking his measure—and his measure was on the increase, with every moment that passed.

'Well,' she said at last, 'if you will not give up the chase, then I suppose I must, for weariness alone . . . And I would not waste my errand.' She advanced a step upwards, until she was touching him, and their arms went out to each other at the same moment. Her breath was warm on his lips, and her body as hot and urgent as his own. 'You said, "Not this side of my bed",' she murmured. 'But I will cheat you yet. It shall be *your* bed.'

He was still hesitant, even at this peak of his desire. 'What is this, Lucy?' he whispered, as he mounted the stairs to the attic. 'Why such a change of mind?'

'As to *why*, I know not. It must be the planet Venus in conjugation with Mars.' She was repeating something overheard, not to be understood, fit only for soft laughter. 'As to *what*, if you do not know, you are no sailor.'

Was she then a wanton after all—and *for* all?

She was not a wanton of any kind; she was virgin. Matthew had never been more astonished in all his life to find it so. Though she was quick and lively and willing, yet this was her deflowering: the bridal narrowness was there, the gasping surprise, the pain. When all was done, and her sweet body had given its last shudder, and his manhood its pent-up stream, they lay apart in the attic bed, and he was nothing but tenderness.

'I did not know,' he told her. 'I thought you were———'

'A ready whore?'

'Never that. But a woman, a full woman.' He could not forget how, when the first shock of his invasion was over, she had locked him in with her slim legs as close as prison bars, and moved like quicksilver for his delight. 'Where did you learn, Lucy?'

'Talk.'

'And the master?'

'What of the master?'

'Did he never want more?'

'What he wanted, I gave him. . . . He seemed a-feared of more. I know not why. He is a dear man. A good man. But all he wished from me was playing hands.'

Matthew leaned over to stroke her breast, and then her flanks. 'What did *you* wish?'

'What I wished then is different from what I wish now.'

'*Now?* A man must sleep!'

'A man must earn it . . . When you are man again, Matthew, I will be woman. That *full* woman.' She laughed, her lips close to his ear. 'And yours the darling spout that fills me.'

So presently they went to again, with stormy appetite, and the sweet evil of the day attained its crown and promised everlasting more, and love seemed set in its ways. As long as a man had his strength to pour, and this girl her wakeful honey-jar, their attic bliss must blessedly endure.

Then all the world, from roof-tree down, began to crumble about their heads.

V

The year of 1688 brought to England the worst turmoil since 'the late disturbances', as Mr Pepys had once described a brutal civil war and a royal beheading. This time it was revolution against another king, James, who had become mad for Popery and his subjects equally mad against it, and against their sovereign for that reason. He was vilified, abused, opposed, and threatened, until his kingdom grew ungovernable and the realm was in peril.

Certain prudent or ambitious men, with the power to make such decisions, ruled that this king must go. To replace English James they sent for a Dutchman, as a husband might send for a trustworthy nurse if his wife died in lunacy and left him with a brood too quarrelsome for comfort.

William Prince of Orange was a good safe man, with a pedigree to reassure the doubtful: grandson of Charles the First, husband to King James' daughter, and a brave Protestant soldier. Receiving his secret invitation, he landed in force at the head of 14,000 soldiers, having first outwitted the whole English fleet—forged by Mr Pepys into a great sea-weapon and now tossed away by a Lord High Admiral, Dartmouth, so timid and irresolute that his matchless squadrons were still trapped by the sands off Thames-Mouth while the invader was landing 300 miles away at Torbay in Devon.

As, from this sleepy, surprised haven, Dutch William stormed eastwards towards London, men rallied to his standard as if they were obedient subjects already. King James, briefly opposing the advance at Salisbury, took to his bed with a violent nose-bleed, then retreated to London: then fled, throwing the Great Seal of the Realm into the Thames as he passed.

He was captured in disguise by some Kentish fishermen who found him trapped in good Kentish mud, rudely searched for gold down to his royal under-drawers, and imprisoned; then allowed to escape to France—and into limbo, as William of Orange drove down Piccadilly to Whitehall Palace and his throne.

James had put, indeed, a clown's end to an ill-judged reign. It was enough for England: King Billy was their man! It was also enough, in bitter anguish of spirit, for Mr Secretary Pepys.

He had done his loyal best, but now wind and tide must set against him, beyond any swimmer's strength. Once he had said, of a newly-hatched place-seeker: 'By God, he can sniff a wind!' Mr Pepys could do so himself, as well as any in the land. Friend and servant of two kings, he knew that he would not suit this third, even if he were allowed to try; and that was doubtful.

He was a marked man for all that he had done: the wind of change had not caught him by surprise, but as a loyal and loving Crown servant he had never responded to it, nor ever seen a dilemma of allegiance.

It was known before all the world that he clung closely to King James, the Navy's staunchest friend: known also that, being bound in honour and duty to this lawful sovereign, he had laboured mightily to mobilize the fleet against the Dutch invasion, endeavoured to steel the reluctant Lord Dartmouth to fight, and supported King James to the very end, as his oath of office commanded him.

He was now bare to that biting wind, and defenceless. While others had trimmed their sails, or lain low, or, like Lord Churchill the commander of the Army, entered into secret treason to support the invader, Pepys had remained steadfast at his desk. He could remain there no longer. Having broken the law of the Psalmist: 'Put not your trust in Princes, in whom there is no help', he must pay the price.

If a king could be toppled, any man smaller could be crushed in the mire.

There came a dreadful day, after a cold audience with his new sovereign which had left him tired and ill-feeling, when he called Matthew into his sanctum and motioned him to close the door between it and the clerks in the Long Room. He spent some time in sombre thought as Matthew waited. Then without preamble he began:

'A letter to Lord Dartmouth. The Right Honourable the Lord Dartmouth, *et cetera*. My dear Lord, At this time of uncertainty it is beyond my skill to foresee what further letters I shall have the honour of addressing to your Lordship. This is to commend to your gracious favour all the Crown servants at present working in the Office of the Admiralty, and in particular one known to you, Matthew Lawe, who has served as my confidential clerk with a zeal and industry not less than those I have endeavoured to accord to your Lordship and his late Majesty. When you have time to consider, I pray that you will bear this name in mind. I have the honour to be, my dear Lord, with every truth and regard, your Lordship's most humble and obedient servant.'

It was Samuel Pepys' way—a most moving way—of breaking the news to his friend.

Matthew was thunderstruck. He could scarcely complete his writing, which at the end scrawled and wandered like a child's.

'For God's sake, sir—what is this?'

'There is no need to tell you more . . . I do not know if it will serve any purpose.'

'But *why*? Are they all mad?'

'They are new, and strong. I am not.'

Indeed, he did not look so, at that forlorn moment. Matthew had never seen the great Secretary so downcast, nor so careless of showing his wounds. His face, in the pale light of a February day, seemed as grey as the London sky, and as ready to dissolve into misty tears. Matthew, bitterly mourning already, could only think: The end of the world is not some awesome thunderbolt, it is a child's cry in a grown man's mouth. Humbly he asked:

'Sir, must you go?'

Mr Pepys stirred himself, as ready to deal with this question as with a thousand others, from the price of clinching-irons to the deference due to a broad pennant from some lesser piece of bunting.

'As to that, I have not yet been told. But today there was proclaimed a cessation of all public offices—' he pronounced the phrase with delicate irony, 'until their holders are given new authority . . . I have it in mind that they will not give it to me.'

'Then they are beyond reason!'

'No passion, Matthew,' Mr Pepys reproved him. 'The world turns, and some are left behind. . . . It is better to walk away than to be led in chains.'

'*Chains?* For God's sake——'

But gallant Mr Pepys was gaining in spirit, even as Matthew Lawe gave way to despair. Suddenly he was busy with a paper which he had drawn from his breast pocket, and then with his laborious reading-tube.

'Think not of chains,' he said. 'Think of freedom and love.' He tapped the page before him. 'Here is something I have carried next to my heart these two months. It is a letter from Will Hewer, written at the first moment when he feared my downfall. I ask you to remember it—here speaks a tender spirit at a time of trial.' He bent to his leather spy-glass, and read slowly: ' "You may rest assured that I am wholly yours, and that you shall never want the utmost of my constant, faithful and personal service, the utmost I can do being inconsiderable to what your kindness and favour to me has and does oblige me to. And therefore, as all I have proceeded from you, so all I have and am is and shall be at your service".' Mr Pepys raised his head again, and his eyes were moist with more than painful straining. 'With such friends,' he murmured, 'who cares a fig for enemies?'

Now they were both near to tears. 'I would write you such a letter, this very hour,' Matthew said.

'That I know. And so I have writ you one of mine.' Then Mr Pepys recollected himself, and took command again. 'Cheerly, Matthew! Our cause will survive. Make one last note of *why*!' His voice changed to the dry dictating tone. '*Summa magna* of the figures for tonnage of the Royal Navy during my tenure of office. At the beginning: 62,600. At this end, 101,000.' He caught his breath, and sat back. 'Let them milk *that* from me!'

Then the first Secretary of the Admiralty put his desk in order, and—a noble heart drained, within a man untarnished—walked ashore for ever.

FIVE

FISHERMAN

1720

'Give up the fishery, and you lose your breed of seamen.'

Admiral Sir Charles Saunders, in the
House of Commons, London, 1774

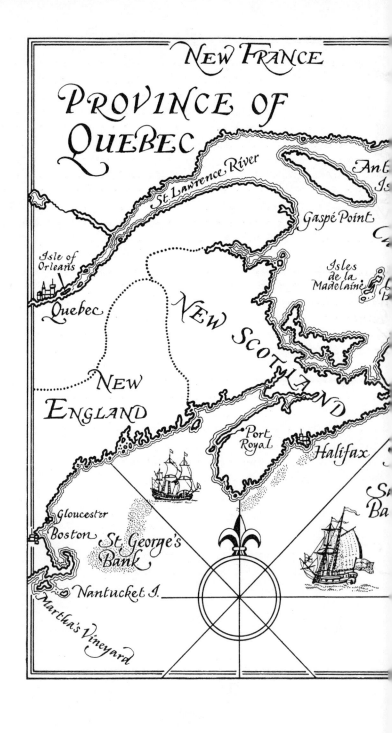

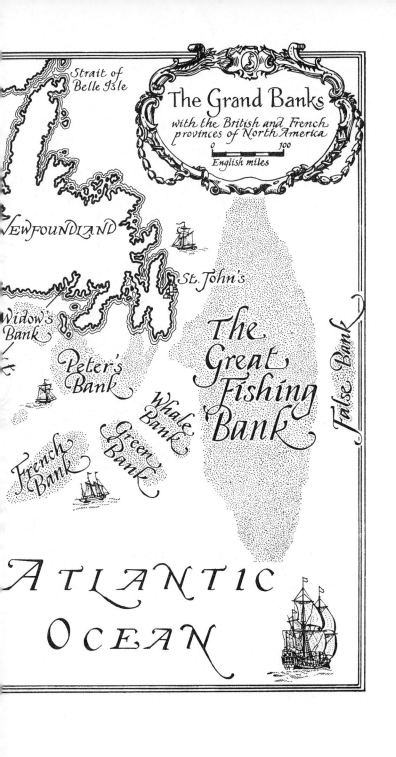

Strait of
Belle Isle

The Grand Banks

with the British and French
provinces of North America

0   100

English miles

NEWFOUNDLAND

St. John's

Widow's
Bank

Peter's
Bank

The
Great
Fishing
Bank

Whale
Bank

False Bank

Green
Bank

French
Bank

ATLANTIC

OCEAN

MATTHEW, waking in the noisome filth of the Fleet Prison for the one-hundred-and-twenty-first time, had no need to wonder where he was. Unlike some confused dreamers, he knew for certain; and it was this certainty which distinguished one foul place, Bedlam, from another fouler still. Most of the Fleet inmates had all their wits about them, and felt their misery to the last bitter drop of despair.

Here, none was better than his neighbour; though one of them might be wicked, and the other misfortunate, both were doomed. The dregs of criminal London might share their berths with fine folk who had come down in the world, by accident or foolishness. But where was the distinction? In the cut of a coat? In the height of a nose?

They all lay together in the new Fleet—debtors, thieves, runaways, murderers, beggars, pickpockets, sellers of false gold and sellers of flesh—and they all woke to the same hopeless dawn.

Was it better to be a debtor, Matthew Lawe sometimes wondered, rather than one of his fellow-captives, born in the image of Christ, who had been branded with his transgression in livid lettering: 'R' for Rogue on the shoulder, 'M' for Man-slayer on the right hand, 'T' for Thief on the left hand, 'F' for Felon on the cheek? If branded he must be, Matthew would have been forced to choose 'F' for Fool. . . .

But whether distinguished or not, they were all clipped in the same cage: starving in misery, beaten by their captors, fouled by their own dung-hill, waiting for the release of death or the end of a harsh sentence, or—as with debtors—for the moment when some magical friend would wave a golden wand and secure their release.

Such was the fate which could befall a man who had lost his guardian angel.

When Samuel Pepys, that proud and humble servant who would not bend to the stiffest of winds nor succumb to the most delicate of lures, quit his office of Admiralty, much that was private dissolved with him. The household of Buckingham Street ceased to be, either as a family or a fabric; it was a measure of his velvet grasp that as the firm hand within was withdrawn, all that remained was a puppet-show struck lifeless at the cold hour of midnight.

Mr Pepys, losing his quarters, withdrew to the country: the servants melted away to other masters, other places: little Lucy was here one day, gone the next: Matthew himself was without a bed, nor anyone to share it. Having found lodgings close by, he worked on at the Admiralty for a man he could not love, and it was not the same, nor ever would be.

He owed his continuing berth to Mr Pepys' kind intervention, and was thus under suspicion from all the new breed of office-holders. Though the decline of trust was slow, it could be felt in every breath, every sidelong glance. A fall from grace need not be as swift as Adam's, in order to be fatal.

The years withered down like leaves, and the people with them. Henry Morgan the pirate-king had died, full of honours and evil, in the year of Revolution: Port Royal itself, that sink of all iniquity, perished entirely, by earthquake and the wrath of God, some four years later. So one black page was turned and the world was made a cleaner place, though not more comforting for those who were left behind.

Old Sam Pepys himself passed to his great reward, at the allotted span of three score years and ten, in 1703. A legend in his own time, casting the longest shadow in the service of the sea, he was fit to be mourned by any sailor—and by any king also. His life-long friend and deep admirer, John Evelyn, noted in his famous Diary for the benefit of later years:

'This day died Mr Samuel Pepys, a very worthy, industrious, and curious person, none in England exceeding him in knowledge of the Navy.'

Thus this shield of England, and of countless smaller men, failed by one year to witness something which would have delighted his sailor's heart—the capture of Gibraltar, one of those Pillars of Hercules standing guard at the western end of the Middle Sea, from the combined might of France and Spain. But as one giant flag was hoisted, another tiny pennant fluttered and fell. Clerk Lawe, deemed 'unwanted in the present state of the Admiralty service', received his dismissal.

There could be no argument. The 'shield of England' had been

Matthew's breast-plate also. A man who lacked such protection was a man in peril.

As if by a last echo from the past, he was soon befriended by Edward Lloyd of the coffee-house, and took service with him as book-keeper—or was it cellar-man?—or pot-boy? It did not matter greatly; it was a modest berth in a mortal storm.

'Lloyd's'—the name was now in common currency—had prospered, and was crowded at all hours by men who held the great sea-commerce of England within their grasp, and other men who were hopeful of sharing this spoil, and yet others—the rats and mice of a nation's vast dock-side—who snatched at fallen crumbs or nibbled at bulging sacks, and scuttled for cover when observed.

There Matthew toiled every day at every task, whether it was the entering-up of the daily profits, or the assembling of *Lloyd's News*, or the swabbing down of a table ring-stained by the coffee-mugs. He might have prospered himself if common greed had not poisoned him. But once more a giant pendulum swung, and brushed him off this perch as well.

The South Sea Bubble, that great ball of fatuous madness which infected England for ten fevered years, and then burst like the rank abscess which it was, had its origin in the second curse of mankind, which was debt. But this was no mere husband's worry-wart, like the slate at a tavern or an account overdue for ribbons. It was the Great National Debt, creeping up on a nation encouraged by its dung-headed masters to believe that the best way to save was to spend.

Any prudent householder knew that this could not be true. From his private life he had learned beyond question that if he overran his income he must make do with less—less of everything, from cakes and ale to a new carriage, until the balance had been restored. But when private life was enlarged to public, the rules were magically changed, by fools or scoundrels, to persuade this prudent citizen, multiplied by six million, that a debtor could borrow his way to riches.

Hence the National Debt. When Mr Pepys quit his office, it stood at £700,000—a sum shameful enough, if a man were thinking in terms of honest house-keeping. By 1714, a mere twenty-five years later, plagued by wars which none could afford, it had risen to a staggering thirty-six million pounds.

Already this was regarded almost with pride, as if it were the mark of a great nation to produce children with two heads or four feet or twenty toes—or thirty-six million millstones round their necks.

Then a man, a fool who was also the Chancellor of the Exchequer and soon to be an earl, came forward with a splendid scheme. Since the nation was deep in debt, this must be repaid by public subscription. Money—let us say, a fair round fifty million pounds— would be borrowed on some promise of gain from the very people— that is, the nation—who themselves owed the National Debt already.

Thus all would be in sweet balance again, the lion would lie down with the lamb, and none would owe a penny.

So was born the South Sea Company. It started small, it grew enormous, and it spawned a monstrous brood of children, to whom everyone ran with money spilling out from their fists, avid to make investment in schemes of any sort. Then it withered overnight to nothing, bringing down with it half the families of England, rocking a throne (since King George the First was the august Governor of the enterprise), ruining a government, and dishonouring the slippery club of ministers who had been bribed to give it licence.

The lion did indeed lie down with the lamb, but only one rose up to greet the dawn.

It started small but saucy, with a state-trading company which would sell exclusive permits to trade in the South Sea—that part of South America which was, in fact, the province of Spain. In return, the Company would have an assured revenue from import duties, to be levied in perpetuity on all things from vinegar to whale-fins. It would also assume the burden of the National Debt.

Then it began to grow, out of a contrived, mendacious prosperity which hoisted the stock from £100 to £500. The author of this conjuring trick was one John Blount, the Great Projector, the most energetic and fertile of the Board of Directors. Son of a shoemaker in Rochester, he had risen from petty scrivener, who would write you a love-letter at sixpence a page, to solid money-merchant, and then to the most arrogant swindler in the kingdom.

As the stock mounted, despite growls from Spain that the South Sea and all its riches were theirs, whatever fraudulent claims might be swallowed by the gullible, so men mounted with it. John

Blount was dubbed Baronet, while his fellow directors grew regal in their contempt for non-believers. But there were very few of these, as South Sea stock reached £850, and fortunes were made, and all England, from dukes to door-keepers, listened to rumours and flocked to buy.

Then other bubbles began to float and whirl, as other men moved into a mad market-place which had set the country ablaze.

These were the stock-jobbers, the first of their breed, who set up business in Change Alley, near enough to Lombard Street and the Bank of England to give a spurious countenance to their endeavours, which were rapacious and dishonest beyond anything imagined.

Their favourite haunts were two coffee-houses, Jonathan's and Garraway's, where they set up pavement tables; and soon, as a kind of lunacy gripped the country, and men mortgaged their great estates or pawned their shoes to raise money for investment, one might wait hours in a milling throng to come near enough to these man-traps in order to thrust out gold and beg for it to be taken.

At the peak of the turmoil, when South Sea £100 stock stood at £1050, these other bubbles were also swelling to the size of monsters. Their schemes, when soberly judged, were past belief, but never beyond the greedy aspiration of fools.

They invited men—grown men, not children or parish boobies— to buy stock in such enterprises as making oil out of radishes, butter out of beech-nuts, fresh water out of salt, and silver plate from fluid mercury.

A wheel for perpetual motion was all the vogue: so also was coral fishing in the Middle Sea, trading in coarse hair from foreign climes, curing venereal infection at its source, and importing large Spanish jack-asses to improve the British mule.

A contrivance for emptying every Necessary House in the City of London had all the sniff of success: so had funeral furnishing by a new and secret method of disposal, carrying live fish tanks to reach every hamlet in England, and Puckle's Machine Gun for discharging round and square balls.

All, no matter how foolish or brazen, found ready buyers. Most invited subscriptions of two million pounds each, and were filled within hours. During the last few months of Bubble Time, the capital sum garnered from all the live fish in this whirlpool was three hundred million pounds. Though there was not so much fool's gold in all Europe, yet by some stolid English alchemy it was to be found in Change Alley.

Lloyd's Coffee House, and Matthew Lawe with it, had long resisted the lures of bubbles great and small. The men who gathered there every morning were of more sober stock: nothing that was fly-by-night, nothing with the taint of gambling, nothing vulgar, nothing too close to the wind of the law, had any currency or repute in this clean, industrious, and respectable hive.

Not for them was investment in foreign hair or Puckle's round and square balls. The solid bottom of a ship was the very least fortress in which they put their trust.

A man, a stock-jobber, a foolish fellow whose tongue ran like the Fleet Sewer, and who had tried to set up a table outside Lloyd's and there sell stock in a company 'for dyeing Calicut to resemble the finest silk', had been overturned, table and all, and then bundled beyond the chaste confines of Lombard Street. Matthew's share had been an arm and a leg, and he had done clawing justice to both.

Yet the Sirens' sweet song was to be heard daily all around them, and was hard to deny, even with the wooden ear of rectitude. Men, though careful to say that they would never indulge in this bubbling nonsense themselves, reported constantly that this fellow or that, a wife's cousin or a duke's body-servant or the confidential clerk to the Lord Mayor of London—or the Lord Mayor himself—had made a fortune in South Sea stock, invested his gains in a second venture, and made a second fortune so vast that he now dressed his thirty servants in liveries of cloth-of-gold.

Matthew, waiting at table and keeping alert as he dispensed his wares, overheard some ripe morsels which made honest toil seem as dull as ditch-water.

'They say South Sea stock will top two thousand pounds, and then there will be another subscription.'

'There is a man at Garraway's who will warrant you a hundred pounds a year for life, for every hundred you subscribe.'

'They say the German mistresses have sent half a million out of the country—all from the King's own bubbles.'

'They say Thomas Guy the bible-seller will build a hospital from his profit.'

'Sir John Blount eats off gold plate—and voids guineas every morning!'

'There was a woman of quality at Jonathan's who stripped off her diamond necklace and changed it for Fish Pool stock.'

'They say Squire Boone bought Marvell's Magic Ointment to the tune of six thousand pounds, sold it next week for eleven, and went back to the West Country with his daughter's dowry—enough to buy an earl for a son-in-law!'

'There are more than a thousand new carriages ordered. God save us when we try to cross Piccadilly!'

'They say it will continue for ever—till we all have a million pounds each.'

'What then?'

What then, indeed ... Matthew, listening hungrily, tempted mightily, setting his humble wage against the glitter of wealth which appeared ripe for anyone's grasp, could not see beyond a million pounds. 'They say' seemed to ring as true as guinea-gold itself. There were times when he set down a pot of coffee, and smelled the fragrance, not of the roasted beans of far-off Araby, but the new-minted coin of Change Alley, a mere fifty paces away.

A million pounds a-piece? How could he forbear, at this one great moment when he might transform his whole life?

At last, in a quiet time when trade was slack, he sought advice from a man he could trust. This was Meyer Mond, young son of the man who had been Samuel Pepys' banker. Mond, whose black hair above an olive face was already vanishing—as the wits said, 'Grass does not grow on a busy street'—sat alone at his customary table. His left hand, idle, was drumming a steady tune on the table-top; his right was busy with pen, ink, paper, and a column of figures thrice as long as his nose.

'Mr Mond, do I disturb you?'

'Yes,' Meyer Mond replied instantly. Then he peered upwards, and smiled with the ageless charm of his race, and said: 'No, Matt, you do not. By chance I came at this moment to the end of calculation. So you want to ask me a question? I will tell you the question, and the answer, in one breath. You wish to know, are these tales of prosperity true, and how can you have a share in them?'

'Yes, sir.' Matthew, marvelling but set at ease, stood boldly by his side. 'I would value advice.'

'*Quaere*,' Mond said—and he spoke like old Mr Pepys, which was

comforting and moving at the same time, 'are these tales true?
Answer, Yes. *Quaere*, is the prosperity true? Answer, No, not in
harsh fact. We have only planted a money-tree. It blooms now
beyond belief, but autumn will come, or the woodsman with his
axe. . . . Yet an old French philosopher once said, *Cogito, ergo sum*—
I think, therefore I am—meaning, if a thought is present in the
mind of man, then it is a fact. So prosperity is true, as long as we
think it so. *Quaere*, how can you take advantage of this madness?
Answer: Grow mad yourself.' Mond sat back on his bench, keeping his hand on the piece of paper in front of him, patting it as if
it were cherished gold itself. Then he smiled again, the smile of a
man who knows all of himself, from weakness to strength, and can
still survive such knowledge, and added: 'As I have.'

Matthew was astonished. He had thought that young Meyer
Mond, a Jew of repute as good as his father's, would not have
dipped a single toe in these waters. 'Sir, you have ventured
yourself?'

Mr Mond gave him a stare and then a smile. 'Like all that you
see here.'

'Then it must be safe!'

'It is a wager,' Mond replied, 'call it by whatever fancy name
you wish. It is a wager, even for those who know the cockpit.' He
was still looking closely at Matthew. 'But I can cite you another
philosopher, Matt, who might be just as wise as old Master
Descartes. Mr William Shakespeare, who told us: "Poor and
content is rich enough." Think on it.'

But Matthew was too hot for such pale advice. If men like
Meyer Mond, and others of solid substance at Lloyd's, were in
this game, then only a fool or a coward would keep aloof. He
said 'Thank you, sir,' and Mr Mond, with meaning, replied:
'Thank me on this day next year,' and the great trick of confidence
was on the slip-way.

That evening, as soon as he was free, he walked down to Change
Alley—and straight to the arms of a rogue.

The noise in the narrow alley had abated somewhat with the
coming of dusk, but the tables were still busy under a flickering
lamplight which cast strange shadows among the bobbing heads,
grasping arms, gleam of gold, rustle of paper, and the constant
wandering drift of men and women in search of gain, or sport, or
release from their money-cares. There were also whores a-plenty,

pushing forward most brazenly, but Matthew wanted no whores. He wanted first to invest, and then swiftly to prosper.

He knew nothing, and remembered nothing save the tenor of his talk with Mond—that all men were in this game, which was a wager. He thirsted to wager, and not to be left astern in this, the greatest gold race in the world.

There was a man sitting somewhat apart from the throng, in an angle between one house and another. He seemed idle, and he seemed not to care: his table spread with papers, and his candle burning on one side, were enough for him. But his eyes, set in a dark hawk-like face, were always busy, and when he caught the gaze of Matthew's, he smiled and gently beckoned with an inclination of the head.

Matthew was lured, and approached. Close to, the man was seen to be richly dressed; his fingers were ringed with gold, and the purse lying on the table-top was as plump as a fowl. Without a word he drew up a stool, and when Matthew had sat down he still kept silent.

Matthew ventured: 'What business are you doing?'

'None,' the man answered carelessly. 'I have finished for the day—and a monstrous fine day it has been! Now I take the night air before I cast up my accounts.' He looked at Matthew. 'I have seen you at Garraway's, I think.'

'No. At Lloyd's, perhaps. My name is Matthew Lawe.'

'Of course! I know it well!' In truth they had never met, nor cast eyes on one another, but a warm welcome never lost its relish. 'I am Finger, the stock-jobber of Cheapside. You will have heard my name, beyond doubt. What brings you to Change Alley?'

Matthew said: 'To invest.'

'Very wise,' said Mr Finger. 'There is no greater moment than this. We are only at the beginning of a vast prosperity. I wish, as a friend, that I had something in hand for you. If you had come only an hour earlier! I have a company founded which makes all other companies seem like games of penny marbles.' The weird candle-light flickered between them, conjuring dreams of profit from an embracing, lavish darkness. 'But my books are full. I have been pressed all day, by people begging to take part. It is the greatest enterprise——' he shook his head, in honest regret. 'One must catch fortune at the flood, Mr Lawe.'

'Is it too late?'

'I could say yes, and I could say no,' Mr Finger answered, most wise and judicious. 'I have one portion of stock certificates, numbered—' he consulted a paper lying before him, 'numbered one-thousand-nine-hundred-and-one to two thousand. I have promised it to my Lord—well, we will name no names. But he must have other business, because he has not passed by this evening. However, I am a man of my word, so to you I must say no.'

'But if he is not here,' said Matthew, already desperate, 'then he must have changed his mind.'

'That is true,' said Mr Finger. 'That is very true. I see that you are a man of perception. If only I could be certain. What sum have you in mind, to invest?'

'Twelve hundred pounds.' It was all his wealth in the world, scrimped and saved in more than twenty years. It seemed a large sum, as he named it. But Mr Finger pursed his lips at the figure.

'It is rare for me to deal in these small amounts,' he said. 'A great enterprise such as mine . . . Can you not increase it? The price of these last one hundred certificates is twenty pounds a-piece. I must tell you that I might have sold them ten times over in the last hour.'

'But may I not buy less than one hundred?'

'No. That is not my practice.'

'Then the price would be two thousand pounds?'

'Two thousand. *I cannot promise you,*' Mr Finger became confidential, 'that you would sell your stock for twenty thousand within a month, but I have my own opinions.'

'I have no more than twelve hundred pounds.'

'Well,' said Mr Finger, and began to collect his papers.

'Perhaps I might borrow,' said Matthew, alarmed.

'A man of substance can always borrow.'

'But what is this enterprise?'

Mr Finger dropped his voice lower. 'That is the great secret,' he said, 'and your greatest chance of fortune.' He picked up from the table a piece of paper, so handsomely engraved that it might have conferred the freedom of the city, and passed it across to Matthew. 'Read it,' he commanded. 'But not a word to a living soul!'

Matthew, his brain on fire, read: 'A Company for carrying on an undertaking of Great Advantage, but no one to know what it is.'

After a long, mysterious pause Mr Finger said: 'Enough for any man of spirit, think you not?'

'But what is it?' Matthew asked, confused.

'Great God! Do you expect me to tell you, before you have bought?'

'I beg pardon,' Matthew said humbly. 'I am not skilled in this. But is it—is it based upon the sea?'

Mr Finger looked this way and that as if wary of being overheard. Then he winked, very slowly, and whispered: 'All brave things are based upon the sea.'

It was enough for Matthew Lawe. 'I cannot raise the full sum now,' he said—and found that he also was whispering. 'Can the matter wait until noon tomorrow?'

'I will not swear to that,' Mr Finger replied. 'If my Lord comes by before noon, what am I to say to him? That I have given his share to another *who has not yet paid me*? He would have me clapped in the stocks! But I like your spirit, Mr Lawe,' he went on generously. 'Come as early as you may, with your two thousand pounds, and if you are not forestalled then you are a partner in our enterprise. But not a word more to anyone. You must promise me.'

'I promise,' Matthew answered.

Mr Finger leaned forward and, as if completing some holy ceremony, blew out his candle. In the sudden darkness his voice, full of this great conspiracy, echoed like the Oracle of Delphi: 'I swear—if you seize your chance *now*, you will be riding a gold coach by Christmas!'

Now was six months past, and *now* he lay in that golden coach—fashioned of straw and iron bars, carpeted with filth, motionless on the low-road of life. He had been duped, robbed, and condemned in debt.

When he had returned to Mr Finger's table with his two thousand pounds—so easy was it to borrow money in those fatal days, though at a foolish interest—the stock-jobber, all smiles, had first taken his cash, counted it, locked it within his safe-chest, and then bid him come back that same evening to receive his stock certificates.

When he did return, full of bursting hope and pride, Mr Finger was nowhere to be seen, and his narrow corner was occupied by a

stranger, a surly fellow who could only smile at another's misfortune.

Their converse had been a bitter prelude to his damnation.

'Are you with Finger?'

'No. Nor with child either.'

'Mr Finger the stock-jobber. Do you know him?'

'I know ten fingers. Five on each hand.' The man raised two of
them, in a gesture of insult. 'Here's two for you, if you plague me
when I am busy.'

'But I was to meet him here! He was to hand me my stock
certificates!'

'What's that to me?'

'But where would he be?'

'France, I would wager. Or Holland, if you allowed him enough
gold . . . Do you wish to invest?'

'No.'

'Then give place to those who do.'

Thus, between one sunset and another, Matthew had been
ruined: one small fool among a hundred thousand others. For he
was not alone, nor was his ruin more than a speck of dirt on a
mountain of ordure.

The rout had been led by the greatest bubble of all, the South
Sea Company, whose stock, in its last six months of life, fell from
£1050 to £150. The noble directors were the first to sell, though
secretly; then the humbler dupes, in open panic. It seemed that
there were no riches for England in the South Sea: no magic
repayment of the Great National Debt: nothing but empty coffers,
and the sour taste of a vast swindle.

All other enterprises had followed it to oblivion.

Families by the thousand, great and small, were picked clean.
Noblemen wept while the mob roared. Suicide was the order of the
day. Though the principal villains, among them Crown ministers,
were stripped of their gains, the greater part of the money had
vanished, and the National Debt shone bigger and brighter than
ever, a monument to a nation which, at a depth of greed and folly,
had out-swum the fish in its lust to swallow the bait.

At ten of the clock, with his black bread and foul water swallowed
ravenously as if they were the food of the gods, Matthew still
waited for the morning release. The surly turnkey was always late,

in this corridor of the damned; he had no care for Matthew nor any of his mates in misery, because they had no money, not even for the most petty of bribes, and were thus a burden instead of a source of profit.

In this hell of the Fleet, the poor fed on the poorest, and the poorest of all were starveling dogs whose only freedom would be death.

But it did not greatly matter—so said Giant Despair himself, eating away once more at a shrivelled heart. The morning release was only to the prison's general yard, which he hated as much as his cell.

The lock squeaked, the hinges groaned as his door was at last flung open. 'Out!' the turnkey commanded. His humour as foul as ever, he eyed Matthew as if he were one of the swarming rats which shared this paradise. 'And be present when I return, else you will be lashed for a runaway!'

Matthew had been lashed already. All the poor men had, for anything or nothing, to press home the virtue of solvency. Without a word which might be construed as insolence, he shambled down the passage-way and stepped out to the courtyard.

No sun shone here, nor ever had: perhaps its kindly face was hid in shame. A milling inferno of unfortunates filled this pit, and from them rose a tumult of voices as well as a stench fit to flake the nose. There were quarrelling men, speech-making men, whimpering men, capering and jumping men trying to bring warmth to their imprisoned bodies: men howling in anger, or singing of the joys of heaven, or lying in chains, as mute and useless as abandoned logs.

They all bore the name of prisoner, and some could scarcely bear it, for grief or rage or despair.

There were women also in this cage, harpies who crowded together for their own safety and snarled like rabid dogs if any man approached. They would claw off his face if they chose, or, in another mood, would form a curtain of Venus round him while one of their sisterhood lustily swallowed his privates. But it would be a brave man who tried his luck in this lottery.

Matthew exchanged a word with a good friend in misfortune, one of the Honourable Company of Prick't Bubblers: one Nathan, late of Corn Hill—and if a Jew could be brought to this bondage, what hope was there for simple Gentiles? Nathan had pledged his fortune to an enterprise of great promise: a share of a ship, a stalwart, oak-hearted English privateer which would roam the Spanish Main and return with plunder so great that it would need six months to count it.

The stalwart privateer had come no nearer to the Spanish Main than Rotherhithe Reach, where her rotten timbers shed their only fastening, which was paint, and incontinently sank—and with her, one Nathan.

Then Matthew heard his name called, in a voice almost a shriek, and turned round. Advancing upon him, arms outstretched and flailing, was one of the women from the female barricade.

As he gave ground, fearing she might be the spear-head of some foray, the prudent Nathan did more: he disappeared without trace, like a minnow at sight of a shark in the bay. The woman, a squat slattern with swarthy shoulders and hair like snakes a-coiling, pressed forward, a grinning welcome on her face. But she repeated a pass-word which halted his retreat.

'Matthew! Is it you?'

Astounded, he peered at her. Could he have forgotten such a befouled scarecrow as this? He asked uncertainly:

'Do you know me?'

'Better than your mother,' the woman answered, with a lewd laugh. 'Unless you were at the same tricks with her. I am Lucy! Who else? And you are the same as ever!'

Little Lucy ... Matthew was shocked beyond endurance. She was little no longer. She was hardly Lucy at all, so gross was her person after nigh twenty-five years. The pretty, inviting face was now rotten, over-blown fruit, with sagging jowls and tell-tale pustules round the mouth. The breasts he had fondled with such delight swung like monstrous saddle-bags; the legs which had once enclosed him had grown so gross that they might have enclosed a barrel of ale, with nothing to tell the three apart.

She looked what she was: a wheedling old harlot, long past the trade, who could turn to a Fury when trade was denied. And to think that this might be his own doing.

Aghast, and softening to this thought, he drew her aside, and found a corner where they might be private. The women watched them greedily, while men gave way willingly enough to such a formidable pair. Once there, she laid a fat, veinous claw on his arm.

'Matthew! You look so young. I wish I knew your secret! Did you truly not remember me?'

The coy approach was repulsive—and heartbreaking in the same breath. 'I was jesting,' he said. 'Or the sun was in my eyes. Such blinding sunshine! Of course I remembered. How goes the world with you, Lucy?'

'As you see.' Then she forestalled his own question. 'Why are you here?'

'Debt,' he answered, glad of some neutral ground. 'The bubblers made a Tom Fool of me. I owe near a thousand pounds, and can never pay, if I live to be a thousand myself. . . . But what of you?'

'The same,' said fat Lucy with a wheezing sigh. 'But by a different road. Incorrigible whore—that is what they call me here!' She had pronounced the first word laboriously, the second with ease. 'Did you ever hear such evil language? Incorrigible?— I cannot even get my tongue round it! I ran into debt—since the bait is not so fresh—and at the last they came to take my goods. A few pots and pans, and a good broad bed. . . . I might have 'scaped, but I poxed the bum-bailiff!' She cackled with laughter, revealing teeth so rotted that one could not tell where the gums left off, the while her dangling bubs shook like butter-churns at their last delivery. 'So I am here for ever, 'less I can work my way out. But who has money in a stews like this?'

Little Lucy . . . He could have wept for the whole of woman-kind. His glance showed his guilt and shame as he asked: 'Did I bring you to this?'

'Nay, do not fret.' The scabrous hand pressed his arm again. 'If not you, then another. . . . We had good times, Matthew. None better, in all of life. My first love . . . Maybe I can help you. Do you know, there are ways to get you out of this kennel?'

'What ways?'

'Would you go to sea?'

Matthew stared, his heart leaping. 'Christ, I would go to sea in a miller's sieve!'

Suddenly, above their heads, a warning bell tolled its melancholy signal, and Matthew, forever guilty, started up. He had three minutes to reach his cell. 'I must go,' he said. 'Do what you can, for God's sake!'

'Meet me here tomorrow,' Lucy said. 'I can promise you

nothing, and there may be danger for you, but I will try. And Matthew——'

'What then?'

'Think well of me.'

He could not kiss that putrid mouth, but he touched her shoulder as he lied: 'I could not think better if we were climbing Sam Pepys' stairway.'

Nathan joined him as they were swept into the cattle-drove walk to their cells. 'I always knew you a brave man,' he said. 'But this tops all! Who was that, for the love of God?'

'A friend . . .' Trusting this other friend, he looked sideways, and whispered: 'Is it true, one may escape from here by going for a sailor?'

Nathan glanced at him sharply. 'Aye. So I have heard. But it might be one prison for another.'

Lucy did not meet him next day, nor ever after. But at eventide a man came to his cell, and was admitted by a turnkey whose mood seemed much improved, and who even set a rush-candle—never to be seen by poor debtors—between them before he withdrew.

When they were alone they eyed each other, without concealment. In the flickering gloom of his cell, Matthew saw a small, fat, olive-skinned man, dressed like a gentleman, but yet a nervous watchful man, whose very scalp moved to and fro as the brain beneath it thought and schemed.

This was a man both of confidence and fear, a man of limited command—though God knows this was enough, in the debtors' corridor of the Fleet—who might be overthrown by a greater, and knew it every day of his life.

What such a bum-boat admiral saw in Matthew was not to be known. But he must have satisfied himself on certain aspects of precedence, for immediately he spoke it was in the voice of contempt.

'Do not waste my time,' he said, as if Matthew had delivered himself of a windy speech. 'You are a sailor?'

'Aye.'

'With no money?'

'No.'

'There is nothing new in the world. . . . Listen to me, without interruption. I will not pay your debts. But I can give you freedom, and a berth at sea. At your own risk. Agreed?'

'What is your name?'

'Enough of questions!' the olive admiral barked.

'But what berth is it?' Matthew insisted. 'I cannot sign to sail for China in a hen-coop!'

The visitor sighed, humouring a fool. 'There are Portuguese schooners from the Grand Banks of Newfoundland, sheltering in the West Country. Cod fishers. They lack crews to return to the Banks. You will be bound for five years. Any idleness, any insolence, and you will begin your bond again, from that very day. . . . You must make over your first two seasons' wage and share of profit to me. The captain has a paper signed on board already.' He pulled a scroll from his sleeve. 'You will sign this one.'

'Two years' wage? On what do I live for two years?'

'Cod,' said his benefactor.

'It is slavery!'

The man looked round the cell. 'What is this, then?'

Matthew swallowed. There was no answer, in all the harsh world. 'But how do I escape from here?'

'I will return, two nights from now, and you will walk out with me. There will be a postern gate opened, and a man with his back turned. On the way, you say nothing. *Nothing!* Agreed?'

'Aye.'

'Well enough.' Then the man leant forward, within the feeble candle-glow, and his eyes managed a dull gleam. It was not friendly. 'But mark this. By then I will have paid certain sums to certain men to see that you are free. *But your own debts are not paid.* You are a runaway, from that moment. If you are caught by mischance, you will be brought back here, branded for a rogue, and whipped till you never walk again. If you cheat *me*, you are a dead man. I have a long arm, with a fish-hook at the end. . . . You are safe only in one place, on board the schooner *Consuela* in Bristol Dock.'

After a pause, in which he weighed the chances and found them daunting, Matthew asked: 'How do I reach Bristol?'

'In a farm cart. Caged, along with others. Agreed?'

'Aye.'

'Then sign. And keep your mouth shut.' For the first time the olive man with the power of life or death—or so it seemed—allowed himself a modest jest. '*Do not even smile!*'

Rumbling and lurching westwards in the caged farm cart, with foul companions who, if they were not stupefied with liquor, cursed and groaned without ceasing, Matthew surveyed his fate and found it cruel. If all went well, he would escape from land, which had proved disgusting, and find the sea again. But to what end? He had been freed by a land-shark whose teeth still gripped him.

Full of despair, kicking out at a companion whose twining legs, even in this prison of animals, seemed ready to be amorous, he cursed all things, from the distant past to the gross future:

'A Portugee cod fisherman! Cry stinking fish! I was to be a gentleman!'

II

Sea-time once more was the bliss of the world. Fishing brought eternal hope. Sleep after toil did greatly please.

Out on the Grand Banks, the shoal waters to the south of New-foundland, teeming with fish, blessed by the summer sun, plagued by fog and storm at the beginning and end of every season, the motley fleets assembled, and fanned, and went about their secret business.

There were the English, the lords of this empire, sometimes as strong as two hundred sail: the Portuguese, cunning sailors, twice as cunning at squeezing a bargain out of the last dried fish-head: French ships reputed to be full of spies from their fair province of Quebec, little Breton *pinques* more at home in the creeks of St Malo: Spaniards who still dreamed of an imperial past, but would now settle for a dozen schooners full of salt cod.

It was a world of fish and fortitude, commerce and courage, rare skill and evil luck: navigation by guess and by God, storm by appointment: great catches to match great hopes, bitter waste of time to poison past memories. It bred sailors, and it broke them— the eternal lesson of the sea, which ruled all and was indifferent to all save its own twin masters, the wind aloft and the raging tides below.

Once in position above the Banks, and motionless under storm canvas or idle slatting sails, the schooners dropped their long-boats and left them to their toil. The boats spread out like furtive fingers, pretending indecision or ignorance, searching all the time for a chance sign which might promise a shoal of fish: the flurry on a surface suddenly boiling, the cry of a watchman seagull, the changing colour of the water which betrayed great movement below.

There they fished by hand-line baited with capelin, the little fish that the cod loved, and there they stayed until their boats were loaded. On a happy day or night, it might be for two hours: with starvation luck, it could keep them sea-borne orphans for a week, and fifty miles from their mother ship. There was only one command: 'Do not return without a full catch!' and if the weather turned brutal, or the fish were shy as a girl with the spring blemish, that command was still unchanged.

They were all partners in the same enterprise, and no cowards or faint-hearts, fools or bumblers, were welcome back on board until they had earned their share of partnership.

Matthew, crouched in the bows of the *Consuela*'s Number Six long-boat, had only one thought: somehow to keep warm, somehow to keep awake, until the dawn of their fourth day of fishing brought the solace of the sun. Continually he slapped his breast and shoulders, and chafed first one hand and then the other, while he held between his teeth the two long cod-lines which led over the side and thence fathoms deep below.

His hands were not only cold: they were sore to crying, with every cut and scrape and blister of his work inflamed and crusted by salt. His feet were like lead, numbed and useless; his hair and beard ran with icy droplets. In the dim glow of their riding light, the fog which brought this bitter cold pressed and drifted overhead, like a shroud waiting for a dead man.

He had the watch, the desolate middle watch of the night, while his five companions slept unseen: three cod-liners like himself, and the fat boy who was their bait-cutter, and the old man who, with luck, wrenched the fish off the hooks, tumbled them into the barrels, and dashed some salt on top to seal the bargain.

None of these others was visible, in the swirling wet cold darkness. Matthew knew them to be there because he could smell them, as he could smell every single element in Number Six after three days on the Banks: of fresh fish not enough, of urine and excrement swilling about beneath their foul floorboards, too much: of decayed bait and rotting fish-heads, sufficient for the strongest stomach afloat: of sweat, enough to prove their enduring manhood.

In the depth of a late September night, with all the signals of winter—a touch of ice, a drift of snow—already besetting them: with only two of their ten broad barrels full and topped with salt, while the rest still gaped and mocked them: with the icy water slapping and pounding against their hull, and the prospect of another day of damnation before him—in face of all these hazards, only this manhood could sustain a failing spirit.

He might be proud of that, but of nothing more. Dawn was still a long way off, and they must endure for a fourth day, a fourth night, perhaps a fifth dawn, till duty was done. They were four men with eight hand-lines, carrying thirty baited hooks a-piece: tended by an old exhausted shell-back who had seen too many seasons on the Banks, and a hungry boy who should have been at school: all pitted against the malice of the sea, the merciless foe, the ordained winner, the Almighty.

Now and again the fat boy, lying nearest to Matthew, groaned in his sleep, dreaming of freedom, dreaming of food. Matthew dozed, and woke, and twitched his lines to free the hooks, and dozed again. In the unearthly darkness, there were voices borne on the wind, mingling with the splash and surge of the waves; but whether they came from near or far, from east or west, from friend or foe, could never be determined in this mariner's labyrinth, a Grand Banks fog.

As on a thousand other middle watches in the pit of night, gloomy thoughts, haunted sounds, were the only music.

Then, at his next waking, there came a change, and he lifted his head to savour it. There was now a faint light from the east, creeping across the water like a ghost invading the battlements; but it was a friendly ghost, the most cherished ghost of all—dawn at sea.

As its blessed colours changed, from pale grey to white, from white to yellow, from yellow to the promise of blue, hope sprang eternal in a wakeful breast. Daylight was at hand, with no more than a single hour to sunrise.

When he reached up to douse their riding-light, Matthew moved clumsily, making as much noise as he chose. His thoughts were the thoughts of all middle-watchers. The night-guard was over, and these his comfortable shipmates should know it. They had kept their swinish sleep for long enough.

They awoke one by one, in their varying degrees of humour

and elegance. First was the old man, the hook-man, Carlo the Portuguese, a spitting smelly old fellow long past the prime of anything, whether of life or skill. Too lazy to shift his fat bottom from the stern-sheets, he urinated on the floorboards, and broke wind as if he were favouring the world, and thus returned to the precious gift of a new day.

Then the three cod-liners roused themselves in their turn: Manoel the mate of the *Consuela*, another Portugee who was the king of their small realm, and knew it every hour, and had no other voice save a shout, and few words that were not oaths: Jorgensen the Dane, a strong and supple man bred to the sea, the best hand aloft and the best at the wheel, with the sharpest nose for fish, or a change of weather, or, when ashore, for a woman who would stroke his yellow hair till it crested as a cock's own comb; and then another English bondsman like Matthew, a little slippery eel called Trail, who had lied about his seaman's skills and had proved as bad a bargain afloat as he had been on shore.

Lastly the boy, the fat bait-cutter with the thin name of Bac, joined the living world again. He yawned as if he held the weight of this world on his shoulder, scratched himself like a man, spat like a silly child—into the wind, so that it fell back on his bare knees—and then, drawing a morsel of raw salt cod from his breeches, sucked at it like a baby.

Bac was twelve years old, and had much to learn: a fact brought freely to his attention very many times a day.

Matthew, hauling in his lines to inspect the bait, surveyed his shipmates as he had on countless mornings of a long season, and found them poor company. Only the boy, embarking on his servitude with a willing heart, and Jorgensen the Dane, who knew all and gave all without stint, were worth a moment's confidence or friendship; the rest were the rubbish of the sea, never to be trusted with more than their own precious hides. Manoel the mate was——

Manoel was roaring already, even as he buttoned his breeches. 'Mateus!' Such was Matthew's name in this many-tongued world. 'How many fish in the night?'

'None,' Matthew answered, intent on his virgin hooks.

'*Senhor!*' the mate shouted.

'*Senhor.*'

'You were asleep, curse you!'

'The fish were asleep,' said Matthew, freed from care by the dawn. 'Curse *them*.'

Jorgensen laughed. Though for any other man this would have been unwise, the Dane, both in spirit and skill, carried too much weight for Manoel, and Manoel had known it for many months, and turned his barking guns elsewhere. He dismissed Matthew with a contemptuous 'Fish do not sleep. Only lazy English!' and then chose an easier target, which was old Carlo, nearest to him.

'Wake up, *porco*!' he snarled. 'Do I pay you to sleep all day?'

To be truthful, Manoel the mate did not pay Carlo the hook-man anything, not even on the most generous of days. But Carlo, who feared for his berth, was all servile duty.

'Ready, *Senhor*,' he mumbled. 'Just let me have fish in my hands. They will be salted down in the barrel while they are still swimming.'

'I would have *you* salted down,' Manoel said, unappeased. 'Old dung-hill . . .' It was all part of his niggle-naggle bullying, which had not ceased for three empty days. Now he turned on English Trail. 'What are you grinning at, *Inglês*?'

Trail, toady and sneak, could always turn away wrath with a soft answer—from much practice. 'I was thinking of Carlo head-first in a barrel, with a scoop of salt on his tail. But where would they eat such a dish, eh?'

'Not at a poor-house for the blind!' Pleased with his wit, Manoel roared with laughter. But he must have his victims. 'Bac!' he shouted. 'Sit straight! Look alive! Do you want the rope's end again? Is that what you hunger for, fat piglet? Rope's end pie?'

'No, *Senhor*,' the boy said humbly.

'Well, you will taste it, whether you choose or not . . . Have you fresh bait ready?'

'Yes, *Senhor*.'

'And something to eat?'

'*Senhor*, there is nothing. Well, there is raw fish.'

'I want soup, damn you!'

'But the oven-wood is all finished.'

'I am not eating oven-wood, you fool! There was soup left over from yesternight. Have you stolen it, little robber?'

'No, *Senhor*.' Bac rummaged at his feet, and came up with a crusted pannikin of cold soup. It was no more than water and pressed olive oil, with a few fish-skins floating a-top the scum. 'Is this what you want, *Senhor*?'

'Ask no questions! Serve it out. And give me the first dip, and the best, or you will finish in the soup yourself!'

So they ate of this breakfast banquet, and with it a mouldy bread-crust to chew on. Then they hauled in, and baited their hooks afresh, and lowered, and lapsed into silence: six cold and hungry men in a stinking long-boat, with two full barrels of cod and eight standing empty: waiting in the chill fog and the long-swelling sea, and the slow daylight which gained like a snail and would never warm their bones this side of a far-distant summer.

They sat in three pairs, like the fortunate six on a dice: Manoel and Trail in the stern-sheets, Matthew and Jorgensen under the cramped bows: amidships, in the long-boat's bulging waist, was Carlo, girded about with his barrels and his sack of salt, and Bac crouching over his evil bucket of bait.

It was Jorgensen, the matchless sailor, who first brought them all to the alert.

'Listen!' he called suddenly. 'They are coming!'

Because of his repute they heeded this warning, and listened. But there was nothing to be heard, whether the Dane was talking of fish, or of some bloody Viking horde about to burst out of the fog. After a moment, Manoel said crossly:

'What are we listening to, squarehead? There is nothing.'

'There is everything!' Jorgensen answered harshly. '*Now* listen.'

Then they all heard it: an extraordinary sound, like a slow kettle coming to the boil: like a threshing when the corn-husks began to fall: like a harsh sigh from a giant throat. They looked about them, in doubt and dismay; and then every man saw what was bringing this unearthly message.

At the further limit of what the fog allowed, not more than thirty yards away, the surface of the sea began to move, and then to bubble and boil. The very waves changed colour, from black to grey, from grey to creaming white. The reason was any fisherman's alert: it was a multitude of tiny fish, thrashing and leaping as they broke the crown of their world: a million capelin-fry in desperate fear, forced out of their own element by some cruel foe beneath.

The tumult moved nearer: soon it would engulf the long-boat, and they would have their answer.

'Count to ten!' Jorgensen shouted exultantly. Already he was their natural leader, worth a dozen loud-mouthed mates: the voice of command instead of the wind of nothing. 'Then we will catch all the fish in the sea!'

Matthew, entranced, believing in such sailor's magic above all things, was already counting, his lips moving like a child's. The little fish were all round them now, in a froth of terror. But before he had reached six he was gripped by another turmoil, as were they all.

All their eight lines were suddenly struck, with a tremendous jerk which set the whole long-boat rocking. The stout cod-lines dipped and twanged, torturing their fingers, forcing their shoulders downwards. It was a dawn strike, by a huge shoal of cod more ravenous than all the hungry multitudes in the world.

'Haul in!' Manoel cried, as if he alone had commanded this invasion. But there was no need for any such fatuous order. After the first shock, they were all deft fishermen astonished by good fortune, and all prepared to seize it at the flood.

A flood it was, a mad avalanche which many times threatened to overwhelm them. Jorgensen, forewarned by his own skill and knowledge, was the first to haul his line. Out of thirty hooks, twenty-five were laden with fat cod. Manoel was next, with twenty. Matthew, who was slower—'I was still counting to ten,' he said later to Jorgensen. 'I trusted you!'—made a score of twenty-two, while his companion was already hauling his second line, hand over hand, exclaiming at its marvellous weight.

Trail, whose puny arms were better suited to hoisting a pint pot, was the last to bring his catch into the long-boat. To the chagrin of all, every hook on his line was loaded. Thirty cod at one haul . . . God was good—sometimes to the undeserving.

Then they had no more time to count, nor strength to tend more than one cod-line at a time. As soon as this was baited and dropped, it was struck by the fish, and weighted down, and must be hauled in again.

The grey-white glistening cascade never ceased. Pale sunlight gleamed on the silver scales: leaping fish, tumbling fish, gasping fish, all poured into the boat, as fast as strong arms could pull them up and over the side. Carlo the hook-man was overwhelmed.

'For the love of Christ,' he cried, 'free your own fish and toss them into the barrels! I cannot keep pace with the salt!'

It was the happiest complaint of their voyage, and, in this wild moment of endeavour, it was forgiven.

The cry of 'Bac! Bait here!' also had the fat boy at the end of his endurance. But even Manoel the mate had no spirit to curse him. Bac was doing his best. They were all doing their best, in the richest hour that any could remember.

Matthew was in a private heaven. Though his hands were bleeding from a dozen cuts and scars newly opened, the pain, hunger, and cold were alike forgotten, with all else that had plagued him. He loved this work and its rewards. He was toiling a-top a summit of fish—and it might have been gold, or a couch of women, or all the drink in some tavern of the sky, or the blessed body of Christ.

Could one become drunk on draughts of cod-fish?

Jorgensen was of more practical mind. Soon, with their empty tubs nearly filled, there was no more bait ready for the hooks. But who would stop reaping this harvest until winter struck? Not Jorgensen the Dane.

His arms were slimy to the elbow with glittering fish-scales, and his yellow hair ran with salt sweat. 'No bait?' he growled. 'What matter? They will take anything. They are mad down there! Let us see if they are cannibals.'

He drew a knife from his belt, seized the nearest codling flopping at his feet, and sliced it murderously. Then he baited his own hooks, and Matthew did the same. Together they let down their lines, and together hauled in. The cannibal hordes below were still ravenous, and beyond counting.

Manoel and Trail copied them, and found the same story. Even Bac seized a spare cod-line and caught his first fish—his first ten fish—at one gulp. Now the tubs were brimful, and the cod, spilling over, tumbled about their feet. The long-boat itself was filling— ankle-deep, knee-deep, up to the thwarts and beyond.

Then at last they were full. The boat could take no more without sinking. 'Enough!' Manoel cried, and for once he had spoken some sense. As they rested from their labours, peace fell on an astounding sunrise.

Every quarrel was healed. All they need do now was to boast.

From Trail: 'Did you mark that? Thirty fish on thirty hooks! It has never been matched!'

From Bac the boy: 'I caught ten! Ten of the biggest!'

From Manoel the mate: 'I *knew* this was the right ground. I could smell it!'

From Matthew: 'Did I not say they were asleep? They must have had nightmares!'

From Jorgensen, a more considered judgement: 'If others have the same luck, the *Consuela* will be full.'

As their craft settled low in the water, with sometimes a little wave slopping inboard, the sun came up strong and the fog began to melt. It showed them the best news of all. The only other long-boats which had shared this treasure of the morning were *Consuela*'s own, and there were three of them within half a mile.

The crew-men waved to each other, and signalled their good luck. All were content in pride. *Consuela contra mundum!* Their own ship against the world!

'Thole-pins out!' the mate commanded, when they had enjoyed half an hour of idleness. 'We row home before we sink.'

A chorus of groaning greeted the order, though, on this happy day, good humour still prevailed. It was an old quarrel on the Grand Banks, and had brought many a ship to the point of mutiny, many a fisherman to a beating, and—the last of the beneficiaries—many a widow to desolate tears.

The men of the long-boats always maintained that, since they were heavy laden, their own schooners should come to recover them when their task was done; while the schooners, under-manned by the captain, the cook, the carpenter, and other idlers and waisters, ruled that the mother-ship should stay where she was, and that it was for the long-boats to seek her out.

It was a dispute never to be resolved, and the long-boat men felt it most. In their present case, after three days of separation, they might have to row twenty miles; if the *Consuela* herself had shifted berth, it might be fifty.

'Let them come to fetch us,' Matthew said, voicing the general view. 'With this cargo, we deserve it. Who has lost the most sweat?'

'Thirty fish on thirty hooks,' said Trail, who, after a brief foray into manhood, had turned to whining again. 'I deserve more than *fetching*. I should be carried to St John's Harbour in a litter!'

'Would it were so,' Jorgensen muttered to Matthew. 'Make it a coffin, and I would bear it myself.'

They laughed together, and Manoel the mate, sensing rebellion, also returned to his natural self.

'We row!' he shouted. 'Thole-pins out, I said!'

But they could not forget that, out of all the men in the laden long-boat, the mate himself would *not* row.

'Give us a little more ease, for the love of God,' old Carlo protested. He was still puffing like a grampus from his efforts. 'The schooner might be with us in an hour. Who can row this log of wood?'

'You can,' the mate answered promptly. 'So ship your oar. Mateus! Man the bow. Squarehead, you next. Trail, midships. Bac! When this old man drops dead, which I pray for every day of my life, you take his place. Now give me no more of thirty cod on thirty hooks, or ten cod on one, but keep silence and bend your backs! What if the wind rises while you are chattering? I care not if every mother's son drowns. But I will not lose the fish!'

He had his way, as he must, and soon they were manned and moving, with Manoel sitting at ease, nursing the stern-oar like a favourite child. Well he might do so, seeing that his luckless crew of four had to pull against the thole-pins with hands already worn and blistered to bleeding shreds.

But being sailors, as cunning actors as any in the world, they could still row light, yet give the appearance of monstrous effort: seeming to toil like Titans, while pulling like girls round a may-pole —enough to satisfy the onlooker, never too much to mar a later enjoyment.

In this private by-play *Consuela*'s Number Six, more crammed with fish than a pedlar's pack with fleas, made enough progress to escape comment or to earn praise. She was followed by the other long-boats, glad to leave the leading to the mate.

They steered directly into the wind, which had blown a steady south-west for three days; by the stern rule-book of the sea, this would send light-weighted boats in one certain direction while their schooner held most of her ground. Being loaded to within six inches of their freeboard they moved sluggishly, like jellyfish with a year to mate, while Manoel gave light touches to his steering-oar and conserved his strength.

When, now and again, they shipped a slopping wavelet, Bac the boy—using sponge, scraps of sail-cloth, and scooping hands—

sucked up every intruding drop of it and threw it back to the enemy.

There was no need to urge him: for all the play-acting, their safe progress was still a matter of life or death, within the cruel confines of the sea. The boy knew—they all knew—that they sat on no more than a little tender fortress, a shell upon the water: if the shell sank, they all sank. No gasps nor thrashing strokes nor cries to Heaven would ever make them float again.

So they rowed all that morning, with aching backs, bleeding hands, sore lungs; being sailors as well as actors, they expected nothing less and nothing more. They rowed only towards the vast emptiness of the sea, the horizon which must be there. Lonely as seagulls at the death of the year, they had no useful friends in view, only faith and hope and trust.

Their trust was in other shipmates who—if it were possible in humanity—would not leave them to die.

Then they rowed into fog again.

As so often on this coast it came upon them unawares, like a thief of sight. It was a patchwork, fashioned of thin wisps, morbid tendrils of blindness, trailing rags which became blankets of gloom: as treacherous as a moody woman with love and hate to be mixed at her sweet or sour will.

At one moment they saw and felt the sun, at the next a cold grey blight robbed them of their surety, like children when a flickering nightlight is snuffed, and hope becomes fear. Of one thing they might be sure, as they peered about them: this doom of the sailor was thicker ahead.

They cursed, and sucked in the cold air, and by agreement ceased to row.

Matthew would have been ready to wager which of them would be the first to break a sullen wall of silence. It was Trail.

'Now what is this?' he whined. 'Can we not even carry our cargo home? We have worked like dogs for three days——'

'We have all worked like dogs for three days,' Matthew broke in harshly. He was ashamed of Trail, ashamed of an Englishman who mewed like a kitten at the first denial of milk. 'Some of us work like dogs, others like puppies chasing their own fleas.'

'I only——'

' "Thirty cod on thirty hooks," ' Jorgensen mimicked him savagely. The circling fog had unnerved them all, even this stalwart mariner. 'Is that what you want on your tombstone? Say the word! I will carve it!'

'Silence!' Manoel the mate shouted. 'How can I make my plans?' They had often wondered of this: now they were less sure than ever. But then a chance wind blew, and the ragged fog cleared to reveal a thin disc of sun and a stretch of open sea. 'Pull ahead,' he ordered. 'We cannot sit idle. We must move.'

The mate was the only idle man, their raw hands reminded them as they bent to their oars again. But the command could not be faulted. Action was the sole answer to uncertainty, and its only cure. Somewhere ahead, unless they were befooled, lay their schooner waiting for them. She would never close them now. Even less than the long-boats could she move in fog.

They pressed on again into the wind, but cautiously, sniffing their way, listening for danger, listening for anything which might bring hope. They were children in a great unfathomable world, which might strike without warning or caress without limit. Then, before they had time to settle to their task, the next horizon vanished. Ahead was another fog bank, as thick as the veil of the richest, ugliest woman on earth.

On its very edge they stopped again, like wary animals, and waited. To stop was better than to fall into the unknown.

Water sucked greedily at their thin hull. A dying fish flapped in its last throes, mocking its captors. 'Join me, join me!' a feebly flailing tail invited, and a sighing wind seemed to echo: 'They will, they will.'

It was Jorgensen who put an end to this craven nonsense. He raised his head suddenly, and said: 'Hark to that! They are within a mile of us!'

It was not the wings of death he was announcing: it was the promise of life itself. Borne faintly on the wind the sound of fog signals came thinly down, high and low like the summons of God: trumpet noises, whistle noises, a deep-tongued bell, the wail of conch shells—the very music of the sea hunt.

The long-boats were no longer solitary. Their faithful mates were calling them home.

Now at last they had a link. Yet it rested on no more than the insubstantial air, for the sound came out of nothing—a low-lying fog-bank which might have been fifty miles across. Leaning on their oars, the men of Number Six gazed at it, and wondered. Where to attack? Where to enter? The homing sounds might be cheating sounds also, echoed from within a blind cavern which, once they plunged into it, would blind them also.

Then God, for themselves alone, passed a small miracle.

Some wayward gust of wind sliced the ceiling off the fog, or pressed it down closer to the sea. Jorgensen, the eagle-eyed, was the first to profit from it. Suddenly he jumped up, steadying himself against Matthew's shoulder, and stared straight ahead of him. Then he called out in triumph:

'*Topsails!*'

The roof of the fog had lowered again even as he spoke, and they all saw their target. It was indeed topsails, a small squadron of them hanging in the air like a celestial laundering. They were topsails without ships, strange and marvellous, gleaming in the sunlight as the banners of a distant fortress. Below them, all the rest was hidden in luminous, impenetrable grey.

But it was enough for the long-boat men. They knew their own: the patched canvas by which they earned their bread differed from others like a hoist of signal flags, and could be read at a glance. *Consuela*'s streaky, weather-worn sail, stained red by some misfortune with a paint-pot—'Like a damned side of bacon' was the general verdict—stood out as a beckoning finger directly in the eye of the wind.

They hailed the other long-boats to follow them, and plunged into the gloom. The run of the waves, sharp on their prow, was enough guide. Presently their schooner loomed over them like a black ghost in a grey shroud. They closed, and hailed again, and were hooked on—bursting with men and fish—and hauled up her filthy side; and so, with a cheer for their splendid catch, to the wooden walls of home.

If the schooner *Consuela* was not the dirtiest ship afloat, she had small chance of meeting her equal within five years of a roving search. The Grand Banks schooners were work-boats, and their work was fish: they smelled of fish, they were slippery with fish, they sweated fish in the sun and made fish-icicles in the frost.

Consuela's holds were rank with it, and never lost their savour. Rotting fish-skins poisoned her bilges: fish-scales littered her decks, and blocked her scuppers, and clung to her sails and rigging and running-gear. To this malodorous stew the crew added the smells of their own choice: rancid olive oil and raw garlic, rot-gut wine, the sweat of unwashed bodies, and all the refuse of lazy—and healthy—men who would not use a privy if a corner of the deck would serve.

Consuela had brought Matthew face to face with many things which he had not encountered before: a stinking ship which was also a matchless sailer: frayed and patched gear which would outlast the fiercest storm: and Portuguese seamen who were the finest to be met in all the five oceans—and foul as a ditch-beggar who would spit on any man's foot, tread in filth as if it were rich eiderdown, and sleep with a sow to warm his arse.

At sea she was the pride of the Portuguese fleet, for swift sailing and cunning catching; and in harbour, the despair of any man who nosed her down-wind and lost his breakfast between one gulp and the next.

Yet when Matthew stepped aboard this sea-midden, bearing as a trophy a giant cod with a girth as thick as young Bac's belly, and joined the crew thronging round the long-boats' catch, he was ready to forgive *Consuela* all her faults. This was the best day of their six months' season, and one which might set him free for the winter; and if the schooner was, for him, still a kind of prison, on this happy noon-time he could forget it.

The bound debtors on board were for the most part despised, because they could not escape and, at a single word of insolence or protest, must begin their bondage again. They were also penniless. Thus the meanest Portuguese fish-gutter was their master. But there was another reason for their ill-repute. Many were proven liars and shirkers, mere landsmen lost in a sailor's world, passengers borne on the backs of better men.

Many, like Trail, having boasted of their skills to escape prison, were learning what they lacked in harsh fashion. Many, like Matthew, were Englishmen, and hated because England ruled the colony ashore and put on all the airs of majesty. Here, on board, justice could at last overtake them. . . . But Matthew, at least, had earned some respect during his time of bondage. He could fish and he could fight. He could also reef, hand, and steer with the best of them.

Close quarters bred friendship and hate, both equally fierce. There were quarrels. There was knife-play. There was rough justice, and rougher tyranny. The best a man could do was to stand tall, sit small, give a soft answer or—at the last resort—a hard fist: make a friend like Jorgensen and avoid an enemy like Manoel the mate.

Such outcasts would never prosper, though with luck they would live.

But at this fortunate hour there were no outcasts. The crew was all one. They were happy because they were busy, and could see the rich result of their labours, and were likely to carry it home. Never had there been such a catch, and thus, never such pride and comradeship.

The harvest of fish was already pouring into the hold, with spades full of salt to see them well stowed and protected. The captain, Coleiro, came down from his stern cabin to watch the loading as the three full long-boats, and three others less fortunate, were emptied. Coleiro, a hard hawk-nosed man with more spiky hair on his body than a fretful porcupine, walked barefoot and ragged like the rest of his crew. But he needed no gold tassels nor silver whistles to proclaim his authority.

He had been twenty-five years on the Grand Banks. The *Consuela* was his own vessel. He could curse or he could smile according to his mood: the weight of command was the same. He ruled with skill, wisdom, and success, and none had challenged him—or was now on board to tell the tale.

Now, as he watched, he was smiling, and well he might. When the last of the cod had slithered and flopped into the hold, and been topped with salt, the hold itself was topped. *Consuela* at last was full: a small triumph on the face of the globe, yet huge for his crew. It meant, not only a prosperous voyage, but the last of the year; another week might see them homeward bound, loaded like a treasure-ship, in convoy for Europe.

Captain Coleiro walked forward to inspect the hold more closely, while his crew stared at him and waited. There was still a great doubt to be resolved: a greedy captain, or a hard task-master, might yet find deck-space for more. Must they spend another night on the Banks? God forbid!

Coleiro looked down at his cargo, then up at the sky, then southwards where the wind was freshening, and creaming waves ruffled a blue-grey sea. His men watched him still. He held them all in his hand, and loved it, and disdained it. His only measure was the discharge of his own private duty.

Finally he called to the mate, with the most welcome command of all:

'Batten down, Manoel! Then make sail!'

The wind, now strong enough to wipe the fog from the slate of the sea, was fair for St John's, two hundred and fifty miles north by west. Thus they crowded their canvas and went like an arrow before it, into the evening, into the dusk, into the night, and into a new dawn. Another dusk-fall found them threading their way through the villainous entrance to the harbour, past the drying-hill—a mile of split cod hanging high and shrinking small, like former highwaymen, in the pale sunset; and so to a quiet berth at last.

III

Once through that fearsome approach, rightly named The Narrows, which wound its way past rocky corners, islets like bared teeth, sudden eddies, and swift tidal surges enough to startle a full-grown crab, much more a ship at the mercy of the wind also—once inside, that ship would find St John's a fine harbour.

It was a full mile long, wide, deep, and securely locked against any fury which the Atlantic Ocean might launch for its destruction, and made a worthy anchor for England's oldest colony, New-foundland. Though this was fearfully marked on the ancient maps as 'A Land full of Demons', its harbour settlement at least had now proved a trusted refuge against them.

Whether the 'Demons' had been the voices of the icy wind shrieking down from the North Pole, or ravenous wolves and bears, or the howling Indians now held at bay by the stockade, they need not trouble summer sailors.

The English had opened up this rich dominion, first discovered by old John Cabot more than two hundred years before, and now ceded to her in the eyes of the world. But there were others who still disputed this rule: the fishermen from Brittany and Normandy whose boast it was that they had voyaged here every year before a single English ship was sighted, the French who reigned in Quebec and seemed a thorn forever, the Portuguese who named them-selves 'first in every ocean', the Dutch who would call anything their own if there was a groat to be made out of the claim.

Even the Middle Sea men who wandered this way could not forbear to mention that John Cabot was in truth Giovanni Caboto, of Genoa and Venice, and that thus . . .

The rivalry was a frequent cause of quarrels ashore, when all the fishing fleets came under English command. It even infected so staid a man as Jorgensen the Dane.

'Stupid children!' he had once said to Matthew, as they came back from a tavern after one such dispute. ' "We were here before you." Such lies! By God, *we* were here in our own ships before these Portuguese farmers put a foot in the water!'

'How long before?' Matthew asked.

'Five hundred years. Perhaps a thousand.'

Matthew laughed. 'Is it days you mean?'

Jorgensen rounded on him, as fierce as he had been in the tavern. 'Do not laugh at me, you English bogger! Or we will quarrel!'

'Bogger' was often his term of endearment—but not on this night. Matthew gave the discreet answer of a friend:

'Then tell me.'

'All I can *tell* you,' Jorgensen answered, still savage, 'is that we had sailors who crossed the Atlantic in long-boats no bigger than ours today, and found this land we stand on, and made it prosper, before men could write. So we sang of it . . . We found America also. We used to rule all England——'

'*What?*'

'Did they teach you nothing? Have you never heard of Dane-gold?'

'Aye. It was an old tax levied, to raise soldiers to keep out the Danes.'

Jorgensen laughed. 'What a story to tell little boys! . . . It was money you paid to *us*, to keep us away. But we took your money, and still came . . . There were Danish kings in your land before there were English kings!'

The Danes had been everywhere, it seemed . . . Matthew was not sure of his ground, and changed it swiftly.

'Well, we have England *now*, and we have Newfoundland *now*. We took possession of it in the name of the Queen. Sir Humphrey Gilbert raised our standard here. And I will tell you the year, because I learned it at school. After I learned the story of Dane-gold . . . It was 1583.'

'*Who* raised your standard here?'

'Sir Humphrey Gilbert.'

'And what was his next mighty deed?'

'He was drowned, returning home. Wrecked on the Azores.'

'English sailing! . . .'

It was a dispute not to be won by either side; they both knew it, and let it go in friendship. But Matthew preferred his English story to any fantastic tale of Danes a thousand years before. Newfoundland was English soil, transferred by solemn treaty, with the upstart French surrendering their rights and retreating to their lair in Quebec. Let them try to return . . . Newfoundland remained: the first enduring English colony—and ruled like a mad circus.

It was a fishing empire, whose monarch was the 'Fishing Admiral'—a strange and wonderful creature in that he was, by turns, the first English captain to arrive on the coast at the beginning of each season. His word then became law and he reigned without question, whether he was cruel rogue or gentle sea-scholar, drunken sot or prim Puritan, smart as paint or dull as a frog with the dropsy.

His grasp might be slacker than a soldier's splice, or more tight than a drumhead, but no matter: there he was, enthroned on barrel-top or cushioned stool—the Admiral of all Newfoundland. His duty was to preserve order, administer justice as he saw fit, and keep a daily journal for later submission to the Admiralty Commissioners, and thence, if it was legible and informative on any matter, to the King in Council.

There had been men chosen who could not encompass all these tasks; at the worst, some could do nothing at all, and St John's languished for a season, or became a murder-pit or a thieves' kitchen, disgracing its saintly name. But this year had been more fortunate: their monarch was a mild fellow with a tidy sense of duty and no particular greed.

He had only hanged two miscreants who deserved it—and he could read and write.

There came a week, and then there came the day, when the fishing fleet was due to return, in a safe convoy of more than four hundred sail. All had been done that could be done, to conclude a prosperous year. In fourteen days, with a children's delight in packing up and going home, the fishermen of the coast had laboured like slaves to strip down, load, and secure the produce of a season's exhausting toil.

Everything had been poured into the schooners, and the smaller fishing craft, and the cumbersome ships of burden which would make up their convoy. A mountain of salt fish, pressed flat with stones, had disappeared under taut hatches; so had a huge conglomerate of whale oil, blubber, barrel staves, seasoned timber, kegs of spruce-beer, sacks of pine-needles, Indian weaves, and the pelts of Arctic fox, seal, musk-rat, beaver, white bear, and hides of caribou.

As if by locusts, the hillsides of St John's, and the storing-huts on the water-front, and the houses of their settlement, were devoured of everything portable, everything cherished for a brief season, everything of value which might be worth the transport from this new world to what was now the old.

The homeward convoy would leave behind a mere skeleton of habitation, soon to be snow-bound, and only to be re-fleshed by another spring, another year, another voyage of endeavour. Until then, St John's might return—with thanks—to the Ice Age, for all that the warm-blooded men of Europe knew, or cared for, or would remember.

But before this great desertion, one single ordained skeleton had been interviewed by Captain Coleiro of the *Consuela*.

Coleiro, when he summoned Matthew to his cabin, was in high good humour, and why not?—he had enjoyed a season fit for a king, and he was only waiting for a fortunate wind to carry him home. He was also talking to a man who had served him well and honestly, without a groat to show for it, and with the promise of another year of the same bondage. But what he said came as a prime surprise.

'Sit down, Mateus,' he began, and pushed across the cabin table a glass of his best Portuguese brew. 'Drink,' he commanded. 'Warm your heart, and listen! . . . You have done well for me: I wish there were more English like you, and I wish I might give your season's share to you, instead of to that rogue Da Costa.'

'Who?' Matthew asked, confused by this sudden sunshine. Among its other astonishments was the recollection that he had not sat in a chair for more than a year.

'Da Costa. Your slave-master. Did you not know his name?'

'No, *comandante*.'

'You are no poorer . . . But I *must* pay him. It is in the bond. Yet the bond says nothing of carrying you home. If I leave you on the coast for the winter, will you promise me—*me*—that you will be here when I return?'

'Well,' said Matthew, more confused than ever, and stopped on
that poor reply.

'You cannot escape,' Coleiro said frankly. 'If you try, you will
freeze, or drown, or the Indians will give your scalp to the women,
and the rest of you to the dogs. . . . I want your word that you will
join the *Consuela* next spring.'

'But do I stay here alone?'

'No. There will be a hundred men left behind, for their own
reasons. There will be four from this ship, and so—' he added
with a laugh, 'I will save their winter keep. Jorgensen will stay
because he plans to trade in pelts. Trail because he is a debtor,
and I would not have him anyway. Bac the boy because he has no
family, and no reason to return—and he eats too much for my
comfort.'

'But how will we live?'

'There is dried cod, if you can endure more of it. There is
enough corn left for bread, and sweet potatoes to last till Christmas.
There will be deer-meat—the Indians will bring it, and exchange
it for—for what you will.' His eyes became sly. 'An Indian will do
much for a bottle or two, and for ten musket balls you may enjoy
his wife till you both fall sick. . . . We will leave you arms and axes.
There is wood to be had for the cutting, and enough shelter for all.
And a cask of *O Porto* to cure the cold. You will live.'

'But who will rule?'

Captain Coleiro laughed again. 'No one rules in winter. You
spend all your strength in keeping alive. And if you quarrel and
fight, the Indians will bury you. . . . Well? Will you stay, and give
your word?'

With a single glass of good wine down his throat—like the chair,
it was the first of a comfortless year—Matthew did not hesitate
for long. Coleiro had painted a harsh picture, but no worse than
any other which might come his way. To be quit of the *Consuela*,
and of Manoel the mate . . . To be his own man for a season . . .
If they worked hard, and took careful thought, and kept a roaring
fire for two hundred days and nights, they would be snug enough.

Jorgensen's promised company was the best of good news.
Whining Trail was the least—but even *he* might learn to chop
oven-wood without slicing off his own ear.

He said: 'Yes, *comandante*, I will stay, and meet you next year.
By my oath. And thank you.'

'Then, *vá com Deus*!' said Captain Coleiro, and drained the last of the warming wine. 'By which I mean, thank God if we meet again, and thank me if we do not.'

The four castaways—Jorgensen, Matthew, Trail, and Bac—let go *Consuela*'s mooring lines, to the accompaniment of ribald miming from her crew on board: the pretence of great cold attacking their vital parts, and of God's final blessing on their doomed lives. Then, hand-hauled along the quay until her foresail could draw a fair course for the Narrows, the schooner bid them farewell, and began to join the great throng of departing ships.

It seemed as if the whole harbour of St John's were moving away from the land. Sails sprouted like huge flowers: ropes splashed for the last time, and were drawn on board: flags whipped into brilliant life: ships touched and drew apart as men cursed and then laughed: ragged music from a thousand voices signalled the happy hour of homeward bound.

After this slow turmoil, the formal dance of setting sail, the castaways straggled up the hillside, a fortress-wall between the ocean and the harbour, to watch the convoy take shape for its departure. There was a frigate up from the New York station to shepherd them: a frigate like a peacock among these bedraggled crows, smart and seamanlike beyond any dream of the sea.

She would have command over all—schooners, smaller fishing craft, the laden cargo-consorts carrying their joint wealth—as they rolled homewards under the surge of the eternal westerly gales.

It was the first week of October, and those left behind would not see another rag of canvas, nor hear a new sailor's voice, for six months.

Yet they were a cheerful company, sharing the same happiness even in the bitter cold of this onset of Newfoundland's winter. They were free at last: free in a hard land and a daunting season: but free! Matthew's eyes were drawn especially to the Navy frigate.

'Now there's a ship!' he exclaimed admiringly. 'Twenty-eight guns, and paintwork like a picture! She makes the rest of them look like bum-boats.'

'Paintwork,' Jorgensen repeated, with scant respect. 'Paintwork does not make a ship. Those topsails are a-back already.'

'She is waiting for the laggards.'

'She should lead, not loiter.' But Jorgensen's mood was still content. 'Well, I would rather be safe on our dry land than setting sail for Portugal in this cold. They might be two months before they see the coast . . . I have had enough of the sea for this year. Enough of *Consuela* also.'

'I have had enough of St John's,' said Trail, the only complainant. He was puffing from the uphill walk, and chafing blue hands against the chill. 'By God, when will we see the sun again?'

'In six months,' Matthew told him, without kindliness. 'If you live. If you are not snow-blind. If the Indians do not put an arrow between your ribs. If you work.' The convoy was at last moving away from them, slipping into the dusk like the grey ghost of winter. 'Now—down we go, for our first hot dinner! Trail cuts the wood. Bac tends the fire and the lamp. I will cook. Jorgensen scrapes the pans.'

'But what do we eat?' fat little Bac asked.

'Do not trouble yourself,' Jorgensen told him. 'We will keep you until the last.'

In the final hour of that first winter night, in the blood-chilling hut which was to be their home, Jorgensen the Dane, having tapped too early the fat cask of full-bodied wine and taken a draught too much, perceived a great light. It was that shaft of truth which sometimes shone through such a fog.

'Mateus!'

'Aye?'

'Are you with me?'

'Not to the point of marriage.'

'Wait till you are asked . . . We have a great task. What is it, you say.'

'I said nothing.'

'It is to stay alive for six months.'

Matthew looked round their gloomy refuge. The fire was dying, and with it the smoky whale-oil lamp. Trail was invisible under a mound of evil skins. Bac was snoring. The wind sighed and sucked like an enemy spirit. Such company . . . But Matthew had taken his own share of the wine, and with it his own shaft of truth.

'I have wonderful news for you. We will!'

'How?'

'We will carry the weak, and—if we must—kill the strong.'
'Are you weak?' Jorgensen mumbled after a long silence.
'Not tomorrow.'
'Nor I.'

IV

They were ninety-eight men in fifteen huts and sheds straggling along the harbour's edge, with the stockade at their back for safety and the icy water in front to contain their shrunken world; and by the time of Christmas they had settled down, in reasonable order, and brought their winter village into being.

Already the snow lay thick on the hillside, and on the roofs of their homes, and on the quay which was their doorstep; already a rime of ice at the tide-mark proclaimed as a lie the fable that sea-water did not freeze.

The snow and the bitter cold had firmly divided them all, like children who marry and move to different towns, or friends who grow apart and at the last cannot walk a mile uphill to renew their ties. The differing households, distinguished by the names of their ships, the language of their homeland, their age, their liking for wine or innocent water, for a quiet time or a season of riot, grew more private as winter progressed.

It became true, that no one ruled St John's in winter. It was each to his own, and each had enough toil to keep a roof overhead, and warmth within the hearth, and a store of food, and the wind at bay. They were not against each other: they were for themselves. In one such private domain—as mean as a turf hut half buried in the bogs of Ireland, yet precious as the most noble house of any realm—dwelt the four castaways from the *Consuela*.

Next door was a small colony of Portugal, and on the other side some swaggering French from Gascony, who swaggered less as the ice nipped their parts of speech. Soon it became a rare day when they cast eyes upon each other.

Consuela's men were not ill-equipped. The hut, which had served as a store-room for spare barrels, was sound, and grew sounder as they stuffed the wall-cracks with seaweed, and nailed sailcloth to a leaking roof, and trimmed a bank of snow so that it made a wind-break. A chest of small arms, securely locked, stood guardian at the door, another rampart against the wicked world outside.

Also within, they had amassed a mountain of chopped oven-wood, a store of fish, water and wine, and a barrel of hard biscuit which, its weevils having succumbed to the cold, had been generously left behind by Captain Coleiro. There were also six haunches of deer-meat bought from a French half-breed—a *coureur de bois*—with some of Jorgensen's earnings, and a scarlet Edam cheese won from a Dutchman by Trail, who could cheat him at his own game of *trek*.

These would serve until the demons of this land, aware of their plight, came to trade.

The Indians invaded gradually, first as mice testing the mood of the cat, and then as furtive thieves stripping the hillside bare of anything left behind by the fleet—scraps of net and sailcloth, frayed cod-line, barrel-staves, nails, rusty anchor flukes, bottles, tattered jerkins and breeches. Such shy and cunning visitors, who came in ones and twos, smiling by day, creeping like shadows at night, could not be kept out by the stockade.

Its true purpose was to repel an armed attack in force, and this had not been mounted since the hungry year of 1700, when the demons were starvation-wild and the defenders half dead from the same cause.

There were other ways of skinning the winter prey, whose Christmas carols were sung by wolves, more avid than any starveling chorister, driven down from the hills by hunger as gaunt as Caesar's ghost. Their glittering eyes could be seen on the edge of the darkness, as they raced howling to and fro and pillaged for fragments of skin and bones picked clean.

The men within were in no mood to share more than this beggar's offal, and sometimes killed a wolf in fear or in sport—whereupon its companions fell upon it ravenously, and tore out its vitals, and took their fill with cannibal delight. So dog ate dog, and men helped them to their feast, in this holy season wherein their Saviour's birth was celebrated with such tender gramercy.

On a morning late in January, when the wind screamed eternally and the sun did not top the hillside at their back, and they had eaten the last of the deer-meat pounded into a sour mash with the last of the mouldy biscuit, there was a rustling at the door.

It was less than knocking, yet more than the harsh scrape of the gale. As Matthew came to the alert—the castaways kept a watch, from long habit, and this was his—they all listened. The sound was repeated, and with it a snuffling growl, a sort of salutation, as from a dog which says, in all humility: 'I am here. Let me in.'

Whether wolf, or man, or undiscovered demon, it was at their door.

As Matthew stood up, and stared towards the sound, they all quickened, as they had done in long-boat Number Six, in their varying degrees of humanity. Jorgensen, lying on a bed of skins near the stove, was as alert as Matthew; but since he trusted his shipmate beyond question, all he said was:

'Better take a pistol. If it is those Frenchmen, we'll have none of their tricks! We are as hungry as they.'

Bac the boy—no longer Bac the fat boy, but a skinny relict of better days—was willing as ever, and ready for adventure. 'Shall I open, Matthew?' he asked. 'Then you cover them, and shoot off their heads as they advance!'

Only Trail the indolent did not believe in danger, since he did not wish to. They had been private for so long that no other state was acceptable. From his vantage point between stove and wood-pile—so that he could feed the former without stirring more than one arm and half a leg—he said:

''Tis nothing. Some lazy rogue wants a gift of oven-wood. Let him freeze!'

The knocking or rustling, and the low growling, was repeated. Matthew, standing by the door with his pistol at the ready, called out: 'Who is it?' and then, for good measure: '*Qui va là? Sprechen Sie! Quem é?*'

A perfect silence was his only answer, while the wind howled in mockery. Then the muffled growling began again.

He decided that there could be nothing much to fear. If it were a wolf, it was tamed or sick: if a man, he was trying no surprise. With his free hand Matthew drew back the latch, and flung open the door.

In place of the advancing hordes or the slavering wolf-pack, there was only one small Indian, with a squaw at his back, crouched on the dirty trodden snow of their front doorstep. A blast of bitter wind gave him his sole menace.

They stared at each other. The race of Indians was not given to smiling, and this one was no freak in that respect. Though the eyes were watchful, the brown impassive face betrayed not a single thought, either of love or hate. It did not even show his age, which might have been thirty, or forty, or fifty, or any number within reason on either side. The woman was the same: staring, blank of expression, brown, ageless. She was not well-favoured—except perhaps to the man.

They were both dressed in ragged skins, with a furred beaver bonnet above and snow-scarred moccasins below. But they both had other gifts to commend them: humped packs on their backs, a heavy laden sledge a-piece, and—as to the man—a frozen shoulder of deer-meat in his skinny arms.

Matthew lowered his pistol and, as a further sign of peace, thrust it into his waist-belt. He said: 'Have you come to trade?' but the Indian answered nothing. Instead, he tendered the deer-meat, advancing it until it was almost touching Matthew's stomach. The shoulder was well-fleshed, and looked, to him and to the other castaways at his back, a very gift from Heaven. But there was a look in the Indian's eyes which said, more clearly than speech, that it was not to be a gift of any sort.

Matthew said: 'Come in,' and beckoned also. The Indian signed to the woman, who sat down on her sledge. Then he stepped warily into the hut, and Matthew closed the door, shutting out the biting wind, and the patient woman, and anyone else who might be friend or enemy.

Jorgensen now took command. Having tried their visitor in every tongue he knew, without a word of answer, he descended to the language of signs. The contrast between the huge Dane, yellow-bearded, wild-maned, stalking about the hut like a lion in a forest clearing, and the little Indian who sat as still as a mouse, with only his eyes moving from face to face, from corner to corner of the room, was ripe for merriment.

But they were not merry. They were hungry, and both sides of the bargain knew it.

Jorgensen, who must have played this game before, went to work purposefully. With one hand he took the meat gently from its owner, and set it apart; with the other he drew a silver coin from his breeches and laid it on the floor, within reach of the Indian. It did not, indeed, look like the riches of Croesus nor any part of them, and none was surprised when the Indian turned his head away.

Jorgensen set down another coin, with the same effect: the seller stared at it, then looked briefly at the buyer, then moved his face aside.

Trail, who had been examining the deer-meat and licking his lips at what he saw, said impatiently: 'Why do we not take the meat? He is not armed. We can throw him into the harbour if we choose!'

Jorgensen gave him a glance of contempt. 'Think before you talk such muck! First, we are not thieves or murderers. Second, we want more than one shoulder of meat, to see us fed for the last two months of winter. Can you not see beyond your snuffling nose? Last, this man could be a friend.'

'Fine friend! He looks like a thief himself!'

'So do you . . .' There was little love lost between them, and none earned by Trail, who had been the worst of shipmates for three hard months. Jorgensen took out a third coin, placed it on top of the others, and gave it a firm tap with his finger. 'That is all,' said the finger, and his look confirmed it.

The Indian growled deep within his throat, then reached out and took up the coins. The great prize of meat was theirs.

A whole morning of such studious commerce passed at a slow and steady pace. They exchanged a fine char-fish, fresh caught through the ice, for a bottle of wine: jars of maple syrup for fish-hooks: a second haunch of venison for money again: a sack of corn, brought from the sledge outside, for an axe-head, and maize for a hank of cod-line.

Later Jorgensen fetched one of his own beaver-pelts, and indicated that he would buy more. But the skins with which the squaw's sledge was loaded were poor stuff—mouldy caribou, Arctic fox torn by a trap, half a white bear which had already done hard and greasy service as a couch. When Jorgensen refused them, the Indian showed by signs that he would return with others. Then he packed his wares, and raised his hand in salute, and left them to their treasures.

Only Trail was discontented. 'We gave too much,' he whined, still peevish. 'Why should we be cheated by one man, when we are four? At the least, we should have had the use of the woman to bind the bargain. I know these Indian dogs! They will milk us dry!'

'The woman would not milk me a drop of sweat!' Jorgensen answered. With their cares laid to rest for many days, he was in high spirits. 'Did you mark her rare beauty?'

'Aye.'

'Well, she might serve you. Myself, I am not so prime, nor so desperate. Now—bring some more oven-wood, or I will throw *you* into the harbour!'

They feasted like kings that night, and for many more afterwards. The fish and venison were ambrosia. Under Matthew's hand the corn was made up into loaves of coarse bread, and the maize, pounded with stones and tinctured with molasses, a sort of mealy-porridge which no Scotsman would have used for a poultice. But it was warming and filling none the less. The wind was not so harsh, nor the cold so bitter—nor Trail so odious—when their bellies were full and the morrow was safeguarded.

Ten days later the Indian returned, with fresh food and a load of skins. Jorgensen exchanged a half dozen for powder, musket balls, and wine, though they were not of the quality he wanted. Prime beaver and musk-rat, unmarked fox and coney, shining seal, white bear as rich and soft as the pillows of the East—these were the targets of this Viking trader . . . But the Indian remained their firm friend. They called him Sitting Mouse among themselves, since this was his eternal posture.

With the passing weeks, Sitting Mouse became their pet, and proved their ruin.

The second stranger to knock on their door, at the beginning of March, was revealed—when watch-keeper Jorgensen opened it— as the very picture of a *coureur de bois*: one of the Frenchmen, often half-breeds as was this visitor, who acted as trappers, guides, canoe-men, scouts, and trader-messengers to all the world of Eastern Canada.

He was dressed as an Indian, in accordance with wise custom, though his sense of a rank superior had added certain refinements. His leather jerkin was fringed and tasselled like a tapestry: his bonnet was not the feathered head-dress of the tribe, but a coon-skin cap with the tail hanging down to his waistline. A coloured belt of woven beads set with porcupine quills had a curved dagger to give it purpose. Snowshoes of beaver-thong on a wide frame of spruce completed a uniform of some local splendour.

His head came up smartly as Jorgensen opened the door, and he said with a confident smile:

'*Bon jou*'. I am Pierre Dulac, *coureur de bois, à vot' service!*'

'Perhaps the face was of lesser rank than the clothes. It was alert, bold, and sly, all at one time, like a gipsy pedlar ready always to gain something for nothing. The swarthy cheeks still bore traces of the grease-paint which warded off the black fly—the black fly of summer. Here was a man who might act any part to suit his company.

Jorgensen, surveying him guardedly, said: 'What service is this?'

Dulac came straight to the point, like any honest man. 'You look for skins and furs? I have the very finest!'

It was likely enough, and Jorgensen, whose appetite for these wares was still unslaked, relaxed his guard. 'Come in,' he said. 'No harm to talking.'

Pierre Dulac marched forward into the hut, smiling and bold. '*Bon jou' la compagnie!*' he declared in ringing tones, his eyes already busy with each new face, each corner of the room, each promise of profit or lack of it. 'Pierre Dulac, *coureur de bois.*' But he turned immediately to Jorgensen, as a travelling packman turns to the innkeeper, the man with the money and the power. 'Of skins and furs, I have the best,' he said again. 'What is it you seek?'

Jorgensen did not choose to be hurried. 'How did you learn of my seeking?'

Dulac shrugged, and spread his hands. 'How does one learn of anything? A word *par ci et par là*. . . . If you are serious, come to my storehouse. At one stroke, you may spend a fortune—and make another.'

Matthew, a watcher of this scene, asked: 'Can you not bring them here?'

'They are too many.' Dulac did not dare to be contemptuous, but his tone dismissed Matthew as a man of no account in this transaction. 'We are talking of—perhaps five hundred of your pounds, even a thousand. You must come and choose,' he said to Jorgensen. 'However, I have brought one example for you to see.'

He drew from within his jerkin a splendid pelt of white fox fur, as sleek and shining as the moon at the full. All their eyes were pulled towards it, and Jorgensen himself could hardly contain his interest. He leaned forward and touched the fur. It was matchless, and he knew that this was what he sought.

'If you have more like this,' he said, 'then we may trade.'

'*Bien!*' Dulac folded the pelt, and replaced it. Then he turned to harder business. 'What can you offer?'

'Money, of course.'

'What money? If I may inquire.'

Jorgensen did not relish this. 'Enough.'

'*Mais naturellement!* . . . But—gold coins?'

'If necessary.'

'I would also consider,' Dulac said with a touch of grandeur, 'musket-balls, powder, guns or pistols, wine, fish-hooks, and strips of iron. But the gold is important. I would not say No to silver . . . Will you come with me?'

Though he was tempted, Jorgensen still found the pace of this commerce too brisk. 'I will think of it . . . Return at noon, and you will have an answer.'

Dulac, whose fingers on the balance of a bargain were as delicate as a butterfly's wing, allowed himself no more than two hints, mere whispers of enticement, by way of reply:

'*Entendu* . . . I have others to meet here, in any case . . . I should tell you, there is a girl in my storeroom who is also worth the seeing . . . The touching, shall I say?'

'How far is this storeroom?' Jorgensen asked.

'*Deux milles.* A stroll, *une petite promenade*, nothing more. Till noon, then.'

When he was gone, with a smile for all as wide as a rainbow, they fell to discussing their visitor and his offerings. Jorgensen was all excitement, Matthew counselled caution, Trail claimed to have recognized a rogue 'as cunning as a monkey', and Bac the boy wanted to be a *coureur de bois* tomorrow—and dressed as such.

'Did you mark his dagger?' he exclaimed, his eyes shining. 'Perhaps he has skinned a white bear with it!'

'Perhaps he has skinned a white man,' Trail said tartly. 'I would not trust him with a rusty soup ladle.'

Matthew felt that he should warn his friend more seriously. 'These *coureurs*,' he said, 'we know their repute. Some are good, some are bad. Some are both by turns. A cunning trapper or a skilful guide may turn thief or traitor if he sees his chance. So—no need to trust them. I say, go and see the skins. Go well-armed. Discuss the price. Then come back and think on it.'

'I do not trust this man,' Jorgensen answered, 'and I do not mistrust him. I do not know him! I only know one other—the one who came here first, and sold us deer-meat, six haunches of it, at an honest price. No question of that.'

Trail said: 'Perhaps there might be a question from the owner.'

Jorgensen rounded on him. 'All men are not like you, you—you crook't little bogger! If the meat was stolen, then we heard nothing more of it. So it was an honest trade. So is this one. You saw the pelt he carried? The fur was like velvet! If he has more of the same, I will buy.'

'And the girl?'

'What girl?'

'The girl he offered.'

'What of her? If it will sweeten the bargain, why should I say no? It is the furs I want.'

Trail sniggered. 'I'll warrant she has a fine little fur herself.'

'Enough!' Matthew said. They were growing childish, and this was no child's play. He turned back to Jorgensen. 'There is another way. Take no money with you. See the pelts, and make your choice of them. Then tell him that he should bring them here, and here you will pay.'

But Jorgensen was too hot for such delays. 'No. I want to see them safe under this roof, before he sells them to another. I will pay on the barrel-head, and bring them back.'

'A day lost is no great matter.'

Jorgensen said: 'This is my venture, and I will make it in my own fashion.'

It was the end of dispute.

At noon, with Pierre Dulac waiting at the gate of the stockade, Matthew bid his friend farewell. After an hour or so he had come round to Jorgensen's way of thinking. They were conjuring up hobgoblins where none existed. The Dane might be carrying a pouch of gold, and a pack full of other tradeables. But he also bore a pair of pistols, and could use them readily.

'Keep watch,' Matthew warned him. In the frosty air, their chill breath puffed and mingled like tiny cloudlets. Though the sun now showed above the hilltop, it could not yet warm and wake the world. 'You have some sleep stored up. Sometimes it has deafened us! Spend it now.'

Jorgensen was in the highest humour. 'I'll not sleep nor snore,' he promised. 'That girl should keep me wakeful . . . But if I am not returned by next noon, come and seek me. You promise?'

A nod was enough, and so they parted. Matthew's last sight was of two figures, the big and the small, plodding up the forest track between the gaunt, snow-laden silver birch trees. Then they were gone, leaving behind them cold silence, and a garrison reduced by far more than one man.

Though Matthew was not afraid for his friend, he was bereaved already. They had come to love each other: not as the beast with two backs, but as men in utter trust, sealed comradeship, laughing content.

When Jorgensen had not returned by that same nightfall, Matthew was still unperturbed. Winter-time was a limping traveller, whether for mice or for men. If the bargaining had gone slowly, and the girl later moved like quicksilver, then a day could stretch to a night without surprise to anyone. He slept easy, and even when he awoke to a six-o'clock sunrise, and found the hut still empty of the one man he wished to see, he knew only loneliness, not fear.

Then the morning gained, and with it his unease. He went out of the hut a dozen times, to watch the sun show above the rim of the hill—but nothing else. Outside they had an oar set upright in the snow, and a stone placed to the northward, to mark the time of noon; and as the pale shadow crept towards the peak of this child's sundial, he began to itch with fear.

Jorgensen had said 'Noon', and noon he had meant. He knew better than most how firmly the flight of time—watch-time, time appointed, time wasted, time gone forever—ruled a sailor. He would have carried the word 'noon' inside his skull, just as Matthew did, and he would not fail to keep his tryst.

When midday came, and the shadow passed over the stone and brushed the trodden snow instead, Matthew knew that he must make a move.

He returned to the hut, where Trail sat staring at the stove as if defying it to eat more wood—which would entail more hard labour to himself—and Bac the boy was melting snow in an iron pot to provide their day's water. Rubbing his hands to warm them, Matthew waited. No one shared his anxiety, and he was angered. He said:

"'Tis past noon, and he is not returned.'

Trail's eyes came round to him slowly. 'Did you expect him, to the minute? He has better things to do.'

'I expected him, because he promised. We must go look for him.'

'*You* go,' Trail answered. 'I can wait. Either the trade was slow,

or the woman was not. He has his pleasures, either way. I'll not
stand in his path.'

Bac had ceased his work, and was watching and listening.
Matthew tried to control his anger, but it rose sharply at this
heartless answer.

'None the less, we must go look for him. I promised.'

'You promised.'

'Will you not come with me?'

'If you give me good reason.'

'He is a shipmate!'

Trail sneered. 'So was Manoel the mate. Would you go seek for
him? I would not have taken two steps, if that one had fallen dead
on a dung-hill!'

'Jorgensen is our shipmate, *here*. He is in trouble.'

'Or in heat. Perhaps the girl was worth more than one night, and
he delays for a second bout.'

'You talk like a flesh-monger!'

Pig-eyed Trail answered: 'Are you jealous of her?'

Matthew was beside himself with fury. 'One word more of that,
and you are dead!'

Bac the boy, in alarm or in courage, suddenly said: 'I will go
with you, Mateus!'

Matthew expelled a long breath. 'You are welcome. At least we
shall have one *man*.'

In the sulphurous silence which followed, the conflict still sim-
mered like a pot on a stove. But Matthew, possessed by anger, was
also deep in doubt. He had said: 'We must seek him,' and insisted
on it—but where would they seek? Jorgensen had walked up a
hillside track, and disappeared. In which direction? Dulac had
said that the storeroom was two miles off. Was this true? Was it two
miles, or twenty miles? Was there in truth a storeroom? Who would
know, or who could guess?

In a world of questions without answers, the silence persisted.
But then it was broken, by that same outside world. There came a
knock on their door: a modest knock, but a summons none the less.

Bac exclaimed: 'There he is! He is returned!'

Matthew crossed swiftly to the door, and opened. But it was not
Jorgensen, nor anyone of that stature. It was their small Indian
ally, Sitting Mouse.

Sitting Mouse, alone, stood before him, silent and expressionless.

If he brought news, it was not to be found in his face. If he had come to trade, he came with empty hands. Matthew, knowing that words were useless, spread open his arms and made of them a question.

In his only shared language, Sitting Mouse answered. He pointed a finger at Matthew, then at his own breast, then turned and directed it up the hillside behind their refuge. It could only mean: 'Come with me.'

Matthew nodded agreement. Then came the biggest surprise of all the winter. Sitting Mouse spoke, for the first time ever heard:

'Come with sledge.'

Matthew, astonished, asked: 'Why? For the skins?'

The Indian did not answer directly. It was not to be known if he understood. After a pause, he said again: 'Come with sledge.'

It was not to be known, but—for the first time once again—there was a shadow in his face which proclaimed that he knew all.

Before he left with the boy and the Mouse, Matthew walked across the room to Trail, and kicked him roundly on the rump. As Trail cursed him with his eyes, but said nothing, Matthew glared back at the faint-heart.

'That was to remind you,' he told him harshly. 'If aught is amiss when I return, I will drive you out, dead or alive. Be sure of that!'

Matthew Lawe came to feel that he would never forget that journey, nor its fearful end; once, when telling of it later, he added: 'Not if I live to be a hundred years old,' and was conscious in the same moment of a giant, obscene jest which made him the double plaything of fate.

Some of the gods must have been laughing, while others maintained their watchful vengeance, and would do so for ever. If he lived to be a hundred, he was as yet unborn.

Sitting Mouse, their trusted guide, led the way: little Bac trotted after him, with all the energy of youth and an innocence which had not yet fathomed the grotesque evil of the adult world: Matthew, dragging the sledge, was their rearguard: armed, angry, and alert.

It was the longest walk of the last six months, through woods which stood silent and listless under their burden of ice, waiting for the spring. Treading the crunching snow, they plodded onwards for two hours and more. Nothing changed, nothing stirred, nothing welcomed them nor barred their way, and if they were being watched, by wolves or men, demons or angels, the eyes were invisible and the breath unheard.

Sitting Mouse, he noticed, did not look about him, nor keep watch for enemies. He was following a path which he knew, and his patient progress through this silent world seemed to say: 'Fear nothing. We are alone.' Matthew could only fear a trap, and he strove to be wakeful. But his inner heart told him: 'I am a child in these woods, like Bac, like a fledgeling bird, like the youngest newborn fox or hare or deer. If this is a trap, it must prove stronger than I.'

The dead forest closed about them, pressing in as if testing their courage, then opened at last into a clearing. At its further end, banked about with snow, was an Indian wigwam, the largest he had seen, fashioned of hides and birch-bark stretched over a skeleton of poles. When they drew near, and stopped before it, Sitting Mouse motioned him to enter.

Matthew said to Bac: 'Come with me,' and pushed aside a hanging curtain, and bent his head to go in. His pistol was ready; but nothing else, neither body nor soul, could have been ready for what he found, as soon as his eyes grew accustomed to the darkness within.

The wigwam, floored with trodden earth, was bare as a cave, without a stick of furnishing. If this had been used as a storehouse, then it was now robbed or bankrupt. Only one thing was there, and it *was* a thing—a thing of horror. The load for their sledge was now before their eyes.

Jorgensen the Dane lay in the squalor of death, on a dark-stained couch a-glisten with his congealing blood. He lay on his back, in the last surrender: he was naked to the cold and to his vanished enemies. He had been cut, or hacked, or axed by a thousand wounds great and small, the ritual death of the white man who came into the hands of the demons of this land.

Since he had broken their laws, they had broken his body and drunk his blood.

Matthew, and the trembling child at his side, could hardly take in this picture of man in his last torment. The slit nose, the torn scalp, the plucked muscles of his chest, the greedy bite at his vitals, were all the marks of a careful fury. His executioners must have enjoyed their task, and pursued it most lovingly.

Just as children tore off the wings of flies for their sport, so Indian braves had torn life from this living man, while their women stormed and destroyed a dominion which they hated.

Of his private parts, there only remained a bloody pizzle, and nothing more.

Matthew, looking down at the wreck of his friend, cried: 'Sweet Jesus!' while the boy Bac retched, and then turned aside to vomit on the cruel earth. Matthew, in fear and fury, wanted to shout for aid. But what did a man shout, in a deserted forest, to the sole hearer—an Indian guide who would not, or could not, understand?

When he strode to the entrance of the wigwam, and looked about him, he found both his question and his errand futile. There was nothing to be seen save bare snow, icy trees, and the silence of guilt.

Sitting Mouse had vanished. He had been there, and now he was not. He also was an Indian.

Matthew, returning to the bitter cold of this tent of death, found Bac still crouched and shaking. But the boy had spirit and, at this anguished moment, it sprang to life again.

'I am sorry, Mateus,' he said humbly. 'I could not help it.'

'Think nothing, lad,' Matthew told him. 'You cannot have a man's stomach yet. What *man* could, for this butchery!'

'But what shall we do?'

'Return home. And kill anyone who stands in our way.'

Bac looked towards the dark corner with its terrible burden. 'What of him?'

'We bear him back, and we bury. There is nothing else. He is gone.'

A third voice said: '*You dog—you lie!*'

It was Jorgensen from the dead, or the brave ghost of him. The living sound should have been joyous, but it was more ghastly than anything yet witnessed. This stalwart deep-throated man, now robbed of manhood, could only squeak.

Matthew, near to weeping, knelt and bent over his friend. 'By God, you live!' he whispered. The pale eyelids flickered, but they also had been numbered among the slit parts of Jorgensen's body, and the tormenting pain made him scream, on that same fearful bat's cry.

Matthew hushed him gently, like the babe he had become. It

was clear that he must die, but at least he might die at home, by the warmth of his own fire, if they moved *now*.

He called to Bac: 'Bring in the sledge. Then tear down some of these hides to cover him. We must travel quickly.' As Bac hurried to obey, Matthew looked round for Jorgensen's pack and pouch of gold, but—like Sitting Mouse, like the promised store of furs, like the cursed Dulac himself—they were gone. In a fury he cried: 'And I hope to meet a man I may kill!' before he began tenderly to prepare his friend for his last ordeal.

When at dusk they arrived at the harbour hut, after a jolting journey of such pain that Jorgensen, ravaged by movement, had no strength left to screech his suffering, they found a wilderness as bitter as the wigwam in the forest. Their home was in darkness, the stove a heap of dead ash, and the room more icy than the winter wind itself.

'Trail! Where are you?' Matthew called, peering about him in the gloom. There was no answer. A glance showed him that their store of food might have shrunk also. 'That dog!' he growled in infinite rage. 'He has run!' Then he mastered his fury. There would be time enough to see what they had lost when Jorgensen was warmed and sheltered. At this moment, he was only known to be alive because he shuddered so violently in the cold.

It took a long time to light the fire, as Jorgensen lay gasping on his pallet, and Bac's trembling hands built a woodpile within the cold cavern of the stove. From flint-spark to dry bark tinder, tinder to oily calico, calico to a tiny flame, flame to a leaping wood-fire— the laborious task, which seemed a race against death himself, unwound its course, slow-footed as the winter night.

But before dawn they had their glowing stove, and a snug room, and a thin broth of meat-scraps and bread-crusts which could be gently held to a hideous mouth, and a dying man who would at least be warm before he fell into that cold embrace which was the last of all.

He lingered three days. But he did not suffer increasingly: having stepped beyond the threshold of pain, he was moving slowly down-hill to God's mercy—oblivion. He had given all the blood he could: suffered all the cold, endured all the scars, and formed on them all the crusts which the magic body of man could knit together.

Now he was resigned, and now, in tattered whispers which, for very shame, masked his squeaking and mewing, he told his story.

It had been a trap, a simple trap. Pierre Dulac had indeed a storehouse of furs, but nothing near the quality which he had boasted. When Jorgensen said that he wanted others, Dulac had warranted them for the morrow. But in the meantime . . .

In the meantime the promised girl was brought in, as dusk fell and one trade gave place to another. She was Indian, young, strong, compliant, and well-favoured—at least to a man who had not glimpsed the female form for half a year. She had lain down, with a smile; Jorgensen had stripped, with a will. He had scarcely entered her when their shelter itself was entered.

Suddenly the wigwam was full of men, shouting, chanting, brewing up their courage. They crowded in, with knives and axes and claws at the ready; and as he sought to rise, the girl clipped him round the neck so that he could not lift. His feet were bound, and then his arms; Dulac appeared once more, to collect his sorry bait and to steal all that Jorgensen had brought for their bargaining; and the mortal feast began.

Its poor remains grew cold, on the evening of the third day; and no fire, no gentle words, no mourning love could restore them.

Little Bac wept when all was over, and Matthew found that a grown man could also weep—and then awake to angry action. Together they made a rough coffin of driftwood and staves. But there could be no grave for this pitiful body; the ground was rock-hard still, and would never yield to their only digging-tool, the salting shovel from the *Consuela*.

In his last moments also, Jorgensen had whispered of burial at sea. It might be done, for honour's sake.

So they bore the coffin outside and set it in an angle of the hut, and piled a steep mound of snow over it, against the wolves and the foxes and—who could guess at appetite in this savage starving land?—even men themselves. There Jorgensen must sleep until the spring.

'If only we had a priest,' Bac whispered when they were alone again.

A priest.

V

Spring came late and slowly in that sad year. It was another seven weeks before relentless winter loosed its grip, and the late April sun could warm both earth and blood. In those seven weeks, the shrunken household of man and boy had to withstand imprisonment, want, and fear, all so cold and so cruel that it seemed that God held back the spring till He had culled His flock of everyone save saints—who must also be Titans of strength.

They starved and they lived. Trail the traitor had rifled their store of its best, but there was a limit to what one little thief could carry in one furtive journey, and they were left with a margin of ten days' meagre living.

In this time, the half-sack of corn dwindled to nothing; the single leg of deer-meat was stripped to the bare bone, and the marrow sucked from that; the white-fish which Sitting Mouse had brought, in the simple days of friendship, surrendered its rotting flesh, then the pickings of its stomach-bag, then its very smell, as the spiny backbone was boiled for soup.

'What now, Mateus?' Bac asked, as they warmed the last of this for their next day's breakfast. His pinched face and spindled frame had turned the fat boy into a scarecrow with a gaping mouth. 'How do we find food?'

'I do not know,' Matthew answered—and if truth-telling would transform him into one of the Lord's Anointed, then he had earned his place among the elect. 'We will think on it.'

There was little enough to think, or to do. He had a pistol and a musket, locked away from Trail, but nothing to aim at: if game still lived in the hills round the harbour, then it was asleep, or cowering in the snow till the snow melted. Ever since the murder of Jorgensen there had been no Indian trade: guilt and fear had snuffed it out, or else the wandering band which had enjoyed their brief triumph had prudently moved elsewhere.

Nor was there any use in an appeal to others. In this village of castaways, all were starving, and cowered like the deer and the fox till the torment of winter passed. Not even desperate theft, the storming of another hut, would serve them anything: a man and a boy could have little success against a greater household, and if by a miracle they won their fight, there would be nothing for the victor to enjoy.

So they starved for three days, chewing on scraps of birch-bark, sucking at moss gathered from tree-stumps, making an evening meal of snow-water and seaweed. They grew gaunt, the man and the boy: hollow-eyed, wolfish, linked to life only by each other and the warmth of the stove. Then Matthew, wandering out at dusk, sighted and shot a hare, itself famished, which had strayed into their compound; and they feasted again.

But they feasted small, stretching this starveling creature to six days' meat, and corn-mush, and soup with blood, and soup without, and bones pounded to mud between one stone and another, to make a poor man's poorest mouthful.

Then they starved again, and then, on a morning when Matthew in ravenous despair was ready to go out and shoot a man for the flesh off his buttocks, he went forth to find instead, on their doorstep, a wondrous offering of the night—a great strip of whale-meat, or walrus, or seal: four feet long and a foot wide, frozen, fat, and free.

By the side of this holy miracle lay something which Jorgensen, in the high humour of long ago, had once given to Sitting Mouse to seal a bargain: a copper coin with the image of Saint Anskar stamped on it.

'A great saint of Denmark,' Jorgensen had boasted to his unbelieving friends. 'If you have not heard of Saint Anskar, then I pity you.'

The pity, power, and love of this great saint had never been more movingly shown, in a thousand years of piety.

On a certain joyful morning which was—unknown to these outcast souls—the Feast of Saint George of England, Matthew and the boy awoke to the same sound. It was a fretful sound, a puzzle, and then a summons as loud as if some great clapper-bell had burst into clamour above their heads.

Yet it was the most gentle tune in nature. What had wakened them was the primal song of the earth: the first shy notes of the music of spring: the melody of melting snow dripping down from the eaves of their hut and falling at last on yielding ground.

They had lived to hear it! It could not yet fill their bellies, but it charged their hearts to wildness. If they had conquered winter, they had conquered all.

Three more days established a new world of warmth and hope.

As the sun climbed like a waking giant, the last of the ice vanished from the harbour, and the drifting fog which took its place was burned off by mid-morning. Men, for so long strangers, began to move about, or to sit in the sun and cheer their brittle bones.

A trickling stream became a small torrent, while above it a little ruffled bird—singularly mad, worldly wise—perched on a bare branch and sang a song of praise.

In a small household, Matthew took his knife and cut into the last quarter of the whale-meat, the gift of Saint Anskar and his servant Sitting Mouse, while Bac stood ready with the cooking-pot. The cuts were thin and careful, since it was too early to be prodigal, yet one need not go to prison for dreaming of freedom.

All they awaited now was the return of the fishing fleet, with its promise of beautiful briny pork, of savoury weevilly biscuits, of cheese and wine and little oranges . . . But before this manna fell from heaven they had a duty, a sad observance, a ritual of farewell.

They waited for that moment, easy to judge with a sailor's eye, when the ebb-tide had gained strength enough to make the westerly surge of the Atlantic keep its distance, and to overcome the onshore wind as well; then they pushed the raft of logs, with the weighted coffin on top, out into the stream of the Narrows, and leapt back on shore, and so broke the last tie between the living and the dead.

The raft with its solemn load was slowly gripped by the current and began to drift seawards. It caught on a rock, and hesitated, plucked at by fronds of seaweed; then it swung free to continue its gentle journey—the smallest Viking funeral in the world, and the most sad.

Bac, standing beside Matthew as they watched its progress, was anxious.

'How will it go, Mateus?'

'With the ebb-tide.'

'No, I mean——'

Matthew knew what the boy meant, and, sharing his sorrow, explained with care:

'When the raft is in open sea, it will begin to rock. There is troubled water there: you can see it plainly. So the coffin will tip, and slide, and sink, and that will be all.'

'Poor Jorgensen.'

Matthew sighed. Poor Jorgensen, indeed . . . But the true tears
were shed, and mourning past, and spring was here, even if one
lost friend made it seem a smaller ceremony.

'Come on, lad. Let's climb up, and watch.'

They retreated from the rocky edge of the Narrows, and began
to climb the hillside above, picking their way among the bare
patches between the melting snow. As they toiled upwards, often
turning back to watch the raft, the whole of the broad ocean
opened up before them, rolling grey and green and white across
three thousand miles of its majestic realm.

Soon there was only this wide horizon, the greatest in the world,
and the tiny burden which was just embarking on it, and then,
close at hand, a little wreath of yellow crocuses already pushing
upwards to speed the parting and salute the spring.

From this vantage point there were more signs of hope, more
evidence of the magic of a new year. A rim of brown earth now
surrounded St John's, as the snows melted and gave in to the sun.
On higher ground, the trees were shedding their winter burden;
the harbour was a clear patch of still water, waiting for the throng
of sails. Every valley would be exalted, as that Good Book had
promised . . .

Matthew watched, and wondered. Now there were more men,
many men, moving about on the water-front. Was Trail, that other
yellow flower of the spring, among them? Was he watching *them*?
If so, God's curse on him for a thieving rat, and God's forgiveness
on a weakling. But if they should ever meet . . .

He turned back to the Viking raft, and could not see it, and
then marked it as a mere speck among the tumbling waves. Then
it was gone for ever, and all ceremony vanished with the last long
dive of a sailor, and the silence of living souls matched the eternal
silence of the dead.

What joy it would have been if his shipmate friend still lived,
to outrun winter and greet this wakening world.

Bac, who was often quick to catch his thought, asked: 'Why was
he killed, Mateus?'

'French treachery,' Matthew answered gruffly. 'Those moun-
seers would skin their grandmother for a bed-rug!'

'But it was the *Indians* that killed him. Why?'

'Because they hate us.'

'Why?'

It was all 'Why?' with Bac. 'Well, they claim that this was their land. To hear their story, all Canada was their land. All America also. Then the white man took it. Of course we had the right, if we were stronger! But we spoiled their hunting-grounds, so they say. And spoiled their lives besides.'

'But they are Indians!'

'Even so.'

'Will this always be?'

'Aye. Till they lose their spirit—or perish altogether.'

Bac pondered this, and there was silence on the hillside. Then all their thoughts, and all the hours that remained of the day, were interrupted by the strangest sound. It came from somewhere up in the sky, and might have been the gruff voice of God. But God—unless He had a mortal fever—would never speak such unearthly language.

It was a far-off honking noise, a croaking, a tuneless whisper which became an uproar. Matthew looked up and about him and saw, with a thrilling heart, its cause. It could have been a cloud, it could have been wisps and whirls of smoke in the azure sky. But this cloud lived, and moved by its own determined will.

It was a numberless host of winter exiles: a myriad geese returning to the north-land.

They came winging in their hundreds, their thousands, in great skeins, flying in marvellous order, certain of their home. Their clamour filled the air. The strong leaders—who chose these guides?—flew in a single file of ten or twenty; then the flock thickened behind them, and darkened, and changed shape and then drew together again.

They passed overhead, their wings still flailing the air after hours and days and weeks: northwards, ever northwards, until their voices died and the sky was innocent once more, and only the miracle was remembered.

'God's blood!' Bac sometimes essayed a manly oath, and this was such a moment. 'I have never seen such a thing! Where have they come from, Mateus?'

Matthew had his own share of awe. 'From Africa, they say.'

'Where is Africa?'

Matthew pointed south and east. 'There.'

'But how far?'

'Perhaps three thousand miles.'

'But how do they endure? How do they *know*? How do they steer?'

In all solemnity, to mask his own wonder, Matthew answered: 'They have a little lodestone locked in their heads, and a little compass a-top, and this they follow.'

'*What?*' Bac's eyes were shining, in wild surmise. 'Have you seen it? How is it balanced? How——' Then he saw that Matthew was mocking him, and he surrendered to his elders and betters. 'I wished to learn,' he finished, with a hint of the sulks.

'So you shall, lad. And so shall I, if I live . . . All I know is, these geese fly south with the autumn, to winter in Africa, and they come home again in the spring. So—since home is the north, this must be the spring, and God be thanked for it.'

Bac, the earnest seeker after knowledge, was still put out, and gazed sternly seawards like any affronted mariner. Perhaps it was better so: perhaps God ordained all, and both the geese and the talk of the geese were part of some master-plan which now drew to its appointed end. For suddenly the boy stared, and narrowed up his eyes, and then pointed excitedly:

'Mateus! Look!'

'Where away?'

But before Bac could answer, Matthew saw it himself. A-far off, on the eastward horizon, there was a small clear gleam of white. It was a steady gleam: not a creaming wave-top which vanished as it broke, but a picture which held its image. He stared and stared, and still the gleam remained steadfast.

The geese had been true messengers . . . Matthew drew the deepest breath of seven long months, and said: 'By God's grace! *A sail!*'

Their siege was over.

SIX

NAVIGATOR

1759

'Come, cheer up my lads! 'tis to glory we steer,
To add something more to this wonderful year.'

Song *Heart of Oak* (words by David Garrick,
music by William Boyce from the Drury Lane
Pantomime, 1759.

'I would rather have written Gray's *Elegy*.'

General Wolfe, before his death at the taking
of Quebec, September 1759.

'In Memory Of
CAPTAIN JAMES COOK,
Of the Royal Navy,
One of the most Celebrated Navigators that this
or former ages can boast of;
Who was killed by the natives of Owyhee in the Pacific Ocean
On the 14th day of February, 1779, in the fifty-first
year of his age.'

Inscription on a tablet within the communion rail of
the church of St Andrew the Great, at Cambridge.

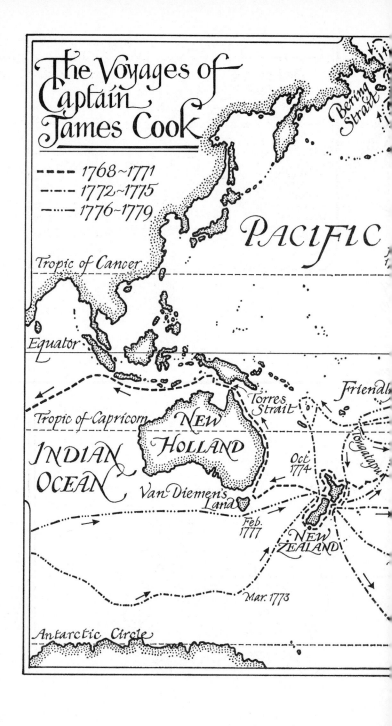

The Voyages of Captain James Cook

---- 1768~1771
-·-· 1772~1775
-··- 1776~1779

PACIFIC

Bering Strait

Tropic of Cancer

Equator

Tropic of Capricorn

INDIAN OCEAN

NEW HOLLAND

Torres Strait

Friendl

Tongatapu

Oct. 1774

Van Diemen's Land

Feb. 1777

NEW ZEALAND

Mar. 1773

Antarctic Circle

The Siege of
QUEBEC 1759

0 1 2 miles

Charlebourg

N

Camp of
Gen. Wolfe

Beauport

Les Batures de Beauport
a shoal, dry at low water

River St. Lawrence

Ile
d'Orléans

QUEBEC

Point
Levis

British
Squadron

Plains of
Abraham

St. Joseph

Batteries

Batteries

Transports

OCEAN

SOUTH
AMERICA

Mar.
1769

Jan. 1774

Feb. 1774

IN THE YEAR OF OUR LORD, 1759, the most singular monster who ever closed the coast of Newfoundland ruled as the Fishing Admiral of St John's. Having arrived the first of all the English fleet, Captain Jasper Bunce had a right which could not be overruled. But Jasper Bunce proved a curse of such evil quality that his reign was the signal for this plan of 'Fishing Admirals' to fall into decay.

Such might be his sole and memorable claim on history. But his claim on men, in that year of '59, was all that the men themselves remembered, and more than enough for those little history books which were men's lives.

Admiral Bunce proclaimed his habits of mind from afar off, and did not care who knew of them. He was a little fat fire-brand, forever bursting out of his finery which inclined to feathers, and belts with buckles bigger than his belly, and polished leather boots with upturned toes. He also burst out of his narrow skull when thwarted, and burst into flames of spitting rage in order to gain his way.

In appearance gross, in behaviour brutal, in ambition greedy, he was above all cunning.

This cunning had been demonstrated from the start. Lately there had been a new rule for the Newfoundland fishery, which decreed that while the captain first arriving was to be the Admiral, the second and third to enter the Narrows were to be named Vice-Admiral and Rear-Admiral respectively, with powers to match their rank.

Bunce had contrived to make his landfall in company with two other captains who were particular cronies, of like mind with his own. They had travelled as a squadron, arriving one, two, three like horses in a race which had been settled before the start, and they now ruled as a cabal.

Thus he had buttressed his reign before it was fairly set up. But of the three, Jasper Bunce was the man whose word was law. There was none to say him nay throughout the fishing season—though there was one other man who might have done so.

By now, as the years unrolled in the colony, Newfoundland boasted a Governor appointed by the Crown. He was, according to custom, a naval officer of Commodore's rank; and, according to another custom—the satisfaction of legal minds three thousand miles away—he was the King's own man and must therefore rule without question. But he was no more than a fair-weather friend, for anyone seeking justice.

The Governor arrived in August, and quit in October: his reign was thus brief, formal, and infirm. He could have no real authority over such as Admiral Bunce, who stayed for six months, usurping all power at the beginning and, with one quick mocking curtsy to a visiting nabob, maintaining it till the end.

Bunce could also rely on another regulation: if any dispute was ruled to be a 'fishing matter', then it was for the Fishing Admiral to settle, and none else. Since this tyrant had unabated power to paralyse the actions of a two-months' figure-head, everything of note became a fishing matter, and the unequal contest was won before it was joined.

Newfoundland, with all its promise and endeavour, was thus governed by a raw-meat pirate, enthroned on a barrel-top with a bottle of spiked spruce-beer in one hand and a pistol in the other: selling justice, or imposing it by whim, and indulging every day a lust for power which either drove men to despair or silenced them for ever.

Jasper Bunce's first 'Court' had set the pattern for all that was to follow.

He held it on the quarter-deck of his swinish ship, a little bark sweetly named the *Lass of Devon* which—since there were more ways of making money than by putting to sea—spent more time alongside in St John's than on the Grand Banks. Men came to fear the very whisper of the word *Lass*, since it stood for all things cruel and corrupt. But on this first occasion it was no more than a pretty name, and Admiral Bunce no more than an ugly face a-top an ugly body.

Both came into prominence as Bunce convened his court. He had a smattering of legal rigmarole, and this also was swiftly brought to bear.

'Now hear ye!' he began, as soon as the company—of his own crew, of passing sailors, of petitioners, prisoners, witnesses, and friends of all—were assembled on the deck of the *Lass of Devon*. 'This is a court of *Oyer and Terminer*, duly constituted by act of Parliament, and if any man knows not what that means, I will tell him—once!' He glared around him, practising his authority before making it bite. '*Oyer* is to hear, and *terminer* is to finish, and that is what I shall do, with no fear and no favour. I will hear and I will finish—and if any man challenge *that*, I will finish *him*. With this, if need be!'

He waved his shining pistol, then sighted upwards and discharged it between the *Lass of Devon*'s drying topsails. The shot rang round the ship and the harbour, startling men, frightening seagulls, and opening Bunce's reign as he meant to close it.

He had doubtless designed this signal to draw more watchers and idlers to the scene, and they came running. Among them was Matthew Lawe, his ship just berthed after its first foray, who listened with a sad heart to an old tale which had now become evil as well as old.

Jasper Bunce began to speak again, with an absurd air of judicial consequence which still had great force to wound:

'The Statute of his late gracious majesty King William the Third, Capitulum 25, Acts 10 and 11, gives me power incontestible— *potestas contra mundum*, as we say—to use or to destroy any huts, fish-stages, or buildings which have been erected on or near the coast during the winter past.'

It was true enough, as Matthew could testify from long experience. There were men in England, West Country merchants of influence, who were still determined that Newfoundland should never be colonized. It was to be kept as a fishing-station for their own use and profit, and otherwise to remain a wilderness unclaimed.

Since they had the ear of Parliament, and money enough to keep this organ attentive, they had clung to their privilege for more than half a century. Though hopeful and hardy settlers crept in, they were always thrust back from the coast as soon as spring returned. They could hold no property in St John's until the fishery had been served. If challenged, their buildings might be appropriated by the fish-monopoly, or levelled to the ground, and new fish-rooms built for the sole use of the fleet.

This noble Statute of William III, which Bunce had spoken of, was sixty-one years old. But it still ruled a sleeping land and, as a weapon, was still in his grasp.

He now said, in the voice of command: 'Richard Bannister, stand forth!'

There was a stir as the luckless Bannister came forward. Compared with Admiral Bunce, he was nothing at all: a shabby hill-farmer in threadbare grey worsted, who had planted hopefully in the autumn, survived a wicked winter, reaped a meagre harvest of corn in the spring, and—in great daring or greater desperation—had set up a fish-room before the English fleet arrived.

Such enterprise must be punished, with all the rigour of the law.

Bunce glared at him. 'Richard Bannister, you have encroached on the beach below the Narrows, built a fish-room there and maintained an anchorage, all to the scandal and detriment of his Majesty's loyal subjects, and practised a trade reserved to the fishing fleet from England. What say you?'

The forlorn Bannister, half starved and wholly at the mercy of the court, still had some spirit left.

'Why should I not fish? We starve if we cannot! What is this Statute? I am English, as well as you. God put the fish into the sea——'

'And the King says who may take them out,' Bunce thundered. 'Would you challenge the King? What are you? Is it treason you are after? I will teach you treason, you dog! *Guilty!* The sentence of this court—' he began to gabble, at a furious rate, 'is that the fish-room be destroyed, any fish within it are forfeit to the court, your boat is claimed for the service of the fishery, and you—' he glared again, summoning his last thunder-bolt, 'you will be put to fish-gutting for a month, if you are not clear of the harbour by sun-down!'

The listening crowd, Matthew among them, drew in its breath at the harsh sentence. Richard Bannister, destroyed at a single stroke, quavered:

'Sir, may I not pay a fine?'

Jasper Bunce softened his voice to an odious piety. 'Certes you may pay a fine, for your great transgressions. This court is merciful, like the God you have blasphemed. Fifty pounds English!'

He might have said a thousand. 'I cannot pay such a sum,' Bannister said, stupefied.

'Then why make pretence? Do you mock the court, you rogue? How much can you pay?'

'Five pounds is all I have, in all this world.'

'You will not need it in the next! . . . . So be it—five pounds, and your boat, and the fish. What say you?'

Bannister hung his head. Without money, he could not buy. Without a boat, he could not fish. Without fish, he was left with an empty fish-room. But he looked at his merciless judge, and saw that he was lucky.

'Agreed.'

'You are wise. But make no mistake! I will remember you—and all others like you.'

There were five such cases of 'encroachment': some of the desperate villains paid their fines, to God-knows-what secret coffers, while others could not and, dispossessed, went their way home in despair. Then it was the turn of the malefactors of the fleet.

These were the drunkards, petty thieves, disturbers of the peace, rioters, laggards, insolents, quarrellers, and late-sleepers, all of whom the Fishing Admiral had decreed must be removed from their own captains' rule-of-thumb and brought within his grasp. For this authority he could cite another chapter from his famous Statute, spoken with the far-off voice of William III, who thus commanded:

'You have power to inflict corporal punishment on all persons profaning the Lord's Day, and all common drunkards, swearers, and lewd persons.'

What sailors could escape this net? What *men*? They were soon to learn that their choice was a fine, or lashes at the capstan-bar, or a little of one and a little of the other.

The onlookers began to murmur as first one, then another of their ship-mates was caught in an iron trap, and paid with his money or his blood to be quit of it. A man reeling out of an alehouse and voiding free into the harbour—his own harbour!—was ruled to be a lewd person, and fined two pounds. He could not or would not pay.

'Two lashes!' Admiral Bunce roared—and two lashes it was, there and then, on a bared back which darkened at the first stroke and bled like a pig's slit throat at the second.

The prudent paid their fines, or promised to pay: the recalcitrants were flogged. Towards the end of this play-bill, when a wretched fellow—or vile criminal, according to the view of the court—had to endure five strokes of the whip for 'Aggravated insolence, in that he did break wind in the face of his superior officer', the murmuring grew to something more menacing.

The fate of the settlers did not matter greatly. They knew the law, and they transgressed it at their peril. But these others were sailors! This was their own world, ruled by their own pattern, set apart from landsmen: a world secure, tolerant, with a give-a-little, take-a-little freedom from prod-nose piety. Breaking wind at the mate? He was lucky if it were only wind!

The Fishing Admiral was more than ready for them. He had been scowling when he passed sentence, and watching with fat satisfaction as it was carried out; now he was scowling again, and prepared to turn a scowl into a stroke, if he could find a victim. Staring round him in baleful anger, his gaze turned out-board, beyond the quarter-deck of the *Lass of Devon*.

By ill-luck, it was Matthew who caught his eye. Standing on the edge of the quay and leaning over the rail to witness this act of justice, Matthew had just murmured to his neighbour: 'By God, this fellow is the biggest rogue of all!' when he found himself gazing directly into Bunce's furious eyes.

'You!' Bunce shouted. 'What is your business here?'

It was time, and more than time, for a soft answer. 'To watch, your Honour.'

The men nearest to him edged away as Matthew was thus distinguished. It was Admiral Bunce who drew closer, goading himself to a fury.

'You were speaking!'

'No, sir.'

'Your lips were moving. Do not lie, you villain! I saw them.'

'I was but chewing cod's roe, your Honour.'

'*Cod's roe?* In my court? I will give you cod's roe till you choke. Come on board!'

After that, it was all swift as a river in raging flood.

'Your name,' Bunce demanded, when Matthew stood before him.

'Lawe, sir.'

Bunce gave him an evil smile. 'I am the only *law* here. . . . *You* say you were chewing cod's roe. *I* say you were speaking some insolence, or worse. In either case, you disturbed the peace and showed contempt for the court. *Guilty!* Fined—what money have you?'

'None, sir.'

'So it is lashes you choose?'

'*What?*' Matthew was stirred beyond all common prudence. 'Lashes for moving my lips?'

'Lashes for blinking an eye, if I say so! . . . So it is one lash for chewing, one lash for talking, and one for your contempt. What say you now?'

Matthew was beside himself with fury. 'And one lash for calling you a son of a whore.'

There was a sudden break of laughter from the watching crowd, and warmth for such bravery. Jasper Bunce in his turn grew icy cold.

'And *five* lashes for calling me a son of a whore. Lay on!'

When it was done, and Matthew was nursing a back alternately shivering and flaming with pain, Jasper Bunce delivered his last magisterial decree:

'The next such case I will treat as mutiny, and the next dog I catch will be hanged.'

But in that year of '59, more things were stirring than such a mountebank ape could comprehend. The Old World and the New were alike in ferment, beyond anything dreamed of: there were wars and rumours of wars: the 'Wonderful Year' was earning its repute. East and west the ships sailed, the cannon roared, the cutlass and the sword flashed—and common rock and turf became English ground.

The French had already been ousted from Cape Breton, their eastern fortress in Canada, never to return. Nearer home, where a man or a nation might be hurt the most, their great Brest fleet was soon to take the thrashing of a lifetime from Admiral Hawke at Quiberon Bay, with six ships-of-the-line burned or wrecked, and nineteen others chased ashore into any creek or river where they might hide their drooping tails—or break their backs.

On the ruins of the Mogul Empire, half-way across the world in India, France was striving to erect her own. She was uprooted and sent packing, by a soldier-statesman who truly earned the title 'Clive of India'. The same fate befell her in the Caribee, where Guadeloupe, the richest jewel of the Leeward Islands, was lost to her and won to British arms.

France had chosen to be the enemy. Now the enemy was on the run. She must be kept at it.

All this mighty surge of power was the child of the brain of one Englishman, William Pitt, the 'Great Commoner' and first minister of Parliament; and one part of his dream of empire, one piece of a huge puzzle, was to drive the French from Canada forever, before they could bind these harsh northern possessions to their languid estates in Louisiana.

Certainly, in English eyes, the colonies of America, growing apart from the mother-land, were rebellious and should be disciplined. That was a pleasure still to come. But first there was the French fortress of Quebec, the thorn which must be plucked out.

To this, Mr Pitt now gave his attention.

High on a hill-side above the Narrows, whence he had watched Jorgensen's funeral voyage so many years ago, Matthew stared long at the town below and the sea beyond it. Both were restless and troubled, but he knew beyond doubt which he preferred. At that moment of loneliness, it was not only the past that he mourned.

As the seasons and the years had reeled out, like a log-line running free, after that first harsh winter on the coast, he had not lost his allegiance for this frontier-harbour, which seemed to cling to Canada as a barnacle to a whale.

He had spent other winters there, and starved, and lived; in soft-seeking years he had sailed south to Floridia, and passed a season among sun-warmed oranges instead of frozen whale-meat, or voyaged deeper still to the Cays of the Bahamas, where rainbow and dolphin and flying fish would jump into an angler's boat, so close to nature was man.

But always in the spring of the year he returned to the Grand Banks of Newfoundland, and commended his soul to cod. Freed from the *Consuela*, whose bones now lay bleaching on Sable Island, he had taken British service and remained within it; the discipline was not less firm, but the food more honest and less devoted to the idolatry of the olive. And here, at least, a man passed his water over the side instead of onto the deck.

His ship this season was a fine new schooner, fresh from a Yankee boat-yard: the *Barnabas*, British bought and manned, graceful as a gull in flight, strong as a bare-fist fighter, built to survive any Atlantic fury. But this very morning the *Barnabas* had sailed without him, and therein lay his mourning mood.

The 'Wonderful Year' of '59 had not been so christened on the the coast of Newfoundland, and never would be. Under the whip of Fishing Admiral Jasper Bunce it had turned sour, and more cruel with every week that passed.

Bunce was enjoying his remorseless rule, without stint and without mercy. He had made a noble fortune from fines, from imposing a lien on men's earnings, from declaring forfeit a half-cargo of fish for any fancied infraction of the fleet's rules: from selling licences-to-trade, from imposing a tax 'for the maintenance of the quays and sea-marks'—which had never been worse maintained.

This was all for his profit; beyond profit there was the sport of cruelty. Already he had hanged his 'mutinous dogs'—seven of them separately, on evidence which would not have nailed a badger-skin to a barn door.

He had locked men in the town stocks, day and night for a week, and set up a 'noon pelting' for the enjoyment of his own hired ruffians. He had whipped them beyond endurance, so that the *Lass of Devon* was now renamed 'The Lash', and Bunce himself 'The Bosun'—the flogger of the fleet.

Though the fishing crews had hoped for relief when the Governor had arrived from England to take up his yearly office, they hoped in vain. Their new ruler arrived early and left earlier yet; and in all case she had pleaded No Jurisdiction: whatever concerned the conduct of the fleet was a fishing matter, and beyond his competence. Let justice be done—and good morrow to you!

Matthew was left with only one sorrowful thought: that he had turned this page before. It was Henry Morgan and Governor Modyford again, the very ghosts of infamy. As in old Jamaica, the Governor held court on his throne of state. But he would have done more business if he had stayed on his privy-stool. It was the pirate who ruled.

Men murmured against this harshness, and were put down. Matthew murmured, and ran straight into the arms of Bunce once more. Bunce had never forgiven him his insolence at the first court, and petty persecution had grown to a threat overhanging him at all times—the danger of death at the hands of this despot. It had now moved closer than ever with the ordeal of next morning, when he must face the same court again, armed with its same butcher's hook, on a charge of 'Disaffection'.

It was for this reason that he had been named 'Chargeable' under Bunce's system of tyranny, and the *Barnabas* ordered to leave him ashore to face the music of his trial.

'Disaffection'—which was no more than complaining of unjust treatment, within the hearing of a spy and toady who had betrayed him to Bunce—disaffection might be stretched to mutiny. Certainly he would be flogged, having been sucked dry of money. But it might be a rope's end of a different sort, if their great Fishing Admiral maintained his spite and had his way.

In such a case, Matthew would be dancing on air by the next sun-down.

There was one other hope of rescue, beyond their lackadaisy Governor who was no more than a resplendent uniform and a deaf ear; and it was for this that Matthew was waiting and watching, on the sunny hill-side above St John's.

The rumours of war had not passed Newfoundland by, set as it was at a cross-roads of the Western Atlantic where so much was stirring. There was talk among the schooner crews, and aboard trading vessels coming up from the south, of a great English fleet carrying five thousand soldiers—some said ten thousand—which was assembling at Halifax, the finest harbour of New Scotland.

Where they were bound was unknown, but they were not in Halifax for their health.

Would some of these ships, wearing the true British flag instead of the Fishing Admiral's soiled banner, put into St John's to water or to send a press-gang ashore in search of useful seamen? It had been done many times before—and now it could not come too soon for certain suffering sailors and settlers. Better to work under honest Navy discipline than to live at the mercy of Bunce's evil eye.

Soon this phantom fleet was more than a rumour. Small squadrons of it had been seen, on the far eastern horizon. Matthew himself, standing a night watch on board the *Barnabas* when she was heading for the Banks, had sighted a great collection of them, moving south-west under light winds. They had ghosted by; only their cluster of stern lanterns, flickering like a giant Christmas tree, was to be remembered. But they were real!

Perhaps there might be more of these staunch invaders. Always there was hope of better things. On one happy day a fighting ship —a frigate, or something more lordly—would put into St John's, on the hunt for water or meat or serviceable men. She would find a welcome, in at least one fearful heart.

Matthew now saw that this great blessing might be at hand.

He had sighted the ship more than an hour before, and took her to be another busy passer-by. Even when she closed the land, it had seemed that she was setting her course south-westward, to round the lower coast of their island and continue on to Halifax. She was a warship, as the tiers of her gun-ports proclaimed, and intent on the business of war.

He now noted that her southerly course had been planned to stem the tide, sluicing northwards at the top of its flood, and that she was heading in for the harbour.

He watched in great admiration and greater hope as she made her run for the entrance. This was a big ship, more than a frigate, and her broad bottom was subject to the pull of the current, while her towering sails, from royals to main courses, were trimmed to push her in an opposite direction. It was all done to perfection, and not slowly, as some masters who were better named mice tried to make the passage of the Narrows.

At the very moment of balance she put her helm hard up and turned as if on a six-pence piece, while the royals and topsails were doused; then she made an arrow-like dart for the entrance. Just as the thrust of the tide lost its power, the ship was pointed directly at the only safe channel: she lost half her speed, and glided inwards like a swimmer paddling the water, knowing it to be an ally instead of an enemy.

Then she began to weave through the inlet, while Matthew, who knew its hazards as he knew the scars on his sea-worn hands, could only say to himself and to the world: 'I swear to God, that fellow could thread a needle with his ship!'

The stranger threaded the needle, turning and twisting through rocks big and small, and sailed out smooth and serene into the broad harbour of St John's. Her cannon boomed forth a salute of nine guns, though whether to wake landsmen from their lethargy, to proclaim that she had powder and shot to spare, to do honour to his Majesty, or to scare the crows, was not clear. Then she backed all sail, dropped her anchor, and lay at peace, with one last cannon-shot to announce: 'I am here.'

There could be no one in St John's harbour who did not know that she was there. As the booming echoed round the heights and valleys, men began to gather on the quays. Others started to run upwards from the harbour to the hills; not all witnesses were ready to welcome this visit, and the doubters were taking to the high ground till its purpose was clear. But this was not the path for Matthew Lawe.

Resolutely he turned, and began to march downhill, into the arms of his friend, his friend unknown but already welcome, a Press-Master of the Royal Navy.

II

The ship was the *Pembroke*, detached from the main British fleet and sent on a swift foray from Halifax to clear the coast of French supply convoys before the attack on Quebec began. On her southerly return, she was to seize any likely sailors from their great breeding-ground of Newfoundland.

The *Pembroke* was a ship-of-the-line, a third-rate of 64 guns, 1250 tons, almost fire-new from Plymouth Dock: Captain John Simcoe in command. The gifted ship-handler who had threaded her through the Narrows with such swift dexterity was one James Cook, master.

Aboard such warships the master bore the sole burden of navigation, pilotage, and any surveys required in uncharted waters. Under the captain, he set the courses and handled the ship; he was also concerned with stores, with masts and spars, sails and rigging; with the working of the ship and the management of the crew. A good master was thus beyond price.

Being without commission, he wore no uniform. Always he was surrounded by glittering officers of superior rank and class: lieutenants, mates, even midshipmen. But he was the trained man among all of them, with parchment to testify to his competence, and responsibility by day and by night to prove it. His quarters showed his true rank, placed just forward of the captain's great cabin in the coach of the quarter-deck.

There, he was at the captain's right hand; there, in truth, he *was* that right hand, and fortunate was the captain who could put all his trust in it.

How James Cook, at thirty, had attained this pinnacle of confidence was a straight tale—straight, laborious, and single-minded. Matthew only learned of it slowly: first from talk among his *Pembroke* ship-mates, later from Cook himself when he had secured the master's friendship. There were elements in it of his own life—the humble beginning, the thirst for the sea above all other provender in the world, the longest of all loves and the fiercest of all hates which ruled a sailor's blood within a sailor's breast.

But there the likeness between the two men was sundered. This farm-boy, son of a Yorkshire day-labourer, had prospered.

The farm-boy became the shop-lad, apprenticed to a life of grocery and ribbons—but on the coast, at Staithes, a fishing-port married to its trade. Here the sea began to exert its pull, like the tide its power. Within a year the grocer's boy of eighteen had bound himself to a ship-owner of Whitby, concerned with that coastal commerce which was then the life-blood of England—coal. Here he found his private heaven, and was slave to it for ever.

Here also the apprentice learned his navigation and seamanship, up and down the east coast from Thames to Tyne, and across to the Baltic ports. It was a nursery of seamen, but the schooling was as harsh as any to be found. These east coast highways, lacking lights or sea-marks or trustable charts, were treacherous, with shifting sands, fierce tides, sudden gales, the fogs of the blind, and always a steep shelving bottom on one side or the other, which could turn a single mounting wave into a score of spiteful curling demons.

But the apprentice boy rose to mate, having been tested in bigger vessels, wider waters: the Channel and the Irish Sea, the coast of Flanders and the ports of Liverpool and Dublin; his range enlarged from the bitter cold of Norway to the softer shores of the Scillies. Step by step he rose, and at the age of twenty-seven was offered command of his own ship.

To this he said 'No', or rather, being both cordial and respectful, 'No, thank you kindly.'

Instead he made his grand decision and took his giant step. Just as Matthew had done, though for more reputable reasons, he turned his back on the past, walked ashore at Wapping Stairs, presented himself to an avid press-gang, and volunteered into the Royal Navy as able seaman.

In James Cook the astonished Press-Master found himself with something of a prize. Instead of the drunken sweepings of London's dockland, the gaol refuse, the lumbering landsmen who would never be turned into sailors if they were flogged for twenty years, here was a skilled and willing man who solemnly wished to serve.

Such a man could only rise again. After four weeks in his first ship, he was rated master's mate; he was appointed master of his second, the *Solebay* frigate, in two years; and master once more, a mere two months later, of his third. This third was the *Pembroke*, and *Pembroke* was destined, like James Cook himself, for great enterprise.

She had spent the winter, before her burst upon St John's, in Halifax harbour, trimming up for the task which was known to lie ahead. It was a hard season, one of the worst of the century: men kept to their ships, sailors to their cramped quarters between decks, officers to their cabins. But Captain John Simcoe, and James Cook his master, did not waste a minute of it.

Simcoe, a man of inquiring mind and open curiosity, had long been convinced that in science and mathematics there lay the true secret of navigation, whether it was to be practised on the broad ocean or the creeks and inlets of such a river as the St Lawrence.

Measurement—that was his great word and his great dream: measurement of moments of time, measurement of the infinite wheeling stars and the dependable sun: measurement of distance, depth, current, tidal height, angles of sight, length of shadow—and all locked within that other most precious measurable element, time itself.

In his own cabin, hard by Cook's quarters under the coach, he set up a small palace of navigation, and all that winter the two of them, captain and master, toiled together to perfect a plan— a picture within the head, then a chart set down plain on paper— which would take *Pembroke* and the fleet up the St Lawrence River to their target, Quebec.

It was only a tiny miniature, when set against the massive surface of the globe: some five hundred miles in all. But if it could be extended to all the world . . . John Simcoe was the teacher and prophet: James Cook the eager pupil, set on fire by new visions so far beyond the lead-line, the dead-reckoning, the tired eyes of the look-out, that it opened up a whole fresh universe. This year would be Quebec's; next year, who but God Almighty could set a limit to it?

For James Cook, this enlargement of mind was his captain's richest gift, and sadly his last. By spring-time, *Pembroke* was poised, with her fleet in company, to lead the way to Quebec and conquest. Beyond all the rest, the master himself was poised for the greatest effort of his life.

Matthew Lawe, when better acquainted, once ventured to ask the master of the *Pembroke* how a man might so rise in the world. How had James Cook, the shining example of such buoyancy, risen?

Cook chanced to be at his ease, which was rare. He was a big man, lean, raw-boned, awkward, yet forever strong. Above a keen and hawkish face his piercing eyes, which could blaze with anger at some poor work which he could have done better in his sleep, were also at rest.

'I followed my star,' James Cook answered. His customary grim smile was tendered to a softer mode. 'There may come a day when I shall follow every star in the night sky. But that is promise, which lies in the future. Performance is for *now*—today and tomorrow. Life is divided small—one hour at a time, one *minute* at a time. Fill them all, put them all together, and you have the whole of life. But forget one hour, forget one minute, and you may have nothing. You may have *failure*!' He pronounced it as the worst word in his private lexicon. 'Think on it! If we run on a rock tomorrow, we lose all. Perhaps as much as twenty working years. If we avoid that rock, we have nineteen years of promise saved. *Do not run on that rock!*'

Do not run on that rock . . . It had forever pleased Matthew that these simple cautionary words, useful alike for mariners in tall ships who were careless of their course and for children playing king-of-the-castle by the sea-side, were of the same tune as had signalled his first link with James Cook, his first acquaintance with this great man, and his first mark in a new world of naval competence.

The link had been forged in so small a moment of time, that measurable gift of God, and was such a chance event in a lifetime of chance, that it was not truly felt until after. It had happened on their passage southwards from St John's to Halifax, when Matthew was standing his first trick at *Pembroke*'s wheel and they were approaching Sable Island.

Matthew, from long experience, knew Sable Island and had cause to curse it. He knew its reputation, he knew its treachery. He had suffered its power to lure and to betray. On one fearful night he had lain exhausted on its beach, surrounded by dead men and the timbers of a dead ship, the old *Consuela*. He still feared it, as did all sailors who had ever coasted south from Newfoundland and given it a wide berth.

Now, planted four-square at the wheel below a billow of canvas, he knew that they were heading into danger.

Master James Cook was busy at his chart-table. He was for ever busy at his chart-table, measuring this, measuring that, confiding to history the fruits of long and patient examination. Meanwhile, the *Pembroke* ploughed on, with nothing on the starboard hand and only the low face of Sable to port. But Sable to port was moving closer than it should.

Though their course had been laid to clear it comfortably, there were certain other hazards attending Sable Island which had trapped a legion of other ships. The great Atlantic tide which now sucked them eastwards had been allowed for, but beneath that tide lay banks of sand and silt and broken rock which changed with the seasons, receding here, stretching out new claws there, adding new perils within the space of any week. The *Pembroke* was standing in too close.

He did not know what to say, nor whether to say it. The master was for him the greatest man on board; Matthew was nothing—a pair of sinewy hands, a pair of bare feet, with 'Aye, sir' on his tongue and simple service in his head. Such walking objects did not speak to masters for any reason save 'Aye, sir.' Above everything, they did not correct his navigation.

But Matthew knew that he must. His hands on the wheel knew it, and his feet on the deck of a ship which was sliding up-wind as well as holding her course. By chance, at that moment, James Cook unbent from the chart-table and walked back to his customary stance at the helmsman's back. Matthew summoned all his spirit, and said, over his shoulder:

'We are carried to port, sir.'

Cook, whose mind had been intent on greater things, answered: 'Aye. 'Tis the tide's pull.' He turned his spy-glass onto Sable Island, lying flat and innocent some two miles ahead on their port hand, and studied it intently. Then he said: 'We are set to clear it, by a long mile.'

It was not true. They were set to clear what they could see, not what might lie underneath. Matthew drew one of the deepest breaths of many a year, and said:

'Sir, may I speak?'

James Cook, surprised, stepped forward until he stood at Matthew's side and could look into his face. 'Speak? Of course you may speak! We breathe the same air . . . What is it?'

Matthew dived headlong in. 'Sir, I know this island. I was cast ashore there, one summer night when we were also set to clear. But there was a long spit of shoal water underneath, and there we grounded, and the foremast went by the board. It was a new arm of land.'

Cook looked at him keenly. 'A *new* arm?'

'Aye, sir. It happens every season. Every tide, for all I know. We had a saying on board, "Sable is black!" It has trapped more ships than——'

The master interrupted. 'What, then?'

Matthew, in new and shoal water himself, was confused. 'Do you mean, what course?'

'Aye. And hurry.'

'Some two points to starboard. Or better yet, come up-wind and leave it on the other hand. We have sea-room, and the tide will serve us better.'

James Cook, when he moved, did so swiftly. 'Down helm!' he ordered. 'Bring her to port.' Then he bawled: 'Watch on deck! Brace your yards back—and jump!'

They jumped, being willing men under a master they trusted, and *Pembroke* stood out to seaward of black Sable Island and presently passed it safely by. When all was clear, and she was back on her course for Halifax, Cook growled to the boatswain: 'Relieve the wheel,' and then to Matthew, more grimly still: 'Come with me.'

But he was not grim in anger, only hungry to learn and to record. At the chart-table he drew a swift picture of Sable Island, which was in the shape of a bean-pod ready to be plucked, with its fat curve towards the south-east. Then he beckoned Matthew with a snap of the fingers, and the questions began.

They were like the strokes of a surgeon's knife, quick and forceful, exact, deep enough to inquire—no more and no less.

'Where did your ship strike? On the ocean side, or the landward?'

'The landward, sir. The side we were set to pass today.'

'Where, precisely? Show me.'

Matthew pointed. 'The northerly end. The new shoal ran out north-west. We had coasted it a month before, and there was nothing to bring us harm.'

'How far off was your course?'

'A half mile.'

James Cook made a mark on his picture. 'What ship was this?'

'A Banks schooner. Portuguese.'

'What was her draught?'

'Some fifteen feet. No more.'

Cook made another note. 'What then?'

'The foremast went, as I said. Then the wind came up fierce, and we were driven hard on, and rolled over. So we jumped and swam ashore. Eight men in all. Eight from thirty.'

'What of the island? Does anything grow?'

Matthew shook his head, remembering a wilderness, a bare refuge for sailors astonished to be alive. 'It was barren land. Some sea-moss. Tufts of grass. Wild pease. A few gulls' nests.'

'But nothing living?'

'Aye. Wild horses.'

Cook's frosty eyes stared at him. 'Do you spin me a tale?'

'Not I, sir. They came at us as if they would eat us. So we leaped and shouted and they ran away. 'Tis said there was a Spanish ship, long ago, that was wrecked on Sable. Her cargo was horses for Peru. They have lived on.'

'Then they do better than men . . .' James Cook's cold face curved to a smile. 'Well enough—I will write "Wild Horses here" on my chart, and if you deceive me, you deceive the world, and wild horses will drag you to your grave . . . How long were you left ashore?'

'No more than three days. The wind fell light, and friends put down a skiff and took us off.'

'God bless all sailors,' said James Cook. 'They look after their own.' Then he made one of his swift, probing changes of mind. 'How old are you?'

'Thirty years, sir.'

'And long on the coast?'

'Ten years.'

'Where have you voyaged, in this region?'

'The fishing grounds. New England. And south to Floridia and the Bahamas.'

'Never in the St Lawrence Gulf?'

'No, sir.'

'Nor I, past Gaspé Point. Nor anyone. But we will make our way . . . Well, thank 'ee.' He smiled again. 'When next we are running on a rock, do not fear to speak out.'

Matthew went forward with his head high. The hot press of St John's had never captured a happier man.

It was a fine advance-guard which set sail from Halifax early in May, rounded Cape Breton, and laid course up Cabot Strait for the mouth of the St Lawrence. They were thirteen ships-of-the-line, charged with one simple duty which was in truth the core of the attack on Quebec: to find the way there in strength, whatever the difficulties and hazards: to mark that highway as plain as it could be made, and to wait for the main fleet and the British Army to follow them and take the city by storm.

The *Pembroke* led the way, since James Cook had voyaged furthest of all in the Gulf and, with Captain Simcoe, had toiled at those winter plans. As well as her manful crew, smartly uniformed in tarred canvas blouses or petticoat trousers, striped waist-coats, and straw hats also tarred to keep out the weather, she carried an advance-guard of another sort: some one hundred soldiers and marines, to give teeth to the ship and see her past the enemies lying in wait on either side of the river.

These soldiers were as proud as Lucifer, and *their* uniforms of scarlet and gold made them seem prouder still. They were disdainful of the marines—'Royal Boiled Lobsters', they called them; and to sailors they would scarcely speak at all. Their greatest pride was in their regiment, the 47th Foot. When asked of this they would answer, with all the arrogance of acknowledged heroes: 'We are Wolfe's Own.'

General James Wolfe, the darling of the British Army, would be in command of all, as soon as battle was joined.

But first, for the *Pembroke*, there must be a sad sacrifice to the gods of war.

Captain John Simcoe, ship-bound for a hard winter and working without respite on the plans for the approach to Quebec, was an exhausted man when the time arrived to put these plans into action. He kept to his cabin from the moment the *Pembroke* cleared Halifax harbour, leaving all to his lieutenant and the master. He could eat nothing, nor rally his strength; and within ten days he took leave of life.

It was a shock and a bitter grief, beyond all expectation. The surgeon spoke of 'ship-fever', rising out of foul bilges which could never be cleansed, from the day that a ship was first launched.

James Cook, mourning a true friend who had opened up a great new window on the world, told himself: 'Care kills.' But if a man's course was run, then his work was done, and others must assume his killing care.

With that heart-breaking sailors' ceremony which could never fail to move, *Pembroke* buried her captain under the lee of Anticosti Island. As the marines beat their muffled drums, all ships in company struck their topsails. The guns began a mournful booming; twenty cannon-shots tolled their salute, with a half-minute pause between each. Then the canvas packet dipped beneath the tide, and at dusk a new captain, Wheelock of the little *Squirrel*, took command.

It was a sea story, as old as the first ship that ever dared the ocean. A man died, another stepped forward to take his place. James Cook's skill was now more precious than ever, as the *Pembroke* pressed on through tumbling ice-floes which, though troublesome, were at least good spring-time news for the fleet. The St Lawrence River was opening to them.

But, like their lost captain, the good news faded swiftly. Though the broad river was indeed opening into a stream of fresh cold water, it was closing as a safe highway. Beyond Gaspé Point, Cook knew nothing at first hand of what lay in store, and must finger his way forward like any other explorer.

The charts which he and John Simcoe had drawn were based upon the hearsay of the coast, on the guess-work of two able mariners, or on captured French maps which were of low esteem— or might even have been doctored for the confusion of later invaders.

For the next two hundred miles the river narrowed inexorably, and their voyage became a multitude of puzzles, each to be solved before the next one could be examined. The St Lawrence began to lapse into a very mill-stream of a river: its currents ran all ways at once, its eddies were sudden and spiteful, its whirlpools fed by waterfalls plunging from the cliffs on either side. They could only make progress with a dependable east wind at their back.

Cannon-fire from the shore attended their passing. They sighted —and were sighted by—Algonquin scouts who kept pace with them, for mile after mile, and at night lit leaping fires which were as much a foretaste of hell as a signal to their masters in Quebec.

Anchored from dusk until dawn within reach of these savages and their treachery, the *Pembroke*'s soldiers mounted a guard, and changed it every hour with much barking and bombast. There were some on board who would rather have risked an attack.

Monstrous military boots rang on the deck like thunder, while the watch below cursed the wearers off the ship. Why could they not go barefoot, like Christian souls?

Though the *Pembroke* captured a French pilot by the oldest trick of warfare, the hoisting of false colours, he proved more of a burden than a benefit. He was a little merry man who could take a jest, and perhaps return it. Well enough—he had been duped by a French pennant. Maybe it would be his own turn next . . .

An armed marine stood sentry over him at all times, and James Cook questioned him fiercely, and even Matthew Lawe, whenever their eyes met, drew a finger across his throat in the general wish to turn this captive into a slave. But the French pilot, smiling like a villain—or a true patriot—never flinched and was never to be trusted.

Yet still they made their slow progress up-river, probing for the deeps and the shallows, marking the channel as they passed; and there came a day—the twenty-third day—when James Cook could say to Captain Wheelock, and to Matthew standing nearby:

'Tomorrow the Traverse!'

III

This Traverse was already a fable—and never more of a fable than when Matthew talked of it to his new-found friend on board the *Pembroke*, who was, wonder of wonders, a soldier.

Matthew had made a friend of Corporal Ned Pym with the best introduction of all, by saving his life. Soldiers, the crew had found on their journey up-river, were prone to falling in the water: thus far *Pembroke* had lost four of them, four land animals who could not comprehend that ships differed from houses and streets: that they moved impudently under the feet, and rolled incessantly with the waves, and had forsaken the habit of standing still.

Corporal Pym, a large and ruddy man with ox-like shoulders who, like an ox, could not swim, came near to bringing the tally of the careless dead to five. On a certain evening when they lay at anchor, though pitching to a brisk swell, he had climbed up to sit on the guard-rail. There he raised, with both hands, a giant portion of bread and beef to his mouth, and promptly fell arse-first into the St Lawrence.

His bottom-weight was enough to take the rest of him down. Forlorn bubbles were already rising to the surface when Matthew, diving headlong into the dark river, found him, forced him upwards again, and secured him with a rope thrown from above.

Corporal Pym, having been drawn up—a task for three stalwart sailors—was rolled to and fro over a barrel until the last gallon of water was squeezed out of him. He then opened his eyes to a world so nearly forsaken—and became a friend for life.

Ned Pym was a simple soul, and no further from God on that account: when he had cause to be thankful he showed it, and when displeased could bear down hard on the offender. He was a soldier because he liked the glory, he was a corporal because he stood six feet tall, and he was now a friend because he had been taught, as a boy, to return favour for favour.

To Matthew his benefactor he was ready to confide all, whether it was pride in his calling or doubt of whatever he could not comprehend.

They fell into the habit of talking together, when the day's work, of sailor and soldier alike, was done, and nothing lay ahead save an anchor-watch for the one and an armed lookout for the other. Corporal Pym, a grenadier, never tired of showing his weapons, and boasting of the skill needed to use them.

As well as a musket with a gleaming bayonet, and a bandolier loaded with musket-balls, and a pouch for powder and flint, he was hung about with the special tools of his grenadier's trade: the small bombs, with fuses attached, to be lighted and thrown ahead as he advanced.

He would demonstrate such an attack with fierce energy and much stamping and shouting. ONE! Set light to the grenade: TWO! Throw it at the foe: THREE! Fire the musket: FOUR! Charge!—with that razor-bayonet aimed anywhere between the breast-bone and the genitals: FIVE! Cry 'Huzzah!' reload, and do it all again.

Confined as he was to such weaponry as an oar, a marline-spike, and a knife suitable only for cutting salt beef off the bone, Matthew felt himself envious. But on the evening when James Cook had declared: 'Tomorrow the Traverse,' it was Corporal Pym who lacked instruction and Matthew Lawe who could give it.

They were at ease in a magical June dusk, warm as the feathers of a nesting bird, peaceful as Paradise. The *Pembroke* rode gently to her anchor as if she loved it; round them was lapping water, the riding-lights of other ships, and a distant view, dissolving moment by moment, of hills on either side holding in their hands the soft bosom of the river. Only the Indian fires ashore, winking with envious eyes, marred a scene of deep content.

But the soldier, who now crouched safely on the deck instead of braving the guard-rail, had thoughts beyond peace, which was not the first word in the manual of his trade. He was quick with a question which must have gone the rounds of the ship before reaching the military ear.

'What is this Traverse?' he asked, as soon as he was settled within the security of wooden walls. 'All are talking of it. What does it mean?'

''Tis the next piece of the river, and the last.' Matthew could answer with certainty, since he had been attending the master and had seen the great chart which might—or might not—give them their clue to what lay ahead. 'From here we have one mile to go, not more, before we reach Quebec. But it is the worst mile of all. Not one big ship has ever passed through. It is a crossing, from one side of the river to the other, and then on through to the pool below the citadel. There were sea-marks laid down by the French, but they have been cut away. Now we have to discover it for ourselves.'

This was sailor's talk, which Corporal Pym could scarcely comprehend. Certainly it was a puzzle not to be solved by the bayonet, nor with hand-grenades and cries of 'Huzzah!'

'But how do we discover it?'

'We row,' Matthew answered, in the tone of a man who would be pulling an oar. 'We take six long-boats, and six men with lead-lines to sound the depths, and twenty buoys with flags, and some soldiers to drive off the Indians while we do the work. We find our way through, and we mark it so that the fleet can follow, and then we return.'

'But is there a way through?'

'That is what we must discover.'

'And if not?'

Matthew grew cautious. He had been entrusted with secrets, and could not forget it. 'Then we think of another plan.'

Corporal Pym, whose frame demanded that he should be forever eating, ate of beef, bread, and raw onion for a comfortable space. Then he asked:

'How long might this take?'

'The master says, three days.'

'And if there is a safe channel, what then?'

'The *Pembroke* and the rest of our ships sail through, and anchor. Then we signal to the main fleet, and they come up-river, and we are all knit together, and ready for the attack.'

Ned Pym asked: 'How do you make this signal?'

'Simple enough. We send it by the fish.'

'The fish?'

'Aye. We whisper into their little ears, and they swim down stream and tell it to the admiral.'

'*What?*' The corporal was a slow man with a jest. 'How——' Then a hand the size of a small ham descended forcibly on Matthew's shoulder. 'You damned dog! You make game of me! Tell me truly.'

'We send a pinnace back, and she will signal "All clear" with a gun-shot or a flag.' Matthew rubbed his shoulder. 'And do not hit me again, or my friend the master will send you in it, with a sizzling bomb stuck up your backside.'

'Did I hurt you?' Ned Pym asked, contrite.

'Only to the maiming point. So keep your fist for those frog-eaters—and may you meet them soon!'

'It cannot be too soon for me!' Corporal Pym was ready to eat fire again. 'But how do we engage?'

'We land with boats. We have them ready in the hold, in pieces.'

'Pieces?'

'Aye. But never fear. There is a carpenter sticks them together, before we hazard your precious lives . . . So we land. There will be a hill above us, or a cliff. We climb it. Or better, *you* climb it while we hold your coats. Then you climb down again, and tell us: "Huzzah! We have taken Quebec."'

'That we will do—never fear! With General Wolfe to lead, we can take anything.'

Matthew had heard this boasting talk before, and could not mock it. In his own life he had been fired with the same feeling and the same certainty. It was the one sure mark of leadership, that it could prompt such words: Give us a man we know and trust, and we will follow him to hell and back, with much pride and pleasure on the way. But the renowned General Wolfe was still a mystery to sailors.

'Have you met this Wolfe?' he asked.

'Aye. Well—I have seen him, and followed him at Louisburg last year, when we trounced the French from eastern Canada. We know he is there. We know he has the king's admiration. We know he has not been beaten in the field. So——' honest Ned Pym could not quite find the phrases, but the urgent belief shone out. 'Yes, I have met him. As I have met Christ at the communion rail.'

It was a long speech, and could only be followed by long silence. The *Pembroke* rocked gently to the river's urge. Night was coming down like the curtain of the world. Jingling stamping soldiers were marshalling in the waist of the ship, ready for the first hour of their vigil.

For sailor and soldier alike, tomorrow might be the testing day.

Corporal Pym said suddenly: 'He reads poems to his officers.'

It was not the easiest recommendation to swallow. 'Why would he do that?'

'God's blood, how can I tell? I have never heard a poem, nor ever shall. But it seems they love him for it.'

'Will he read poems to the French? Will he read them to sleep?'

Corporal Pym of Wolfe's Own returned to proper regimental pride. 'He will do what he chooses, and we will follow him.'

It was the hour for the end of wakefulness. Matthew stretched and flexed his arms, feeling the first dew of a river's night begin its downward drift. 'Well, I to my cot. I am first boat off tomorrow, with the master.'

'Did you say that you take soldiers with you, on this Traverse?'

'Aye. We are all together.' Matthew rose, and prepared to make a prudent withdrawal. 'How can we leave you alone on board? If you do not fall in the water——'

'What else?' Ned Pym asked menacingly.

'You cry Wolfe,' said Matthew, and disappeared into secure darkness.

Though the two days that followed were back-breaking work for common sailors, it was not Matthew's back which was broken. By the first dawn of this Traverse time, he found himself promoted to master's mate, and could sit beside James Cook in the long-boat's thwarts, and happily watch his late comrades, mere galley-slaves, working their arms to death and their rumps to rags, while he had only to exercise his brain.

It had happened as they were about to lower the first long-boat, the master's own, over *Pembroke*'s broad bulwark. Cook had snapped his fingers at the boatswain and, indicating Matthew, told him: 'I need this man at my side. Find another oarsman to take his place.' To Matthew he said: 'Carry these for me into the boat, and drop nothing, or we lose our labour.'

'These' were a spy-glass, a precious chart, a set of pencils of black lead and soft slate, and an ancient back-staff which, cunningly altered to show the distance off-shore, was still the master's particular treasure. As Matthew gathered them, and wrapped them carefully in canvas, James Cook told him: 'You are rated master's mate on our return. *If we do not run on a rock!*'

They did not meet any such disaster, though it was not for lack of rocks.

As soon as the clutch of six long-boats put out from the fleet and began to row up-stream, they were caught in a swift current which harassed them for the rest of the day. James Cook's plan was to lead the procession, taking soundings with the lead-line as he went: at any chosen moment, when the water grew shallow and a turn might be necessary, the boat immediately astern dropped a flagged fairway buoy, while the rest played leap-frog, passing ahead and continuing on their course.

It was as if they were threading one needle with six strands of awkward twine. The power of the broad river seemed determined to deny their progress: little whirlpools became engulfing tides: a boat could be swept sideways, between one oar-stroke and another, and forced off its course into the river-bank, while the coxswain shouted: 'Row, damn you! Bend your backs!' and the oarsmen, as sullen in labour as a woman with her seventh child, stared heavenwards and cursed the day when they had said, whether to priest or Press-Master, 'I will.'

All had to be done swiftly: there were other enemies beside a spiteful current and an unknown course. The boats, in single file, crawling like dung-beetles with the weight of the world as their burden, were a ready target for those on shore. There was musket-fire from a dozen vantage points, and sometimes a volley of cannon-shot which hissed over their heads before, being over-aimed, it plunged into the river.

As yet the gunners on the bank could not depress their weapons deep enough to find a target. But for the rowers there might come a moment when their own heads would hiss, and plunge, and sink.

In the bows of Matthew's long-boat, the spear-point of this enterprise, the leadsman called continually, as fast as he could lower his line and reel in: 'Three fathom and shelving! By the deep, four! No bottom! Dry rock to starboard!' James Cook himself steered, nursing the tiller like the arm of a favourite child, while Matthew at his side marked the chart as he was ordered, and wondered at the intricate secrets of the world, and the skill of men to surmount them.

The sailors arched their backs under the hot sun, and gasped, and toiled without ceasing. They could not even see where they were going. They were simple brawn, and eyeless. Others were brain and sight. They cursed the difference—and prayed that it might be wide and wise.

Thus the whole day passed, with no respite from pulling and hauling, taking soundings exact enough to satisfy James Cook's love of perfection, heaving over the side the anchors which secured the buoys, and seeing that these floated free on the shortest possible length of chain.

But it was deep bliss for exhausted men to turn for home at last, and to float down-stream once more with scarcely a touch on the oars: especially to move through the cool dusk past the evidence of their long toil—fifteen captive buoys topped by their coloured flags, securely set in a deep channel leading through the Traverse.

Even the master, re-living a day's work which had borne such fruit, was satisfied. 'We have done well,' he said to Matthew, as he examined the chart against each buoy as it was passed. 'That is more than half our course plotted . . .' Then, with housewifely care, he added: 'Roll the chart, and cover it carefully. Admirals do not care for dog-ears.'

Next day, new crews for the boats, new buoys for the second half of the journey, and—at Cook's request—more soldiers for their guarding. 'The French will wake up tomorrow, and see what we are after,' he told Captain Wheelock of the *Pembroke*. 'They will not raise their hats as we pass by.'

Nor did they: they raised instead an unholy rumpus. As the

buoy-layers moved up-stream again to complete their task, the musket fire from the shore increased, volley by volley, and found its mark in wounded men and splintered oars. One French cannoneer at least discovered the range, scoring a hit on a long-boat.

Its crew was sucked out of the ruined stern and, if they were lucky, rescued by the next boat in line. But by noon-time the frail convoy came under the lee of Ile d'Orleans, the last island to be passed before they reached their goal which was the pool of Quebec, and in safety they planted their final markers. Now the sea-way was open.

They could be satisfied as they prepared to take their ease on the island's beach before returning home; they could also be surprised in the same posture. It had been thought that the Ile d'Orleans was deserted, and by the French it was, the *habitants* having moved across the river to the shelter of the Heights of Quebec. But it presently appeared that they had left behind certain warriors, a fighting rear-guard who were their Algonquin allies.

Matthew's boat, and two others with it, had scarcely grounded on the shallows of the island before the blow was struck. It came to them as a strangled scream, breaking their companionable silence; it came to their own leadsman who had done such valiant work all day long; and it came in the guise of death, as a toppled body with an arrow through its throat.

Cook, who had been intent on his chart, looked up, saw the murder not six feet away, and shouted 'On guard!' The corporal of marines who commanded their six marksmen bellowed: 'Load! Load and prime!' Then they all turned landwards as a sudden howl assailed them, and a band of Indians in their fiercest war-paint hurled themselves across the beach and into the shallows of the river.

It was swift, brutal, and violent. Though the marines were steady, and fired a volley at the advancing foe, it could not stop such a weight of attack. Hands were laid upon the boats, to be chopped at with sailors' knives and soldiers' musket-butts: tomahawks and clubs flailed down on bare heads and shoulders: all along the tumultuous beach the tide of blood and fury threatened to engulf them.

Matthew, laying about him with a heavy stern-oar which was his only weapon, thought: 'Is this how we are to die? Is this how the master will die?'

Then, as suddenly as it had begun, it ended. The marine corporal, who had a pistol, discharged it at the leading warrior and,
being at close quarters, carried off half his chest. He must have
been their chief, for his men turned aghast and faltered at the fearful sight. Then, cool as ever, the corporal lit two grenades, counted
to five, and threw them into the heart of the mob. As they burst,
screams of agony arose: the screams became panic, and the panic
flight.

Within a minute the boats lay alone again, and the only sounds
were the groans of wounded men and the bubbling death-throes of
the attackers as the shallows drowned them.

'Back away!' Cook commanded. He was gasping, having taken a
heavy clubbing on his shoulder. The oars bit into the bloody froth of
the river as the little fleet, separating from a treacherous land,
drifted down stream into safety.

So, on that second, murderous dusk, they bore their dead and
wounded home. But it could not be a mourning day, in spite of all.
Their work was done, and the great Traverse was subdued.

For sailors, it was indeed subdued. On the morrow, which was
the tenth day of June, the *Pembroke* and the rest of her squadron
sailed majestically through, on a fine easterly wind which drove
them surging past all hazards to an anchorage below Quebec. The
little whispering fish of which Matthew had spoken were sent down
to summon the main fleet; and within two weeks they were all
assembled, without the loss of a single ship, or a scrape of paint or a
trace of mud on the keel.

The newcomers were majestic also: nine ships-of-the-line, led by
the *Neptune* of ninety guns: thirteen frigates: and one hundred and
nineteen transports carrying the soldiery and the stores. With the
first path-finders, who had reason to be proud of their skill, they all
lay at anchor in the pool below Quebec, or behind the western tip
of the Ile d'Orleans.

There they waited. Men and stores were landed on the island:
the artillery set up their batteries: tented camps sprouted like
mushrooms. Out of the holds came the pieces of the landing-craft, to
be put together by ships' carpenters who grumbled at this tinker's
task. All was ready: the huge enterprise stood tip-toe on the brink
of its launching.

But there they waited still, like maidens at the church door, like
blind men on a cross-way, like babes still-born. Presently it became

clear that, for all this great huzzah, the city of Quebec was un-
assailable.

It was a problem for the soldiers, as sceptic sailors were not slow to
point out. They themselves had done all that could be asked of them,
and more. The military had been carried to Quebec most tenderly,
like children to a dame-school: there they had been stood up on the
shore, their beautiful scarlet coats had been dusted, the very sweat
wiped from their brows. Every service had been given them save to
blow their noses.

Now what did they lack? A change of napkins? And where was
their famed General Wolfe, and what was he doing? Sucking a
teat?

Corporal Ned Pym, met on shore when he was guarding a
store-party from the *Pembroke*, answered curtly: 'He takes his time.
And why not? He does not care to lose men. So he is making his
plans. When all is ready, he will move.'

Matthew, with all the confidence of a friend, asked: 'Is there
much poetry read?'

Ned Pym scowled. Though the soldiers had been bearing a
mountain of mockery with stoic fortitude, the regimental pride of
Wolfe's Own was growing tender. 'Matt! Do not provoke me! I
will smash you, else!'

'Pray take your time.'

Certainly General James Wolfe was taking his time. But it was
precious time, and running out like the sand as the summer, his
only ally, advanced and then began to fade. The weeks passed, and
stretched to months. June turned to July, then to August, then to
the first week of September. The season of attack was faltering to
its close without a gainful stroke to show for it, and after that, ice
would do the work of men, and throttle the invaders where they
impotently sat.

Wolfe delayed—the worst misfortune for a commander at the
spear-head of a turbulent, avid mass of men—because he could not
find an answer to a soldier's puzzle which had plagued him for
eleven tormenting weeks. The citadel of Quebec sat upon a rampart of rock. It had been fortified by cunning French engineers
who might have taught Troy to withstand its siege until the crack
of doom.

Its protective lines were alive with troops: its guns were sited to command east and west, north and south: it had food in plenty, and pure water springs and underground cisterns to last until the next snows fell. It had also a noble commander—Louis Joseph, Marquis de Montcalm Gezan de Saint Veran—who, as a Governor of French Canada scandalized by this blasphemy of his empire, would never submit to such a common Anglo-Saxon rabble till his own head fell on the block.

For James Wolfe, there was no way to scale these forbidding, insolent heights. He might thrust his troops forward as a battering-ram, but this would be blunted in such bloody chaos that it would need a team of butchers to sort the meat from the bone. He could not use men like that . . . Then, on a day which was the tenth of September—fearfully late, in a season of mortal doubt—General Wolfe, a remote figure, came touching-close, and remained so for ever.

The party of fifteen cautious men who filtered one by one through the trees onto the heights near Point Levis, across the river from the western outskirts of Quebec, were all dressed alike. They wore the dull grey working coats of private soldiers, so that, if they were sighted by the spying eyes opposite, they would pass for a patrol or a team of wood-gatherers.

But under this drab disguise there was some quite different finery, and under the finery some most un-private men. They were General James Wolfe and five of his aides: Admiral Saunders, commander of the British fleet, and five of his: and Captain Wheelock of the *Pembroke*, the first ship to scale this watery ladder. In addition, Wheelock had brought his master, and James Cook his master's mate.

Of all the fifteen men, there could be no doubt which was their leader.

Wolfe, at thirty-two, was wan, slight, and undeniably frail. He had perhaps the strangest face ever worn by a Major-General: chinless, meagre of forehead, coming to a little ferrety peak with the nose. For weeks he had been wretchedly ill, and still was. He might indeed have been a pale poet himself, in the twin throes of composition and consumption. But none of this mattered a jot.

He was a valiant leader, and all who met him, all who served him, and all who derided him knew it, in their hearts and minds. At some magic moment, God had laid a finger on his brow, and proclaimed: 'This is my beloved warrior, in whom I am well pleased.' It was never disputed—not in the past, and not in this present, when as a general officer he stood motionless before an enemy he could not crack.

He was a commander to whom legends attached like ribbons round a may-pole. The very latest of these was still going the rounds of the regiments, the camps, and the ships. It seemed, according to his staff officers, that he had fallen in love with yet another poem, mournfully called *Elegy Written in a Country Churchyard*, by one Thomas Gray. It was newly published, newly admired by Wolfe, and newly read by him whenever the chance occurred.

One evening on their westward journey, according to this tale, he had read aloud a stanza which ended on a most forlorn note: 'The paths of glory lead but to the grave.' Then he had closed the book, looked round the silent company, and said:

'Gentlemen, I would rather have written that than capture Quebec.'

It was not what his officers had expected to hear, nor wished to. They had come near four thousand miles for that very purpose— Quebec—and a poem, however fine, with burial promised at its end, did not seem a profitable alternative to soldiers marked for battle. Then a young ensign, who must surely have merited promotion, broke an uneasy silence:

'Sir,' he spoke up boldly, 'we are sure you will do both.'

There was a breath-catching pause. Then, to an audience much relieved, Wolfe answered with a smile: 'That, young man, is my intention.'

Now he stood on the windy heights, this small egg-shell of a man, gazing across at that same Quebec, so near to his grasp and so far from it; and they all watched him, and when he spoke they all listened.

He was speaking, as so often, of time and place and action.

'We are here,' he said, within the earshot of all, whether lordly admiral or eager young military sprig—or of Matthew Lawe, so lately promoted from the oar to the black lead pencil, 'we are here because it is the tenth day of September, because this is the last place left for us to try, because *there*—' he pointed across the river, 'is the city we must take unless we are to return in failure, with our duty *not* done. I pray you, all of you, to consider what that means—particularly the word failure, which does not belong in my vocabulary, any more than it is a favourite of our admiral's.'

He smiled briefly towards Admiral Saunders. At this high level, there was nothing of rancour betwen the two arms of service, the naval and the military; nothing save mutual regard and mutual trust. Wolfe had spoken of '*our* admiral' because this was the core of his brotherhood and his belief. Then he continued: 'We are here to find the last way to Quebec. We have tried all places else, and been rebuffed. If it is not to be found here, then it is nowhere. Let us look.'

He suited his action to the word, raised his spy-glass, and stared across the river. All followed his lead, whether they were men of discernment or others avid to learn. There were none present who had not earned their place. Wolfe did not promote courtiers; Admiral Saunders had no liking for young officers who said 'Yes, sir' because this was thought to be the currency of advancement. It was not.

What they were staring at was a cliff, a great bluff of bush and rock leading to the heights near Quebec which they had already learned were called the Plains of Abraham. This was a bare plateau for the military—or, according to other tastes, a pleasant park for honest *bourgeois*, a trysting-place for night-time lovers, and a wilderness in winter. Beyond it lay the city, and below it, this giant staircase. Could it be scaled?

They all knew that, to find out, they must cross the river and try. They were there to discover if such a mad enterprise were possible. On this fair September morning, among confident men, it might seem a prosperous scheme; but there was none among this elect company of watchers who had not learned the difference between such sunshine courage and the dull, brutish, bloody endurance which could alone turn dream into reality.

There would be no soft passage for any of them. Leaders of men, staff officers, *generals*, might be the first to be killed.

General Wolfe said suddenly: 'Gentlemen, there seems to be a path.' He said it without excitement, in the same level tone as he might have observed: 'There seems to be cloud coming.' But then he was quick to lead, and all others to follow. He commanded the searchers: 'Take a line of sight from the church spire on the horizon. Come down to the cliff edge. A little to the left, there is a clump of trees.' They found the clump of trees. 'Now, below that there is a track. Or it seems to be a track. At least it is worn bare. *It has been used by men.* Do you not see? It goes all the way down to the water's edge!' He dropped his spy-glass, and looked round him. 'If that path has been used by men, even agile Indians, it can be used by us.'

One of his aides, of that valuable kind which keeps its feet on the ground and its head free of the clouds, said: 'Sir, it might be a dry water-course.'

'No,' his general answered. 'It takes too many turns. It was men made that path, not tumbling water. And they were climbing from the river's edge to the heights above.'

'It is a great climb,' said the admiral. 'Do we know how high?'

His brother-sailor, Captain Wheelock, answered: 'We can measure, here and now. Master Cook!'

James Cook, who was not present for the pleasure of the sunshine, signed to Matthew, who brought forward his treasured backstaff. Cook sighted it: one arm on the base of the cliff, the other on its peak. He knew already the exact breadth of the river. Swiftly he drew his triangle, and made his calculation. Then he said to Wheelock, and to the listening general:

'Sir, some three hundred forty feet. No more.'

There was silence as this harsh fact was digested. Three hundred and forty feet up a goat-track clinging to a cliff might be no hindrance to that agile Indian whom General Wolfe had cited. But for plodding foot-soldiers with back-packs and muskets? For their ammunition and supplies? For cannon to cover their advance?

Only one voice could break the silence. It came in the cool tones of command: 'I have it in mind to scale that cliff.'

Only one voice would dare to dissuade him. Admiral Saunders said: 'It is a fearful risk.'

'I agree.' From private strength Wolfe could always listen: from greater vision he could answer. 'But there is a worse risk in standing still—we lose the whole enterprise! Look again at the empty Plains and the clump of trees. Where is their defence? The trees may hide a guard-hut or a tent or two, but beyond that there is nothing. Once we are up, we can stroll to Quebec for our Sunday promenade!'

'Once we are up.' Admiral Saunders was not opposing for opposition's sake: he was no more than a sensible man with his sums to do, as James Cook had done his own for all of them. 'How many men?'

Wolfe answered: 'Some five thousand. We must match the French.'

'Five thousand means one hundred of my landing-craft. There will be guns also?'

'Two batteries. Let us say, twenty cannon.'

'Ten more boats. And supplies?'

'Supplies may wait. We are not setting up home for the winter—we are attacking!' Suddenly Wolfe was all a-fire; the pale face took on the true cast of its inmost spirit. 'Sir Charles, I know your mind,' he told Saunders, 'but weigh the prospect fairly. This is our only way. In all others we have made no progress. We have at the most a month before we must leave, or be iced in and starved. I tell you plainly—if *you* must leave in a month, *I* must climb now!'

'And the plan?'

'We cross in force, at night. We scale. We haul up our guns. At dawn we advance. We do or die!'

Every word of Wolfe's ringing challenge was to come true, save for the last of the smallest, where 'Do or die' was translated by malignant fate into 'Do *and* die.'

We cross in force, at night, he had said.

The Armada of small landing-craft which set out, in the deepening dusk of the night of September 13th, had not gathered like a flock of sheep round a chance gap in a hedge-row. With sailors' skill and soldiers' discipline, both men and boats—5,000 of the one and 125 of the other—had been marshalled in darkness, and hidden by daylight behind the Ile d'Orleans, for three successive days.

Each boat was driven by ten pairs of oars, muffled with cotton-waste soaked in tallow: each boat carried two cannon, their polished snouts dulled by lamp-black, or fifty soldiers with muskets clasped between their knees, sitting in rows like men at a play.

Now their play was to commence. But first it must be in the black of night and the mime of silence. Silence was all, secrecy was paramount. If either were betrayed, this hair-raising assault, the like of which had never yet been attempted by British arms, would end at daybreak in blood, misery, and rout.

In the first boat, proudly styled *Pembroke*'s Number One, was a cargo equally proud: James Cook the most trusted pilot of all, Colonel Howard of Wolfe's Own, Captain Wheelock who would serve as beach-master as soon as the fleet landed; and a pale figure in a boat-cloak, shivering with fever, who was the god of this enterprise.

Wolfe would lead them, sick or well, in darkness or in daylight, in cruel times or easy triumph; and his armed guard of forty picked men loved him for it, and wished no other in the world.

Hunched in the stern-sheets, Matthew steered—to glory or to ruin—under the narrow eye of the master; in the bows Corporal Ned Pym crouched his great bulk and nursed his musket. He was to be the first man ashore, the first to find if there were guards posted at the water-level, the first to try the unknown.

Following *Pembroke*'s Number One was a single long file of boats, pushing out from the shallows, presently stretching across the river like a snake closing in to strike: all rowing in utter silence towards the towering darkness of the cliff. There were no voices, no lights, no rasp of steel on steel, no gift to the enemy. The moon was down. Only the black river knew their passing as they crept in under the shadow of the Plains of Abraham.

We scale, he had said.

The moment that his boat grounded, Ned Pym leapt ashore— and into blank silence. His footfall on the sand was the only sound, and his swift advance the only movement. After a long minute's sweaty wrestling with the fiends of doubt, the corporal—and with him a whole army—had their answer. The beach was not guarded, and the cliff above awaited them.

He returned to the boat, and whispered: 'Out! Follow me! Not a word or a sound!' As the forty men of the general's guard slid over the gunwale and waded five steps ashore, the second boat of their Armada came in alongside the first. Soon the whole beach was full of craft, and moving shadows, and well-drilled men forming their platoons. When the first five hundred soldiers, the advance guard, were marshalled, with ten boats' crews laden with ropes for hauling the cannon, the climb began.

The fearful risk of which Admiral Saunders had spoken became true fact, and more fearful than ever.

Corporal Pym and ten men led the way, with General Wolfe, whose keen eyes had discerned the pathway, close at his back. It was pitiless work: for some, the hardest of their lives. Step by labouring step they lifted themselves skywards, as if by their own boot-straps.

Now there *was* some noise, which could not be avoided. Stones dislodged tumbled downwards, falling on other stones—and sometimes upon the heads of men who, with the best will in the world, could not withhold a grunt of pain. The breathing of heavy-laden soldiers and sailors often came near to groaning. Whispered consultation marked those moments of doubt when the path was missed, and a turn had to be made.

Their only guide was pricking star-light, and their only goal the sky itself, and the dark lip of the hill so far above them that it might have joined hands with that same sky.

One hundred feet, two hundred, three hundred: they toiled forever upwards like men doomed to an endless labour of Hercules, like acrobats, like sleep-walkers, like goats. Each shared the same hazard, the same crushing labour, whether he was a dumb ox of a grenadier or a frail general with a dream of conquest within his thudding heart. Matthew Lawe toiled: round his neck was ten fathom of coiled rope, and in each hand a pair of iron grappling-hooks. It was a burden for a giant—or a master's mate.

When he reached the top, astern of the advance guard, he was exhausted beyond all endurance. His sobbing breath could not have summoned a pale curate's 'Amen'. But as he stood at last on the Plains of Abraham, among the trees which had been their marker for the climb, it was to find that a small battle had been joined and won already.

The victor was Corporal Pym and his squadron of ten. They were resting briefly from their labours, but at their feet were three French soldiers who would rest eternally—the men from the guard-post, surprised, overrun, clubbed or strangled, dead to this night and all other nights.

In the darkness, triumphant Ned Pym's first salutation was a monstrous stroke on the shoulder, and a whispered: 'Well met, Matthew! Now we show *you* the way!'

Matthew peered at his friend, and then at the crumpled bodies staring eyeless at a night sky already peopled by their spirits. He whispered back: 'How did it go?'

'Easy as a butter slide.' Corporal Pym had routed the whole French army at one blow, and was not slow to tell it. 'Two were asleep in the tent. One was eating his supper outside. Frogs' guts,

for all I know. His back was to the river, and to me. A jealous woman could not have wished a better target! Then we scratched on
the tent wall and called "*En garde!*" which is the French tongue for
"On guard!" and the two came running out, and they were
frogs' guts themselves before they could say Bon Jew!'

Ned Pym was clearly at his peak of triumph, and might have run
on for ever. But a shrewish officer's voice hissed 'Silence!' and
added, to their great delight: 'Corporal! Take their names!' Then
silence it was, as men from below began to stumble towards the
trees, in their tens and hundreds, and Wolfe's Own became, not a
straggling line of blind mountaineers but an army massing on the
field of battle.

We haul up our guns.

This was work for the sailors, who were better skilled at lifting
things aloft than at stamping the earth with soldiers' beetle-crushing
boots. There were twenty of these guns: they must be hoisted three
hundred feet, from the water's edge to the level plain above, up a
goat-track which wandered through scrub and trees, round stony
corners, past overhanging rocks, onwards and upwards as if on
some mad steeplechase whose last steeple was the one crowning the
roof of the world.

Matthew Lawe, with his grappling irons and coils of rope, was
the anchor-man at the top; below him, James Cook—a master-
rigger who could also solve the puzzles of the universal stars—
calculated the corners, the turns and twists of the path, the linked
machinery which could draw a gun past one hazard, and be released, and lowered again for the next arrival.

Much of it was rope-work, donkey-work, sweating pull-and-haul;
all of it was groaning labour, with aching muscle set against stubborn weight, the law of gravity against the will to conquer it, the
slaves against the Pharaohs. The guns had wheels, but they could
not run free on this rocky traverse; the wheels had spokes, and
these at least could be turned—by hands and arms which cursed
the day the wheel had been invented.

Yet one by one the guns went up, and topped the crest of the
cliff, and took their place in twin batteries whose target was the
fortress of Quebec and the enemy's protective lines outside.
Matthew's realm of haulage was the last thirty feet, which could
only be overcome by iron claws biting into trees, and ropes and
pulleys on which the lives of the men toiling below depended, and a

final thrust upwards which proved—which *must* prove—that human sinew was stronger than the weapons it had fashioned, and blind endeavour a match for ugly, mutinous, immovable lumps of iron.

Two hours before dawn the task was done. The guns stood in their ranks like dumb soldiers waiting for that touch of magic—the wand of the sun—which would bring them to life. Matthew, his duty discharged, coiled his ropes, collected the grappling-hooks of all the hoisting parties, counted them, and set them on one side. In the past, there had always been a harsh man who mustered such stores; now it was himself.

He was dog-tired twice over: once by the ascent, then by the hauling of the guns. He wished that he might sleep for ever, but who could sleep, on the edge of such a battle, within touching distance of the dawn? Bemused, exhausted, he wandered towards the gun battery, and found himself face to face, in the star-lit darkness, with a man who by his bearing and half-seen dress, by his air of busyness, could only be an officer.

'Sir,' he said, when the man had a moment, 'who is it pulls these guns, in open country?'

The officer was tired and testy, with no time for stupid sailors' questions. 'Why—horses! What else?'

'Thank 'ee, sir.'

Then silence fell on the camp. No fires could be lighted. Men dozed with their cold bellies to the hard earth, or stared wakeful at the stars, thinking of the battle which was almost upon them. They had all come so far, soldiers and sailors alike: three thousand miles across a salt ocean, two hundred miles up a river, one mile across it, and three hundred feet up the cliffs of Abraham.

Now they were all *here*. But now it was for the soldiers alone.

At dawn we advance.

At the first hint of that dawn, the soldiers stirred, whether from brief sleep or the canker-worm of wakefulness. The obliging sun, still fathoms deep below the eastern horizon, showed them their goal, the distant lofty towers of the city of Quebec. On the level ground between, a mist still lingered: an autumn morning mist, coming up from the river below, promising them cover and secrecy at the moment when it was most needed.

Under its shelter, they moved forward a full half-mile, and their guns with them: drawn now not by horses or sailors, but by heavy dragoons who had only to lean against the traces in order to move their burdens. There they paused, and set up their batteries, and formed their ranks.

General Wolfe, pale as the dawn, eager and spirited as the youngest unblooded recruit, was now the man most to be seen and most to be obeyed. He had a new plan of battle, and it took shape under his able command: it was intended, as with most true generals, to add strength to his assault and also to spare the lives of his men.

It was to form up in two ranks, instead of a serried mass of infantry which no enemy marksman could miss: two ranks only, the first shoulder to shoulder, the second spaced so that they could fire between the necks of their comrades.

He had named it, to his officers, as his 'thin red line', and now, on the Plains of Abraham, it became real, replacing a commander's dream.

While these ranks formed in their six battalions, and moved about, and settled, the light was growing: still misty, always mysterious, but firmly established as another day. Presently it brought them a cold sunrise, that hardest moment for courage in all the world of fear. As daylight gained and the mist melted, the gunners sighted their cannon at the enemy lines before the city.

James Wolfe, a master of patience and perfection, waited until he was fully satisfied with the light and with his dispositions. Then he took his place, as a fighting general should, on the right of the line and in front of his troops. As the last of the mist vanished, he gave his signal.

The battery commander called 'Fire!', and Wolfe himself, raising his sword, called 'Advance!'

As the cannon boomed and roared, announcing this fearful visitation, the thin red line began to walk and then to trot. The city of Quebec awoke to its dawn of terror: the sight of five thousand red-coated men advancing on them, where no men ought to be.

There was one short mile between them.

We do or die!

Soldiers started to fall, and they were Wolfe's Own; yet more soldiers, of another colour, started to run, and they were Frenchmen, surprised, appalled by their danger, ready to be routed. As the weight of the British army began to press, they fell back from their protecting lines and made for the only gate open, the western portal of the citadel.

But they did not pass through. The Marquis de Montcalm, wakened from simple sleep by a wild alarm, collected his wits, remembered his nobility, summoned his undoubted bravery, and sallied out to stiffen his fleeing compatriots and to stem the foe. It was a mistake: better by far to have let the runners in, slammed shut the gate, and waited for winter itself to starve the British force and trap the British fleet.

But for once in his life, Montcalm was all noble courage and no clear thought. It was to cost him that life. It was also the turn of the battle.

The far-away watchers—the camp-followers and those staff officers left behind to plan, if need be, for a second stroke—saw amid the clamour and billowing smoke of battle swift pictures of victory. Soon it was red coats everywhere, a curling wave which broke over the blue and tossed it back against the rocky beach of the citadel walls. Now the city gate *was* shut—but caught outside were its main defenders, steadily chased and harried and cut down. For them there was no escape from brutal bloody defeat.

Presently men, red-coated men, began to drop from the battle-field, like worn-out slaves who were beyond the lash of punishment. They straggled back the way they had come: limping, swaying, holding on to each other, pausing to staunch their wounds or perhaps to die. Some were borne on hurdles, some on the broad backs of comrades.

They were the victors who had paid much, and sometimes all.

There was a litter coming slowly towards the men in the trees, with four bearers as tender as children. Ominously, it seemed to be set apart. Fatally, it was preceded by a drum-major who beat a slow, mournful roll as he marched. It was a rhythm which Matthew had never heard before: a sort of whisper on the drum-head, then a single solid beat, then another whisper, then a beat, all repeated at a pace so dolorous that marchers could take eight steps between one heavy thud and the next.

It was Colonel Howard who broke a fearful silence. 'Sweet Jesus!' he murmured—and for once this soldier's oath was a prayer from a stricken heart. 'He beats "The Warrior Home".'

Wolfe's shattered body was laid gently down at the edge of the trees. His splendid uniform—a red coat lined with blue satin, and white knee breeches—was a torn ruin of grime and gore. Three gross wounds to head and chest bespoke the agony of this frail remnant of man. His face was beyond paleness, sinking down to a greenish mask of death.

On the instant, Matthew was surrounded by anguished, weeping men. Soldiers gave one glance to the litter, and turned aside in speechless grief. They were all Wolfe's Own, and suddenly they were all orphaned of a loving father.

'Soft toads,' Matthew thought unworthily, and found his own cheeks wet.

Colonel Howard, his face working, bent to the litter which could only be a funeral bier. 'They run, sir!' he said—he almost shouted —striving to pierce the most terrible deafness of all. 'See how they run!'

James Wolfe did not see how they ran. He had already closed his mortal eyes. But the soldier who spoke to him lived out his life in the hope, and then the holy belief, that his general had heard.

Preparing to drop down river in the dusk of the year, with the sailors' work all done and the soldiers triumphant, James Cook had news of great moment to impart to his master's mate.

'I am bound for a new ship,' he told Matthew. He spoke in the quiet of the quarter-deck, when the first watch had been set and the *Pembroke*, half alert, half sleepy, was ready for the night. 'She is the *Northumberland*, the flag-ship of our squadron. We winter in Halifax again, and then I am set to survey the coast for the next season. Perhaps longer. They require every league and fathom, from Newfoundland to New England, examined and described, and that is my task.'

After a moment Matthew asked: 'Sir, who is "they", that want this survey?'

'Great ministers of state in England. But do not ask me more, for I do not know. Sailors do as they are bid. If we do it well, we escape censure. If not, we bear the blame, and starve.' But he did not sound too downcast. He would do it well, and he knew it. 'If our masters require charts of half the American coast-line, that is what will be provided.'

Matthew waited. There was no pleasure to match the confidence of a great commander, and this, on this easy night, was his.

'For myself,' said James Cook, 'Admiral Saunders speaks of a commission. I might be captain soon!' The dry sailor with the hawkish nose was fired with the promise of the future. Fortunate man, Matthew thought, with envy and admiration mixed: he follows his star, and it has only begun to climb the sky. But the next words were still an astonishment. 'And you? What would you be, in a perfect world?'

Matthew was not proud. He chose the lowest rank of officer. 'A midshipman?'

'Why not?' Cook echoed a phrase of long ago: 'A man can rise if he wills.' But who had first said this? Old Sam Pepys? Old Drake? Old God? All the turning universe was old . . . 'If you have that will, then I can carry you with me. You should stay with me, Matthew. I shall not always be on this coast, sniffing out the St Lawrence or the tides of Plymouth Rock.' From a great desire, he spoke a great ambition: 'I can warrant you this: as soon as I am set, I plan to survey the world!'

IV

In twenty years he had surveyed that world, from the cold Russian seas to the gentle South Pacific: from that North American coast (where he spent eight full years) to the ice-line of the Antarctic: from Cape Horn to the Sandwich Islands, Bering Strait to Botany Bay, Plymouth Sound to the Great Barrier Reef.

He had measured all, and charted it, and tied it to the sun and the moon and the stars, with exact bonds which were never to be matched in his own century. Orion's Belt was his girdle, and Sirius the Dog-star his faithful guide, and the Southern Cross his altar-piece. Beneath their friendly gaze, land at last took shape—not on the face of the globe where it had always lain, but on countless treasured maps which were its faithful reflection.

His instruments were mechanical, and marvellous. He knew what he needed from the scientists, the mathematicians, the clock-makers: he asked for it, and received it, and used it to his own perfection. From one such gifted man he won a sextant, much improved, for measuring the celestial sky: a true son of the octant which was Isaac Newton's, and the great-grand-child of the ancient astrolabe itself.

From another came tables of time and tide which could at last be trusted; and from a third a compass better by far than anything yet devised, whose bowl floated in a bath of whale-oil to protect it against the shocks of the sea.

But the greatest gift of all was given by his Yorkshire friend John Harrison, the village carpenter turned clock-maker who laboured a lifetime to make, for mariners, a trustworthy seafaring clock.

The worst enemy of such a time-piece, apart from the movement of the ship, was the variation of temperature as it was carried from one end of the world to the other: from the Arctic to the tropics, through the cold of night and the heat of the day. John Harrison, who had made his change from woodwork to metal-blending as simply as a child progressed from a crawl to a walk, solved it after long experiment: his answer was to balance this variation by the use of different metals which would, in their sum, annul the rigours of heat and cold.

At his fourth attempt, he fashioned a large watch which could make such compensations. James Cook took a copy of this marvel on his second great voyage. It served to such effect that Cook, returning to Plymouth after a three-year voyage round the world, found its error in longitude to be less than eight miles—or one-tenth of one second *per diem*.

This was triumph—and freedom at last. At sea, the measurement of latitude—the distance north or south of the equator—had been easy, since the days of that first astrolabe. It was governed by the noon-day sun. But longitude—the east and west measurement—had long eluded all mariners.

Now, with Harrison's clock, which James Cook called 'my trusty watch', it had been solved. At any moment, by night or day, in foul weather or fair, he knew the exact time at the meridian line which ran through Greenwich Observatory in London: at that same moment, by star measurement, he knew the time in the place where he sat, whether this was ten miles or ten thousand miles from home.

Then he put the two together, and made a simple calculation. Since twenty-four hours was the sun's circle of the earth, one hour of difference, east or west, measured fifteen degrees on the face of the globe. Fifteen degrees was some 900 sea-miles at the equator, and less as the ship moved north or south of it. But by how much? Here the tables gave the answer—and here was longitude at last!

From that moment forward, each mile of heaving surging sea was a measurable square on a chart. Man need no longer go in search of a coast, or in fear of it. He knew his patch of ocean, and he knew his next plot of land.

Tapping his finger on such a chart, Cook could say: 'I am *here*. The land is *there*. If God holds the wind, I shall reach that coast at noon on Christmas Day.'

His other instruments were not mechanical, but marvellous also: his eye, his brain, his fierce genius for challenging the sea and wringing victory from it. He was a sea-commander first; if a ship would be safer in any other hands, then (his men believed) they could only be God's.

He was also a man of science and inquiry. He was able to draw towards him, and then to command, any like-minded voyager—a mathematician, a naturalist, a hydrographer, an astronomer, a surgeon—who would recognize in him a humane and resolute leader, just as did his common sailors who knew that when Captain Cook said: 'The starboard clew-line on the main is chafed—reeve another!' his keen eyes had found the flaw before anyone else had woken to the fact that it might put their lives at risk.

He cared for these his sailors in other ways. Principally, he was determined to find ways of ridding his ships of those sea-diseases which seemed to have settled on the Navy like a curse. The West Indies station, for example—so fair, so benign, so blessed by God with sun and fine sand and coral-clear water—had a record of distress and of wasted men so appalling that it had become a national scandal, well concealed by lazy or indifferent officers, but now a-flame.

In a single Caribbean year, while 59 men had died of battle-wounds, 715 had perished of disease on board the squadron's ships, and a further 862 in hospital ashore. The bald and cruel truth was that every seventh man who joined one of His Majesty's ships-of-war could expect to die, not by any exertion of the enemy, but of ship-fever (which was typhus), scurvy, or dysentery.

James Cook would have none of such self-inflicted slaughter . . . He began, first of all, to cleanse his ships and to keep them so. The middle-decks were smoked out with gun-powder mixed with vinegar, the noisome lower holds made sweet (or sweeter than poison) by fire-pots burning deep within the hull.

He opened gun-ports whenever this was safe; he designed, fashioned, and installed certain vents which would admit air on one side of the ship and suck it out of the other. He rigged windscoops of canvas, to press a fresh breeze below. He gave his men the means of drying their clothes as soon as they came off watch—and made them do so.

Then he turned his attention to those strange diseases of the sea which, on a long voyage, might strike down enough men to bring the whole enterprise to ruin. Of these, scurvy was the primal curse, the Killer Maritime.

The symptoms of this scourge of the ocean grew so horrible and so clear that many a ship-mate, at the first light of day, feared to look into a comrade's face, in terror lest one or other of them might have been blighted.

It would begin with weakness—a deadly lassitude which made a man unable to stand, unable to walk, unable to work: so invariable was this feebleness that for a long time it was said that scurvy was no more than a disease of shirkers, and that lively men with spirit need have no fear of it.

The ancient, scornful cure had been to chase the sufferer from his berth and send him aloft—whence he fell into the sea or upon the harsh deck, and saved himself much misery by this prompt execution.

After lassitude came the onset of disease unmistakable. A foetid breath, enough to stun a bird in flight, was the forerunner of a skin spotted red or black or green, like a many-hued pox, like a leopard which had forsaken its tribe. Then came the moment of appalling truth, when the appetite failed, not from any surfeit but because the gums grew spongy and swollen, and shrank back from the teeth, which presently began to fall and might be found upon the pillow in the morning.

Fluxes of blood from a stomach starved were mocked by inflated, dropsical flesh into which a man might press his thumb and leave a pit as deep as a pipe-bowl. Legs grew so ulcered that they were turned to foul fungus. It was the time to die, amid the fear and hatred of ship-mates still wholesome; and this was signalled, at the last, by a raging fever which consumed the sufferer's wits.

In the delirium of this he would imagine the sea to be a green pasture, and try to walk out on it into blessed freedom. If he were prevented, the wrecked skeleton which had been a man would be thrown upon it within half a day, for this was the last torment of all, not to be survived by any living soul in any ship under the sun.

James Cook, who had witnessed this horror many times, was
determined to find its cure. He had long thought, in common with
naval surgeons far in advance of their age, that a constant supply
of fresh fruit and vegetables, to counter the morbid salt beef and
mouldy bread, must be the answer. If these rarities were not to be
found, save when a ship touched from time to time on a friendly and
fertile coast, then something else must take their place, something
which ships could carry with them.

Summoning all the cunning and authority at his command, he
set to work.

As his first, and second, and third voyages unfolded, strange
foods, mysterious liquors, forever crept into the innocent bill-of-fare.
Words stranger still, such as anti-scorbutic, twisted men's tongues
and baffled their brains. 'Anti-scorbutic' was summoned to excuse a
substance such as sourcrout, which was nothing more than pickled
cabbage, as sharp to the tongue as vinegar kept too long in captivity: well-suited to Germans, no doubt, but for stalwart British
sailors, as heathen as the Pope.

There might follow sticks of celery, tough enough to plug a leak
in an old bucket: a marmalade of carrots which stained the teeth to
a brilliance beyond the sunset; or a dish of boiled weeds which,
even if christened 'scurvy grass', was still boiled weeds.

To wash down this magnificent repast, honest ale from an honest
barrel was no longer offered. In its stead, a concoction of malt and
yeast, boiled to a thick syrup, then watered till it might be forced
past the gullet, was the best to be had; or, on feast days, the rind
and juice of lemons and oranges, mistreated in the same way. Or
the milk of coconuts. *Coconuts!*

The sailors grumbled till the air grew thick. They had never been
called upon to eat such stuff before, nor to drink it. They had never
heard such names on the lips of men. That alone was enough to
damn it all to perdition, and then to turn them stubborn. Had they
been carried tens of thousands of miles, in order to try something
new? A fine dialogue ensued, echoing round the world—which was
more than most scholars could ever boast.

It was delivered, on the one side, from the corner of a mutinous
mouth, and on the other, in crisp command.

From the sailors: 'What is this on our platters? Carrot-marma-
lade? Are we donkeys, to be led by the nose? Boiled grass? Would
he poison us? Sour fermented cabbage—what banquet is this? And

coconut milk is for monkeys! We are men, and we were weaned long ago. Monkey milk? We'll have none of it!'

'Eat it and drink it,' Cook told them, without dwelling on argument. 'It is good for you.'

They grumbled still, but since there was no other choice save short commons and the threat of punishment, they presently came to bear their lot. The hardest tussle of all concerned onions—raw onions. The order had gone out at the beginning of the second voyage: each man was to eat twenty pounds of onions within a week, and half that amount later on.

Twenty pounds of onions? This was truly beyond belief! He must be jesting! Twenty pounds was likely to be eighty onions. Were they expected to work the ship, while stuffed with onions like a duck in an oven?

'Eat them,' Cook again commanded. 'They are good for you.'

And a little later: 'Eat them, and all the rest, or you will be lashed.'

He was a determined man, and he kept his word. One marine who refused his burden of onions, and then turned insolent, received twelve lashes with the promise of more.

The grumbling, like the onions, disappeared promptly.

'Our captain,' said one irreverent surgeon, 'will safeguard his men's health, if they have to lose their lives for it.'

But the answer lay in the happy result, and long before Cook's course was run every man knew it. On that second voyage, with all its hazards and its three-year span, only one man out of two ships' companies died of disease—and he the cook!

The captain had the acclaim of countless grateful sailors; his own were the foremost, but never the last. They had only to remind themselves that another famous commander, Commodore Anson, sailing his squadron round the world some thirty years earlier and achieving great renown on his return, made that return with only one-third of his 960 men left alive.

For James Cook, the prized gold Copley Medal from that most august and critical body of judges, the Royal Society of London, bespoke his matchless exercise of a commander's care. Another expert jury, his own men, awarded their own medallion and thus engraved it in speech: that on board Cook's ships every day was a Sunday—the day of cleanliness and order.

And now, on this last voyage, what wonders he had led those men to see! For Midshipman Lawe they surpassed all the ancient memories of his first great wandering with Drake, in the very springtime of the world.

At the heart of this journey was still the search for that North-West Passage which for ever eluded sailors; and it took them, from Plymouth, round the Cape of Good Hope, across to Tasmania and New Zealand, and through the islands of the Pacific Ocean—one-third of the surface of the globe—to the west coast of North America.

There the search led northwards, to the Bering Straits of Alaschka; and there they were turned back, defeated as ever by the swirling fogs and the ice-walls of that cruel fortress, and found a softer haven in Hawaii.

On their way they had encountered everything: everything that made a sailor's life its mixture of foul and fair, gentle and harsh, lazy and dog-weary, ugly and entrancing, real and dream-like. Off Tenerife they sprang a severe leak, and worked the pumps like galley-slaves till they could put in for repairs. At Table Bay, in the shadow of its brooding flat-topped mountain, Boer and Briton alike gave them a welcome of such magnificence that it seemed a hardship ever to put to sea again.

In Tasman's Land all their faithful marines, sent to guard their landing against the naked aborigines, made themselves dead drunk on stolen wine—and woke to find these same savages tenderly bathing their brows. On New Zealand's coast they gazed at gushing spouts of hot water, and harbours fit to equal Australia's Botany Bay, so christened by Cook, when first he had charted it nearly ten years earlier, for its lush and beautiful vegetation.

They also encountered visitors who, eager for a change of diet, ate any candles and lamp-wicks left unguarded; and complaisant husbands ready to exchange, for a single nail or spike, the freedom of their wives and daughters. They had been taught commerce, James Cook considered, but insufficient morality.

But it was in the strange paradise of the Pacific that they could all revel in a new language of discovery.

Here were coral atolls riotous with taro and yams, edible land crabs, holy-water clams, pearl-oyster beds, beaked rainbow-fish, the soldier crabs of the Marquesas which cleaned the beaches at night, and the fondly trustful turtles which would lay a hundred eggs in the sand at the voyagers' feet before being turned over and clubbed to death.

Here, as if allies in revenge, were lurking fish with poisonous spines, which delighted the naturalists more than any sufferer could appreciate.

Here were people—as in Tahiti—also fondly trustful, who sometimes fared as ill as the turtles: chiefs who would welcome Cook as a god and then be robbed by his sailors: humbler natives whose wives were stolen from them and returned diseased at the hands—or the fouler parts—of these invaders of their Eden.

Here was a world of gentle, abundant, innocent nature: God-given, sun-blessed, man-measured by gifted inquirers, man-soiled by animals of the same breed. Here was all the generous wonder of creation, and its admirers, and its children, and its despoilers.

Here voyaged Captain James Cook, skilful and benevolent explorer, prince of sailors; and here he came at last, in his ship the *Resolution*, to the place of his death.

V

By the time he returned from that cold forbidding northland whose ice-walls had shattered his plans, and with his consort-ship the *Discovery* had anchored off Hawaii, James Cook was growing weary. He was near fifty-one; he had held a world-encircling sea captaincy, with all its burdens and cares, for nine long years without respite; and now his fine and feeling mind was becoming overwhelmed by a voyage which seemed to be stretching to infinity.

Even now, after two years and eight months of the longest journey of his life, he was still half the globe and some fifteen thousand miles from home.

He had squandered his strength and skill on all the multiplicity of command: controlling the progress of two ships of unequal speed and sea-keeping quality, directing two crews of men who were as weary as himself, without any secret dream to inspire them; maintaining his meticulous observation of coasts and unknown islands, and preserving the perfection of his navigating and his search of the heavens.

Below these same heavens his tired *Resolution* was continually leaking, or falling foul of shoals and sand-bars and murderous coral reefs, or losing spars and rigging, and forever demanding new masts and patched-up sails.

When at last he came to anchor, under January skies, in the fine harbour of Kealakekua Bay which his master William Bligh had found for him after a probing search, Cook had little left to sustain his spirit and much to make it bleed away.

There could be no doubt that, for a particular reason, his men were full of discontent, and mutinous at the harshness of a captain who would not let them enjoy this paradise but never ceased urging them to live cleanly and work until they dropped.

'A connection with women I allow, because I cannot prevent it,' he had once told his second lieutenant, James King, a handsome and engaging young man so close to his commander that the island-ers took him to be James Cook's own son. 'But I will prevent the spreading of foulness, by whatever means I can!'

'A connection with women' was, for his subjects, the only meritorious phrase of this decree; the rest was pious fudge. 'Is he jealous?' his men murmured spitefully, when acquainted with their captain's resolve. 'If he does not want a woman, we do! Look at this new book of rules! Does he think we are monks?'

These rules, which Cook set out as soon as they were settled in Hawaii, were as strict as they were clear, and made necessary (as he thought) because of certain excesses on the late northern journey to the coast of Alaschka.

Here they had found Indians and Eskimo, wild uncouth tribes of men and women scarcely distinguishable: short, filthy, smelling of fish-oil and wood-smoke, their faces daubed red or white or black. The men were bow-legged from their long sitting in canoes, the women likewise from God-knew-what endeavours. But to balance these outlandish savages there was a Russian settlement of fur-traders, some from their own home across the Bering Strait, others from as far off as Moscow, who ruled their little empire in the firmest grip that James Cook had ever noted.

They were strange men, and they greeted their visitors from the *Resolution* in a manner not to be foreseen: at times full of roaring good-fellowship while the fierce wine flowed and eternal friendship was sworn, at others darkly suspicious, as if Cook and his sailors were nothing but spies and thieves of their land.

It was soon discovered that these Russians had treated the women of Alaschka with the contempt of overlords, and crowned it with a venereal blessing which was as firmly planted as it was universal. The men of the *Resolution* and the *Discovery* were its unfortunate heirs. All the way south to Hawaii, the surgeons were busy and the men despondent. To take such infection from such unlovely flesh! One might as well drink whale oil because it slipped in so easy!

Yet, being sailors, when they anchored in Kealakekua Bay they were more than ready for further encounters. At least the dusky island women were beautiful and free—and be damned to the fiery foreskin!

Cook was determined that they should not so damage their innocent hosts. He had formed a special affection for these Hawaiian islands; when he had touched there earlier, for a brief two weeks before the bitter northland foray, he had been astonished and moved by his reception. Several hundred islanders had prostrated themselves, face downwards, as his foot touched the beach: he was received as a divine king, and there had been little to mar a fabulous welcome.

Canoes, decorated with the green fronds or white cloths of peace, had daily flocked round the *Resolution*. Trade had been brisk and profitable: hogs, yams, and fruit had poured on board in such generous profusion that a thrifty sailor could complain that he had only been offered two large fish for a single yellow bead. The mouldy, briny contents of their harness-casks were forgotten in a daily surfeit of the food of paradise.

Ashore, feasting and entertainment had followed. On the last night of their stay both had been princely. Huge piles of yams, thirty feet high and crowned with roasted hogs, were ready for them at the edge of the beach. Under flaring yellow torches paddle dances, singing dances, dances which mimed the act of love in surging bursting vitality, delighted all their senses. Stalwart men fought in boxing matches, and women, naked to the waist, wrestled with each other and (so said the lewd among the onlookers) let their best talents go to waste.

In all such things their poor hosts had been generous beyond belief, and when *Resolution* sailed away were struck down with sorrow. Now, on his return, Cook knew in his heart that he could not repay them, nor betray them, with the tainted gifts his men might provide; and thus, as the trustful canoes began once more to crowd their anchorage, he set out his 'book of rules'. They were tailored to saints, and fitted no one else.

No woman was to be allowed to board either ship, for any reason, without the captain's permission. Any sentry who connived at such a boarding would be punished. Any man suffering from the venereal, or suspected of it, who lay with a woman would likewise be punished. No man with this venereal touch would be allowed ashore on any pretext. No man, whether clean or foul, was to be left on shore at night.

The men were scandalized. Had they sailed more than half-way round the world, to take Holy Orders at the end? Was their last taste of amorous joy to be from a squat, square, fat, fish-smelling Alaschkan female, who peeled off her small-clothes of rancid walrus-hide to confer her heavenly charms—and Muscovite gonorrhea?

The island women of Hawaii, rebuffed, insulted, were more angry than any. 'Are we not good enough?' they screamed as they were fended off.

They could never have understood James Cook's gentle mind. It was his own men who were not good enough. It was his own men also who, whenever the chance offered, set these rules at naught, fomenting a spirit of desertion which made fools of their officers and, it seemed, a beleaguered tyrant of their captain.

Their captain had other cares to plague him. Most damaging to his spirit was the thieving of the islanders, which he had noted before but which now grew so bold and impudent that it was like a disease—matching that other disease which so pressed upon his conscience.

Thieving, for these Hawaiian innocents who now flocked to Kealakekua Bay, was a kind of merriment: a deed of daring, a saucy retribution for favours already granted, whether woman, hog, or top-mast tree: and also a jest against these almighty gods whose powers, perhaps, were not absolute. Thus men would slip into the water from their canoes, and sneak on board—sometimes in broad day, sometimes by night: there they would seize anything which lay loose, and dive overboard again, like slippery eels with grasping hands.

The strangest objects were spirited away: a single oar, a nail, a hank of twine; a bolt from the windlass, a cat, a ramrod from the sentry's own musket. Cook's gold watch disappeared from his cabin, but was returned after threats of fierce punishment. Yet on the next night a determined effort was made to cut loose and tow away one of the *Resolution*'s boats. There was no limit to this in-solence, nor any rest from it.

Midshipman Matthew Lawe, who lost the bronze cap of his spy-

glass—useless to anyone but himself—and glumly set it off as misfortune, found greater interest in observing how those on board the
Resolution treated their constant harassment. The men, as always
with those whose own modest property was not at risk, could afford
to be tolerant. Let these Indians steal the ship, if they could! And
good fortune to them!

A few of the officers showed disapproval, especially the younger,
who had never met such ingratitude. 'These are the greatest thieves
in the world,' the youngest midshipman of all announced, dismissing the whole Pacific Ocean from his private atlas. He was in
mourning for his Book of Common Prayer, bound in Suffolk sheep-
skin, a farewell gift from his mother. 'These jolly full-faced fellows—
they are naught but horrid cheats!'

Other officers, like the men they commanded, were more
generous. So the natives were light-fingered? So they were envious?
So they plundered touring visitors? What else was to be expected?
If thieves discovered could laugh it away as a joke, then rich men—
and they were all rich, in this simple world—rich men could laugh
also, and forget.

But there was one among them who, astonishingly, could forget
nothing. The worst effect of all this thieving was upon James Cook
himself. First he grew irritated, then angry; then he fumed, and
then he began to take revenge.

He was angry, first for the affront to his pride, and then because
he loved this ship of resolution which had carried him so far and so
faithfully on two great voyages. He could not bear to see her robbed
. . . He had chosen her himself, and bought her for the Board of
Admiralty: a Whitby cat of 460 tons, bred to the collier's trade,
from that Northumbrian cradle of ships and men which had first
nursed James Cook the apprentice-boy of long ago.

Resolution had been the ideal for what he had in mind: deep-
waisted enough to carry three years' stores, strong enough to take
the ground without damage, sound enough to endure, and armed
with twelve guns for any race of men who might dispute her passage. This fine purchase had been made eight years earlier, and
from that moment on she had served him and the whole world well.

Now, she was not to be plundered and set at naught by men
he had welcomed on board, men he had befriended, doctored,
rescued from storm and famine: men on whom he had already
showered gifts which might benefit their whole lives—bulls, heifers,

horses, mares, sheep, goats, rabbits, poultry: breeding sows, new plants, new trees, new seed, new means of catching fish and growing food.

He would take whatever means were necessary to prevent or discourage such insult.

His officers could not ignore a new severity in their commander, as soon as the men of Kealakekua Bay returned to their old tricks. While thieves who smiled merrily when they were caught were triced up and flogged: while other cheerful malefactors had their heads shaved in derision and then their ears cropped off in brutal punishment, these officers could not hold their tongues. The James Cook of these harsh latter days was scarcely to be recognized.

In the cramped mess-room where they all ate their noon-time meal together—save for the captain, who kept to his own cabin— their talk had the freedom of men who, living side by side for years and for sea-leagues uncounted, had loved and revered their commander without stint, and now saw a man humane no longer, but sullen in spite.

They were a company of seven, the rest being ashore on wood and water parties. They were John Gore the first lieutenant, the oldest officer after Cook himself: James King, the second, that literate and attractive young man believed by romantic islanders to be Cook's heir: John Williamson the third lieutenant, a fiery, raging Irishman much given to violence, who excited only two feelings among his companions—either dislike or hatred: and Midshipman James Trevenen, a lively gifted lad of Cornish blood who worshipped Cook but could still call him 'The Despot'.

Next in rank were George Vancouver, visiting from the *Discovery* on some business of the fleet: another midshipman risen from able seaman, like Matthew Lawe himself, and already a navigator of brilliance, as intent as was James Cook on opening up the undiscovered world: and David Nelson, a gardener of Kew (but a gardener with friends among the great, who had sent him out as a 'botanical collector'): a devoted naturalist who would hardly leave his cherished seeds and plants and flowers to eat a meal of any kind.

Last of all, Midshipman Lawe, now bearing the proud badge of a 'three-voyage' ship-mate of Captain Cook, who had advanced among men and minds of great achievement in a world which could still astonish him.

It astonished him now to hear Lieutenant Williamson the Irish fire-eater, talking of the punishment witnessed that morning, thunder out: 'Let him crop off their ears, I say! He may crop off their privates, for all I care! These natives understand only one language, and if they choose to rob us they should hear it, loud and clear!'

Matthew, knowing his ship-mates, was ready to wager that it was Williamson alone who held to this harsh view, and that all others would disagree. By now it was enough for this Irish ranter to say 'Aye' to anything, for every listener in earshot to answer 'No'—and so it proved.

After a reflective, munching pause—today's fare was roasted piglet, not to be wasted in argument—James King, his senior, put the contrary view:

'Cropping ears and privates? What are we—cannibal kings? This desperate robber we have just seen bleeding and crying was young Mourea, a chief's nephew. We know him. And what did he steal?—the lid off the water-cask! What does a cask-lid matter to us? Reuben the carpenter can make another out of drift-wood, in a half-hour.'

'It was not Mourea's first theft,' Williamson countered harshly. 'We know that also.'

'So?'

Williamson smirked. 'I think it will be his last. And the same for some other worthless rogues too.'

'Then the price is too high.'

Others joined Lieutenant King in his objections. They all agreed, in their own manner, that Cook had borne down too cruelly on a young rascal, freely welcomed on board, who wished to show a trophy or two to his friends. Even David Nelson the gardener, who usually sat wordless—dreaming of propagation, his companions jested—agreed that the price was too high.

'An ear for a cask-lid,' he said in his slow voice. 'What a Jew's bargain! What a revenge! I would not revenge *plants* with blood!'

'Except with your own,' said young Midshipman Trevenen.

Nelson grinned. He owned to his consuming passion. 'That is different.'

George Vancouver, their visitor, said: 'Certainly we are not so stern in the *Discovery*.'

'So what have you lost to these thieves?' Williamson asked contemptuously.

A courteous guest returned a mild answer: 'Nothing we cannot afford.'

'There's what I maintain,' Midshipman Trevenen supported him, in his soft Cornish brogue which seemed to speak of yellow cream, country matters, and dairy-maids all in a row. 'If we cannot afford the lid off a cask, then we should all be asleep in our paupers' graves! It is monstrous! Only the Despot would exact such punishment.'

A cautionary cough from the head of the table halted him. It was First Lieutenant John Gore, a heavy grey man, a plodder, a workhorse to be trusted, and a respecter of rank. To the young sparks on board he was always 'Old Gore'—one foot in the grave (this was the irreverent Trevenen again), and the other on a peeled banana. But as a three-voyage mariner under Cook, there was no mistaking his authority.

'Mr Trevenen.'

'Sir?'

'Despot he may seem to be, in this morning's business. But he is no despot to you.'

It was a lesser rebuke than young Trevenen might have expected, and he bounced upwards like the simple cheerful soul he was. 'Sir, with all respect, I have the scars to show for it.'

Lieutenant Gore eyed him closely. He knew the midshipman's quality: it had never been his own, even thirty years ago, but he could read the signs and portents of youthful mockery advancing to the limits of prudence. 'Well—keep us not in torment. What scars are these?'

Midshipman Trevenen loosed his clasp on knife and fork, for the first time since they had sat down, and spread open his hands. They were indeed calloused, and four times cracked and blistered within that hard-pressed curve where fingers and hands gripped an oar. 'Sir, we rowed twenty-eight miles up and down the coast, two days ago, taking soundings with the captain. We were nigh to dropping—'

'But you did not drop?'

'Why—no, sir!'

'Well done,' said old Gore. 'We must test you again, to the very limits of endurance.'

There was a laugh around the table. The First Lieutenant was not so slow . . . The midshipman, balked of any advantage, looked about him for allies, and caught Matthew's eye. 'You suffered with us, Matthew. We were Turkish galley slaves together. Show them your blisters!'

Matthew, aware how skilfully John Gore had turned their talk from a captain's harshness to a midshipman's fearful ordeal, sought to continue the disengagement. 'For certain I have blisters on my hands, from all that rowing.' He spread his own claws, and they saw that this was true. 'I have bigger ones on my bottom, if you must know it, but I'll not show them to the company . . . What are blisters, when all is said? We are sailors! If I had not blisters where blisters should be, I would think myself a clerk or a counting-house man.'

James Trevenen, balked again, made a last attempt to rescue his foundering argument. Swiftly he changed his ground. 'Come, Matt!' he said, as if he were faced with a slippery lawyer. 'We talk not of blisters on your hands—nor on your buttocks—and you must own to it.'

'*You* talk of them.'

'In parenthesis.'

Lieutenant King, a midshipman not too many years ago, could not resist a swift interruption. 'What ship was that? Parenthesis? I mind the old *Parthenon*——'

Trevenen laboured on above the mockery, with Cornish persistence. 'I meant, we were talking of other men's sufferings, not our own.' He addressed himself to Matthew again. 'Forget the rest. This is not blisters, it is bloodshed! The thieving, and the punishment? What do you think of it?'

It was the corner-stone of all their talk, and could not be ignored. Matthew could only give his quiet verdict, which he believed to be their own:

'I think he has changed.'

It might have been enough, and they could have moved from foul ground to safe. But Lieutenant Williamson, that man of choleric impulse, was only too ready to re-open the affray.

'He has *not* changed,' he said loudly, and glared at Matthew across the table. 'It is men about him who have changed.' The unmistakable charge of disloyalty brought a chill to the mess-room, and Williamson was not loath to turn it more chilly still. 'Cook is our captain. He can set any rules, whether for us or for these thieving Indians. He can set any punishment also.' Now he was glaring, not at Matthew solely, but all round him at his mess-mates one by one. 'I promise you this,' he ended, on a note of threat. 'If he should get wind of this talk, he might set some punishments for back-biters which would make a cropped ear seem like a kiss on the cheek!'

Matthew did not dare a word of answer: not only the rank of this accuser, but his furious eyes, in a face bursting red, commanded silence and submission. But there was at least one in their company who did not have to suffer slander, nor would do so for a moment. It was Lieutenant King, the Irishman's superior by one blessed rung of the ladder.

'Mr Williamson!' he said, in a voice none had heard before. 'Since you are so free, I will take the same licence . . . You used the word back-biter. There are none here. We have been loyal to the captain, and we will ever be so—perhaps more loyal than some other loud-mouth might proclaim . . . We have talked of this cruel punishment because it shocked us—and it shocked us since it came from a man we love and respect.'

Williamson was not cowed. 'Strange respect!' he declared viciously. 'With such *respect*, what captain needs disloyalty to undermine him!'

There was no knowing where this gross insult might have led—to a mighty round-table rebuke from the First Lieutenant, to further tumultuous quarrelling, even to blows—but there came a fortunate interruption.

Their ship the *Resolution*—their fortress-home for so long—began to tilt, so that the shaft of sunlight piercing the hatchway moved from one corner of the table to another. There was a confused noise from outside: a shouting, a buzz of voices, a hammering, a pattering, a scrabbling on the ship's side. The *Resolution* tilted further, as more weight was added to her loaded hull.

They knew, from long experience, what this weight was. It was an assault! Yet it happened every day. It was the burden of three or four or five hundred bodies, leaving their canoes and climbing the side of the ship on their afternoon visit.

They had come to trade, or to stare, or to steal. But not to kill. There was, at that smoking moment, more spirit of killing within the officers' own mess-room than on all the bulwarks and top-sides of their private ark.

In the silence below, a firm voice which was their captain's sounded from the quarter-deck:

'Mr Bligh!'

'Sir?'

'I'll thank 'ee to turn up the hands.'

'Aye, sir.'

There was no man on board more ready to do so than Master William Bligh.

Bligh, the master, was a man who mystified all, and enraged the unfortunate beings who came directly under the lash, either of his tongue or of his boatswain's biting cat-of-nine-tails. The officers, safe behind their small rampart of rank, could not divine how a young man of one-and-twenty had advanced to the post of master: the crew, bearing the fearsome cross of his authority, could not imagine how the Devil had not taken this his own man to his fiery bosom, long ago, and left them in peace.

On the one hand Bligh was an inspired navigator, ship-handler, star-searcher, haven-finder: in the company of old James Cook and young George Vancouver he need not cover his head. But at one-and-twenty, surely a longer apprenticeship would not have come amiss? . . . On the other, he was born to offend, like Lieutenant Williamson: to quarrel as far as he dared with his superiors, to despise his equals, and to treat every sailor under his command with all the grotesque violence of a Roman bully and tyrant.

All about him were fools or knaves . . . James Cook, who prized Bligh's skill as a most precious part of *Resolution*'s armament, once rebuked him: 'You do not make friends of men by insulting them.' If William Bligh ever took this lesson to heart, it went unnoticed, either in song or story, from the day he first lived till the day when he did not.

David Nelson, the gardener of Kew, had the best song and story of all to tell about Bligh; and he once told it to Matthew Lawe in a quiet anchor-watch off Tongatapu in the Friendly Islands, when Matthew had command of the deck, Nelson with his plants asleep had nothing to do, and William Bligh was ashore with a broken ankle-bone.

'I could wish it were his neck!' Nelson exclaimed, when this news reached the ship, and did not plunge it into immediate mourning. Coming from so quiet and amiable a man, the words were astonishing. But there were many more to follow.

As Matthew turned from seeing the skiff hoisted, and the ship had settled down to peace again, embraced in the warm magic of a Tongan dusk, he had only to ask 'Why so fierce, David?' to receive a most generous ear-full of reasons.

'His brutal treatment of the crew one must let pass,' Nelson answered. 'I suppose it is thought necessary, though there must surely be better ways of setting a willing sailor to work than by starting him up the rigging with a rattan cane across his back. I am not harsh with my plants, and yet they grow for me . . . Perhaps sailors do not grow on love.'

'Every master must keep a firm discipline,' Matthew said.

'This master should not so enjoy it . . . Come, Midshipman Lawe—' they had grown to be firm friends in the course of their voyaging, and raillery now came easy. 'I know you are soft-hearted as a lotus-bud, in spite of those horny fists. *You* do not need a starting-cane to set a man to work, nor the threat of the cat either. But let that go. I will leave discipline to sailors. Did you know that Master William Bligh is concerned with plants?'

'Yes. He takes great interest.'

'He takes the plants too, if he can! Did you know he is a plant collector?'

'No.'

'Some would call it a thief . . . I found him, one night, collecting some of mine. Only they are not *mine*. They are destined for the Botanical Gardens, and thus for the King. Yet there he was, in my garden, as light-fingered as any gipsy filching turnip-tops!'

David Nelson had his own Eden on board, as befitted an ardent gardener of Kew sent round the globe to pursue his trade. It was a fragrant corner, tenderly cared for, nursed like the first cradle of the world. Fruit and flowers grew in profusion, waiting for the moment when they might be tested elsewhere, in the interest of science or—more truly—of love of mankind.

There were bees which had voyaged ten thousand miles, birds which had chorused their greeting to a thousand dawns, and gently taken farewell in matching even-song.

In hot and cold frames, little shoots which would be great trees, vegetables which might one day make the mouth water or the starveling live in England, or the Pacific Ocean, or in the far-off islands of the Caribbean if they could survive so wondrous a voyage, were ranged like soldiers of hope—only they were sailors, carried in the *Resolution*, and, if they could talk, must have been mighty proud of it.

In this Eden, Nelson had one night found a snake.

'I came below,' he told Matthew, 'to give a drop of water to some bread-fruit. It had been a dry hot day for us all. And there was Master William Bligh, putting aside some of my choicest plants!'

'Putting aside?'

'Ready for stealing!'

Nelson's story was now a lyric ode to strife.

'What is this?' he had challenged Bligh. 'Don't touch those! Put them down!'

Bligh had turned at his voice. But he was not at all put out. 'I'll have these,' he said, as bold as brass. 'I wish to try them in a different soil.'

'You will *not* have them!'

Nelson knew his place on board, and he had always kept it. He also knew his duty, which was the same as any other passenger's on any ship in the seven seas: to be of good behaviour, and to obey the lawful commands of the captain. Nothing less, and, by God, nothing more!

He said again: 'You shall *not* have them. They are not your concern!'

'I am the master,' said young Bligh, swelling like a turkey-cock, ready to fly into a passion. 'All that goes on in this ship is my concern.'

'The garden is mine.'

'It is a part of ship.'

'It is *my* part of ship.'

The furious argument, in the shattered peace of the garden, under Bligh's single nefarious lantern-light which flickered across rampant greenery and baleful human faces alike, enlarged to a tirade from the master on his authority over all, and a scandalized defence of his own by Nelson. It had ended thus:

'By God, Mr Nelson, I'll have you at the capstan bar for insolence!'

'By God, Mr Bligh, I'll have you at the Old Bailey Bar for stealing Crown property!'

Now Matthew waited for more, but it seemed that the tale was done. He prompted the silent Nelson: 'What then? What did the master do?'

'Nothing. Just as I would have wagered when he began his impudence. I know such men, and so do you. They are only harsh to those who cannot answer back. All young cock-sparrows are the same. But when they grow to manhood——'

'Did he truly say no more?'

'Not a word. However, I receive a haughty glare when we meet . . . I am no sailor to be put down, or flogged at the capstan bar, and he knows it. Captain Cook knows it best of all, and would have checked him forcibly. That was my strength. I am *less* than a sailor, and so free of Bligh's cruelty. But——' he spoke the cautionary word again, 'if that man ever gains command, with none to overrule him, he will drive his men beyond endurance. Some day, in some ship, on some ocean, he would go too far.'

'What then?'

David Nelson shrugged. 'Only one thing can happen when——' But his tale was truly done, and he was looking away from Matthew and towards the distant shore-line. Tongatapu was withdrawing into the velvet darkness: beyond the pale beach was a purple line which was tree-shadow and hill-slope, and above them only the night sky with its first shy peep-show of stars. 'The Friendly Islands,' he murmured. 'Who can spoil them? Who can spoil an evening such as this?' He turned. 'Would you like to see a yam newly grafted on a banana stem?'

It was not Matthew's prime choice of pastime, but he had a friendship to keep. 'Why—yes.'

'Then leave the deck to the boatswain, and come with me.'

When the ship's company was assembled, with Bligh keeping taut orders over them—and over the invading islanders, who were confined to the forepart of the ship under strong marine guard until the captain's business was done—James Cook addressed his crew.

As was his daunting custom, he stared down at them from the quarter-deck for a full minute before he spoke, while they waited in patience for whatever might befall, whether the proclamation of a joyful public holiday or a call to witness punishment. What they saw above them was a man they still trusted, even in their discontent: a hard gaunt grey man, tall as an unbending tree, who had not eased his grip on ships or men, courses sailed or landfalls made, for three long years.

He was formidable in his captain's blue tail-coat with its blaze of gold braid and buttons, his tricorne hat and white knee-breeches, and his symbol of office—the brass-bound telescope with scarlet leather hand-piece: a man dressed to suit his unquestioned authority, yet a man grown secret and solitary under the weight of command.

What James Cook saw was locked within his breast. What he gave them now was a further lecture on behaviour, though this was more than a lecture, since it told them for the first time the news of their departure.

'We sail tomorrow,' he announced sternly, without preamble. 'So this is our last night in these friendly islands. I wish it to be unmarred by any excess. There will be a farewell feast ashore, and some ceremony. Your conduct must do honour to your ship. If it does not, the usual punishments will follow . . . *And I will sail tomorrow morning with a full ship's company.* I want no deserters. You are sailors of the *Resolution*, and I am leaving no man ashore for a soft life as a landsman. Any laggards will be hunted down, and delivered on board by my friend the chief Palea. So much, he has promised me, and so much, I promise you. Remember it!'

He paused, for the effect of this threat, which was nothing new—and nothing to be disregarded either. Then he turned swiftly to a succession of orders, as crisp and clear as any in his life:

'Mr Bligh! Dismiss the hands. There is to be no trade on the ship's account with the natives: we shall be given presents at to-night's ceremonial, and the ship is near to fullness already . . . Mr King! All boats ready to go ashore at six o'clock. Half the marines to stay on board, the rest to come as our guard.' Then, more formally as rank demanded: 'Lieutenant Gore!'

'Sir?' the old First Lieutenant answered.

'Attend me in my cabin with Mr Bligh for the order of sailing tomorrow . . . Midshipman Lawe! Take the skiff, go on board the *Discovery*, and acquaint Captain Clerke with my intentions. His boats should join ours at six o'clock.'

Matthew, who was standing just below the quarter-deck rail, bethought himself of a time-saving plan. 'Sir, Mr Vancouver of the *Discovery* is with us.'

'Is he so?' Cook answered, in the voice which all dreaded. 'I hope he has been well cared for . . .' Then, as if he had pressed the spring of a magic watch, and recovered thirty seconds of vanished time, he said: 'Midshipman Lawe! Take the skiff, go on board the

Discovery, and acquaint Captain Clerke with my intentions. His boats should join ours at six o'clock.'

James Cook could be stiff, very stiff, when he chose.

The crew dispersed, murmuring of plans gone awry and loves lost forever; the officers, more discreet, went about their duties with faces suitably blank. But at the core of a ship's unease was the eternal question of all sailors: *Where are we going, tomorrow?*

They knew nothing: they knew less than the sea-birds which would follow their wake next morning, and croak to each other: This wooden thing of the strange smell is bound for China, or Peru, or the sunrise or the sunset, or somewhere unknown even to us.

There had been rumours that the *Resolution* was to stay another season in this ocean, and go north again into the forbidding Arctic land as soon as the ice allowed. Could it be true? Were they doomed to a *fourth* year on their endless journey?

Matthew, smarting under his public rebuke, put off in the skiff as mutinous as any flogged marine. He had meant no particle of harm by his question to Cook: only that, if a message must go to Captain Clerke of the *Discovery*, then Vancouver his own midshipman might carry it as well as anyone else. But no—Cook the despot could only send such an order by one of his own officers, and to hint otherwise was insolence.

He, Matthew, should have divined this. He should have kept his mouth shut, and thus saved his face. In these doubtful days it was best to know nothing, and think nothing, and say nothing. James Cook, the changed commander, told his officers little, and his crew even less. No man now dared to question him, and no wise man should think of it any longer.

The feast of their second farewell was of a magnificence even greater than the first, and must have cost the Hawaiian islanders dear. Whatever stern pressures had been placed upon them by their chiefs, by the priests, by their own sense of hospitality to strangers, or by the simple wish to excel their neighbours, the outcome was a display unparalleled. It seemed that they would say goodbye to James Cook as a god, even if they were thereby made poor for ever after.

It might also be that they were at long last glad to see the *Resolution* sail, and that this entertainment, by its very richness, signalled that it was to be the last.

Once more the chief Palea, who was Cook's especial friend, and the high priest Kao, always ready to put a holy *tabu* on their storing-parties ashore so that they might work unmolested, led the ceremonies. Once more the yams and the hogs were piled up for their delight, in smoking mountains of largesse.

Once more there was singing, and dancing, and wrestling; once more the men who slipped away into the shadows of the beach were sure to find a pair of willing lips, willing arms, willing legs, willing all.

If this were in truth the last time for such delights, let them remember it!

Once more the presents, collected from the common people, were poured out upon them. The hogs, the green turtles, the sugar-cane, the fish, the coconuts, the garlands of flowers—all multiplied till they could not be counted. One gift of singular magnificence, from Chief Palea, outshone all others.

It was for Captain James Cook, their visiting god, their benefactor: a cloak beyond price, beyond compare, of feathers plucked two by two from single birds now dying out: an orange, white, and purple cloak, six feet long and a hundred years in the making.

It was for himself alone, and could never be matched again. It was blessed, like all the rest, by Kao the high priest, and presented, like all the rest, with the divine honours due to a departing Deity.

At last, by flickering torch-light, the errant crew—and certain amorous young officers—were rounded up, and the final farewells spoken. One forlorn appeal was made to Cook by Chief Palea. Was it possible that his 'son', Lieutenant James King the beautiful, might stay behind, to bless the islands and give them everlasting good fortune?

Captain Cook could only answer: 'Perhaps next time,' and step into his waiting boat, and be borne off to his ship.

In the morning they sailed away, attended for many miles by a great fleet of canoes, decked out with all the signals of peace and goodwill, dropping away astern like leaves from the very tree of homage. They sailed away—and they returned.

Within two days, a season of violent squalls and ceaseless pounding in a nearby island-bay sprung the *Resolution*'s foremast at the head; it also proved to be rotted at the heel. Such a spar must be replaced, or firmly fished with stout splints, before they could face any further voyaging. It could only be done with certainty at

Kealakekua, where the timber was seasoned and the welcome warm; and there, a week later, *Resolution* and her consort dropped their anchors, in a bay silent and deserted.

Thus James Cook returned winged and in need of help again, and it was the worst thing that could befall, because it should not happen to a god.

VI

For no reason that a working sailor could divine, the humour of the islanders had changed. This was Matthew Lawe's mournful thought, after two days spent ashore with the wood and watering parties, and in overlooking the carpenters working on the *Resolution*'s foremast.

There was little or no trade, and scant ceremonial from chiefs or priests. The natives were cool to their guests: they were sullen when spoken to, and insolent when they were checked for thieving. It seemed that they had fallen out of love with the *Resolution* and, worse still, out of respect for her commander.

Were they perhaps asking: 'Have we not given enough? Enough of our defiled women, enough of our wood and water, enough of yams and hogs and fish and fruit? *Enough of our worship?*' Was this at the core of all, the great question which could now be freely spoken? If Cook were a god, if his men were all-powerful and his ships so magic in their potency, how could their plans come to nothing? How could their fleet be damaged, and turned about?

Perhaps these were false gods. Perhaps they were only men, frail men—even great Cook.

Perhaps it could be put to the test.

The stealing increased, and became more daring, more impudent. The islanders did not scruple to attack the ships themselves: diving from their canoes to cut the fishing lines and make off with the hooks, prising out with flint chisels the very nails from the sheathed hulls. Then the armourer's tongs—another useless trophy for a daring rascal—were stolen from the *Discovery*'s forge; and Captain Clerke, fired by Cook's new harshness, had the culprit triced up to the main shrouds and given forty lashes.

Forty, it was thought, was brutal beyond measure. At least he should have been given the tongs!

Matthew himself, ashore on the second day, found at first hand the measure of the islanders' rebellion.

He was watching the progress of the foremast, set up on trestles near the edge of the beach, with carpenters busy at each end of it, trying to make a wholesome spar out of a weak member. There was some stirring in the trees, and mocking cries, and then a twittering of foolish laughter as a carpenter hammered his thumb and told the world of his misfortune.

Then the throwing began.

It started with a coconut: then a stone: then many stones. None of the sailors was hit, but they had to retreat from their work while the marine sentries took up a closer guard. Then a bold warrior encased in his war-mat of plaited leaves and fibrous stems came out from the trees, raised his spear, and flung it at a sentry.

Perhaps not at the sentry: perhaps at the palm-tree above and behind him, where it struck and quivered. But it was enough for an alarm, and an answering show of force. The sentry discharged his musket, which was loaded with small shot. This did not pierce the war-mat, but pattered down like raindrops onto the sand. The hero then retreated at his own pace into his forest cover, whence shrill cries of welcome soon gave way to more mocking laughter at this triumph over the gods.

Matthew, reporting all this to Cook at noon-time, found him fiercely calm, like a man who knows he must soon decide on some great step, to one side or another. But at the end, the captain only said:

'Very well. I will go see for myself . . . What think you of all these troubles, Mr Lawe?'

Matthew, made confident, spoke from his heart: 'Sir, I say, God damn the foremast!'

James Cook's brow, which was first as black as thunder at this blasphemy—he never swore any oath, and he constantly reproved it in others—turned suddenly to a rare sweet smile which none had seen for many months.

'Matthew,' he said, 'you are all mariner!'

It was the last shaft of sunshine in all their fateful voyage.

True to his resolve, Cook went ashore that same afternoon to inspect the carpenters' work, and ran headlong into trouble. For the first time he was openly insulted and mocked, by a huge crowd which followed him to the water's edge and tried to lay hands on the pinnace.

In the *mêlée*, the like of which the *Resolution* had never met in all her travels, Chief Palea—acting with God-knew-what intent—was struck on the head by an oar, whereupon he seized it out of the oarsman's hand and, in a rage, snapped it in two across his own thigh-bone.

It was possible that, at this moment of anger and tumult, Palea became an enemy instead of a friend. James Cook, returning with great difficulty to his ship, seemed to have matched this mood of fury. His first response was a curt order:

'All marines and sentries on shore-duty will load with ball.'

There could be little doubt that someone must now be killed.

His officers—save for fierce Lieutenant Williamson, to whose hot Irish thinking a musket-ball was the best passport to a Hawaiian heart—were appalled, alike at the order and the risks attending it. Matthew, in mournful brooding on the quarter-deck as dusk fell on Kealakekua Bay, could only see it as a self-inflicted wound, by a man drawn fatally to his own destruction.

He could even be envious, as well as aghast at this movement of the finger of fate. What should a man do, who had worked hard enough, voyaged far enough, and lived long enough? What *did* a man do, who was weary of travel and thus of life?

Only one thing, which now needed one more signal.

The signal was the theft of the *Discovery*'s cutter, that same night. The ship lay nearer to the shore, and was thus an easier target; none the less, it was a most daring assault. Her boat was cut loose from the buoy in the darkest hour, before the moon was up, and towed ashore by swimmers who had captured the greatest prize of their lives.

She was the biggest and best of all the *Discovery*'s little chickens, and for reasons both of pride and plain good ship-keeping the loss was not to be accepted.

At first light on that holy Sunday morn which was the fourteenth day of February, Cook in a quiet rage set in train his counter-blow. Three *Resolution* boats were to be manned and armed. They would go ashore in force, and take as a hostage a chief, greater than Palea, whom they knew as Kerreeoboo. He would be held in captivity against the return of the *Discovery*'s cutter.

James Cook, with one Lieutenant Phillips and nine of his marines, would lead this foray himself in the pinnace. There was also a small cutter under Lieutenant King, and a launch with John

Williamson in command and Matthew Lawe to aid him. No limit
was set to the means whereby this plan would be crowned with
success.

They landed unopposed: there were even signs of respect as Cook
was recognized. The islanders knew well enough what he had come
to recover . . . The small cutter stood on guard at the head of the
bay, while Williamson's launch waited in the shallows, with enough
water under her keel to give her freedom. Cook and his marines
marched uphill towards the village, and the house of Kerreeoboo.

The old man, whose addiction to a brew of the fiery *kava* root
had rendered him a devoted slug-a-bed, was still asleep, and, it
seemed, knew nothing of the theft from the *Discovery*. Roused at
last, he received Cook with every courtesy, and when invited to
return with him to the *Resolution*, agreed readily. Time enough to
dress for such a pleasurable visit was all he asked . . . Presently Cook
and the marines, and Kerreeoboo with his wife and the two sons
of his old age, began their return journey to the bay and the waiting
pinnace. At first they were alone; soon they were not.

They were being followed by an increasing throng, joining
from the huts and the trees as they made their way down. Many of
them were armed: all were murmuring freely, either among themselves or in messages addressed to the back of their chief's head. The
chief did not like what he heard. Nor did his wife.

When they came in sight of those waiting in the pinnace and the
launch, they seemed like a dark stream of lava flowing down to the
sea. Tall Cook stood out, like a forest giant among bushes. The
marines at his back could still be seen. But all were becoming engulfed in a black tide pressing closer, lapping and snarling at their
heels.

The view from the launch lacked nothing of fear, for any man
who had Cook's safety at his heart. Matthew, sitting in the stern-
sheets opposite Lieutenant Williamson, with their four marines beside them, stared through his glass at a scene of great peril.

'They are armed!' he said suddenly. He had glimpsed spears and
daggers, clubs and stones, all openly displayed. The iron daggers
were gifts of friendship from Cook . . . 'They seem ready to—Sir,
they are too close to him!'

Williamson, whose bold Irish colour seemed to have taken on a
touch of pallor, did not answer him. Instead, he said to his coxswain: 'Back water! Come astern!'

Matthew, thinking he had been misunderstood, spoke again urgently: 'Sir, I meant, the *natives* are too close.'

Williamson, with his usual arrogance, disregarded this. '*We* are too close. We must be ready to move.'

To move?

The launch drew further off, and then stopped. As if in answer to this shameless dance, Cook and the great crowd at his heels started to cross the beach. Then and there was played out the beginning of the last act of tragedy.

Chief Kerreeoboo, with one of his young sons already in the pinnace and dancing to be off, began to falter. His wife was holding on to him, pleading, arguing, screaming at him, twining her arms around his neck. Suddenly he sat down in the sand, and began to hang his head and to tremble. He would go no further. Straightway his islanders surged forward and surrounded him, so that he could not be reached—not by nine marines, not by ninety.

Cook and his guard were now alone, near the water's edge, while close about them a vast throng began to shout, and howl, and shake their weapons in the air.

After that, there was nothing but shame and horror.

Making his last prudent decision, Cook ordered Lieutenant Phillips: 'Into the pinnace! We shall never take him now without too much killing.' But before any of the party could make a move, Cook was threatened with a dagger-thrust from behind, and he turned and raised his double-barrelled musket.

It was perhaps a measure of this changed man that before the landing he had loaded his piece with two different charges. One barrel of it held small-shot, a gentler persuasive. The other was charged with ball, which inevitably killed.

He raised, he sighted, and he fired the small-shot barrel. His assailant, a bearded warrior near as tall as Cook, laughed as the pellets pattered against his woven war-mat and fell harmless; and the laugh seemed to spur the crowd to furious assault, just as the sound of the shot had enraged them. They fell upon the marines: Phillips was stabbed, others of his men were knocked down, and, with the line about to be overrun, Cook fired his second barrel, and killed his man.

Then he shouted again: 'To the pinnace!' and at the same time lifted his long arm and waved to the launch.

The view from the launch was now so terrible that the men on

board could not contain themselves. They called out to Williamson
for some action, and to the marines—who were their closer shipmates—to use their muskets. But the Irishman seemed to see all
things with different eyes. He ordered:

'Hold your fire!' and to his coxswain: 'Back water! Come astern
again!'

'Astern' was cowardly retreat, however it was disguised. Matthew
could support this no longer. Were they to be mere spectators of
this slaughter? . . . He spoke his scandalized mind: 'Sir, we *must*
fire. We *must* go in! Did you not see the captain calling us in?'

Williamson turned and roared at him: 'We will not fire! It is too
dangerous. I will shoot the first man who fires!'

Matthew, stupefied at this craven nonsense, repeated: 'But he
called us in. Look there! They will have him in a moment!'

'Not so. He meant, that we should stand off and protect his
pinnace when it is clear of the beach.'

The great fire-eater was dribbling out fear instead. Matthew
beseeched him for the last time, more forcefully: 'It will never be
clear of the beach, if we sit here like dummies.'

Williamson shouted: 'Do you argue with me again, midshipman?'

'I only beg you to move, sir,' Matthew answered.

'Then keep silence, or I will *move* you to the court martial, and
then to the forecastle where you belong!'

At this pause in a shameful dispute, one of the bow-men, looking
landwards towards the tumult, screamed out: 'For God's sake, he is
down!' and Matthew, thinking: *Sweet Christ, I have played this scene
before!* could only watch, with all the rest of a useless, spineless
boat's-crew, the murder of their captain.

Amid a breaking wave of dark bodies and flailing arms, Cook
was struck from behind with a club, then stabbed in the neck with
a dagger. He fell, and villainous hands held him face downwards
in a rock-pool, while he was repeatedly stabbed, and the natives
snatched daggers from each other to partake of this bloody feast.
Hauled high on a rock, the body suffered further wild mutilation
as the pinnace and the launch made their retreat.

They left behind their captain, and four dead marines, on a beach
so swiftly deserted that the islanders must now have been struck
with terror at what they had done. The pinnace, full of wounded
men, could do nothing to recover the dead: the hale and hearty

Williamson, with an unblemished launch, might easily have done so, but neglected it.

He would only mutter, as a crumpled braggart: 'Later, later. You will see!' while his men, in rage and contempt, pulled on their sick-hearted oars.

It was all concluded by eight of the morning, as the boats rowed back to the ships, and on the ships a deathly silence fell.

The first wild grief was for James Cook their captain; and the first great outburst of anger broke against Lieutenant Williamson, that great faithful warrior, who—according to every man's tongue—might have saved him, and did not, and might have recovered his honoured body, and did not do so either.

There were moments of such shaking rage among the crew of the *Resolution* that Williamson could well have forfeited his own life, by some unfortunate night-time accident: men spat on the deck as he passed, and dared this fiery loyalist to check them, or even to take notice of their insolence.

He was a very quiet man, among friend and foe alike, for many days thereafter.

The burden of all this fell upon Captain George Clerke of the *Discovery*, who now took command of their enterprise and—though exhausted with care, and carrying the seeds of his own death within his breast—played a very cool hand of cards for as long as was necessary.

Clerke, as befitted a man of his honourable name, had that kind of mind which, when such were needed, prepared lists of what he would label 'Preferred Matters'; and his officers, after muttering against his caution, came to admire him for the calm appraisal which enabled him to draw a pen-stroke through those 'preferences' which might be forgotten, and leave only those which could not.

He dismissed any immediate idea of action against Williamson, though he went so far as to take written depositions from witnesses. Such a Crowner's Quest could come later. He thought, and then abandoned the thought, of vengeance against the islanders. It could easily be exacted: the whole compass of Kealakekua Bay might be turned into a froth of blood and corpses. But it would prove nothing, and bring back nothing.

All that was left on his list was the foremast of his new ship

Resolution, without which they could not sail, and the body of her captain, without which they should not. For a sailor newly in command of a beleaguered fleet which still had thousands of miles to travel, the foremast must come first. As a friend in deepest mourning, he could not leave James Cook to the kites, whether human or otherwise. On these requirements he bent a mind calm beyond all common quality—and secured them both.

The seaman's task was finished first. A strong armed party, ready to fight—and more than willing—was put ashore on the fatal beach, with Master William Bligh present to see that no harm came to his essential spar. This was recovered without trouble, floated off, towed back to the ship, and hoisted to the *Resolution*'s upper deck for the carpenters to complete their work.

Then began the bargaining, for it was nothing less, over Cook's remains. Natives, creeping towards them fearfully, whispered differing stories. Some said that the body had been carried deep into the country, and would take time to fetch and to deliver; others told how it had been cut up and its parts distributed already, as trophies of war, to all the island chiefs. Both stories were true.

Captain Clerke, and James King his new First Lieutenant, threatened constantly and looked as if they meant it. If the remains were not returned, they thundered, fearful punishment, so far withheld, would be visited upon the whole island. When they *were* returned, with grudging delays, they were remains indeed.

First, after two days, a young priest delivered on board a piece of flesh, nine pounds in weight, from Cook's thigh. Then, five days later, with much ceremonial music and thudding of drums, a chief presented himself on the quarter-deck with a bundle wrapped and covered by a feathered cloak spotted black and white. He indicated that this contained what they most desired and then, with prudent alacrity, withdrew.

The bundle was opened by Clerke, supported by his officers, in Cook's own cabin, and their horrified eyes had to dwell on its contents. They were: both the hands entire, though salted and pickled: the skull *sans* jaw-bone and lacking its scalp: the scalp with the ears adhering: and the longest bones of thigh and leg and arm, which were the private trophies of great Chief Kerreeoboo.

Next day came the final accounting: the missing jaw-bone, the feet, the shoes, and the bruised barrels of that musket which had brought its owner to this pass.

These poor remnants were decently coffined, and buried that evening in the deep of Kealakekua Bay; and on the morrow's morn the *Resolution* and the *Discovery* weighed their anchors, and quit the accursed place for ever.

It was a public print, long afterwards, which showed Matthew Lawe, always mourning a cherished friend and patron, that there still dwelt in England a certain private grief far deeper than any that Cook's ship-mates could know. It lay locked in the heart of their captain's wife.

Mrs Elizabeth Cook, a dutiful yet adoring consort, had borne him six children. Of these, three had died during his lifetime: an only daughter at four years, a son at one month, a second son at four. She was thus left, on her widowhood, with her two eldest sons who were both midshipmen, and her youngest.

Nathaniel, the second, perished within a year of his father, in a West Indian hurricane which sent his ship down with all hands. The youngest, Hugh, destined for the Church, died of a fever when he was seventeen and studying at Cambridge.

One month later Commander James Cook, the darling namesake, the eldest, the pride and joy, the last hostage, was drowned in an open boat off Portsmouth when trying, in fierce weather, to rejoin his ship the *Spitfire* sloop.

Matthew Lawe, reading the cold print of the *Gazette*, could only think, in savage hatred of all fate:

'When God eats, He cleans the platter!'

SEVEN

NAVY CAPTAIN

1790

'Their sayings and doings stir English blood like the sound of a trumpet; and if the Indian Empire, the trade of London, and all the outward and visible ensigns of our greatness should pass away, we should still leave behind us a durable monument of what we were in these sayings and doings of the English admirals.'

Virginibus Puerisque
by Robert Louis Stevenson, 1881

I N THE BLUSTERY MONTH OF MARCH, the most choleric officer in all the five oceans stamped ashore on English soil, after surviving fearful hazards which filled the gazettes and the magazines, and displaced all other coffee-house talk, for many months thereafter.

This was Captain William Bligh, Royal Navy, commander of His Majesty's Ship *Bounty*, who had been subjected to atrocious mutiny on board his vessel, had been cast adrift in an open boat in the midst of the Pacific, and, with scant provisions, had been left to sink or swim by heartless rogues. Yet he had somehow overcome all difficulties, and lived to tell the tale.

Now he was telling his tale, and the tale was enough to startle every ear, and set every ardent reader a-fuming at such gross indiscipline.

The scandalous story lost nothing in the account. HMS *Bounty* —'a happy ship', declared Captain Bligh, when interrogated by 'Our Own Correspondent' on a score of occasions—had been going about her lawful business of charting and exploration in the Pacific Ocean when a gang of miscreant mutineers among the crew took forcible possession of her. A certain Fletcher Christian— what a name for such a monster!—had led the revolt, fomenting trouble out of petty spite, and proclaiming himself the new captain.

He had gone so far as to lay hands on Captain William Bligh— 'Breadfruit Bligh', as he was affectionately known to his admiring crew—and to cast him off in one of the *Bounty*'s boats, thousands of miles from the nearest safe refuge. With him had gone eighteen others of his loyal men, forced into this frail craft with the following provisions to sustain them:

Water	28 gallons
Bread	150 pounds
Pork	32 pounds
Rum	6 quarts
Wine	6 bottles.

To be continued in our next enthralling chapter: 'Captain Bligh and the Cruel Sea'.

Avid readers could hardly endure until next day; among them was Matthew Lawe, sitting in a coffee-house in the shadow of St Paul's and, according to his frugal custom, making his cup last. But he at least did not have to wait: he was reading the old gazettes of last month which were all that this mean house provided, and the next one, dog-eared and dusty, was close to hand. He sipped, and thought, and picked it up.

Cruel was the sea, and frail the craft—no more than twenty-three feet long—in which Captain Bligh and his band of faithful were set adrift. This wicked crime against a noble commander had taken place near Tofua, off the Friendly Islands—another name of evil irony, since when they tried to land there one of their men was murdered by treacherous Indians. There was nothing for Captain Bligh to do but to sail on into the unknown.

Such was the malice of these villains that they had not even spared him a chart. All he had to aid him was the chart within his head, and his own dauntless skill. Thus he had sailed on and on, under lowering skies, burning sun, storm and tempest, fair winds and foul; often afraid to touch at an island for fear of the lustful savages. By the end he had steered his gaunt cockleshell 3600 miles to Timor Island in the Java Seas, where at last a Dutch settlement furnished a haven.

The fearsome journey lasted forty-six days, during which time he and his men subsisted on a ration of one-eighth of a pound of bread and three-quarters of a pint of water *per diem*, eked out with sea-birds' blood and the soup of clams. When they landed they were like skeletons—but undaunted! The astounding voyage was scarcely to be matched in the annals of navigation—nor of endurance, nor bravery.

Matthew set down the gazette on his knee. I will give him that last salvo, he thought, as one sailor for another: this *was* a triumph of men against the sea, and only Bligh's superb skill could have made it possible. But for the rest of the story? His admiring crew? . . . A happy ship? . . . Their affectionate name for William Bligh? . . . The overthrow of a noble commander by one man fomenting trouble? . . . A bucket of whitewash could not have served better!

Then his eye, peeping across his diminished coffee-cup, was caught by the last line of 'This Day's Stirring Chapter'. It contained a name which had not been noted before, and it sprang from the page like a meteor. The name was David Nelson. Matthew read with astonishment:

'Apart from the man who was killed on those most *Unfriendly* Isles, Captain Bligh, in all his multiple conquest of the sea, lost only one other from natural causes. This was Mr David Nelson, the celebrated botanist of Kew, who unfortunately perished of exhaustion as soon as he set foot on the Island of Timor. He, having joined the faithful supporters of Captain Bligh in his tiny craft, rendered invaluable service . . .'

Matthew could read no further. This could only be his gardener-friend from the old *Resolution*, and was thus a moment for weeping. So David Nelson, despite all his talk, had remained steadfast in his undertaking to obey all lawful commands of his captain, even though that captain was William Bligh, who must have tested that loyalty to the very limit—and beyond.

What a creature was man! How like an angel. And poor David Nelson, the ever-willing, ever-faithful servant—how like a fallen angel, who knew not what god had struck him down, but perhaps gazed upon the Island of Timor—so weirdly different from Kew—and wondered for all eternity what precious fruits and flowers might flourish above his grave.

Tomorrow, the concluding chapter of this, the greatest sea-story of all time, with Captain Bligh's own verdict on the Bounty *mutineers: 'I will hunt them down like dogs! I am astonished at their ingratitude!'*

He alone.

Moving to and fro on his errands of service, the coffee-house waiter had begun to stare at Matthew's empty cup each time he passed his table. Matthew knew the signs, from long and mortifying experience, and he knew the reason too. London was full of half-pay officers such as he, hoping for a berth, yearning to get to sea again, yet fatally barred by peacetime doldrums which had put thousands of them high and dry on the beach.

Since they languished in genteel penury, they were no longer favoured customers, in any house which lived by trade; and it would need the skin of an African elephant to be unaware of this disfavour.

Matthew was one such orphan, with no such skin, and had been counting his pennies, day after day, for more than three years of threadbare idleness. He had sworn solemnly that he would accept nothing less than a Navy berth, even if it meant waiting and hoping until the heart grew sick: after rising so high, he could not fall again. So for him it was boots down-at-heel and tarnished braid on a jacket which barely held together: it was poor lodgings, scant food, walking by the river-bank to while away the day, one cup of coffee in the morning and a second to mark the first watch of the night.

It was the shame of occupying a coffee-house chair on sufferance, long after his fair rental was gone: it was the shame *now* of a waiter who finally stopped before him and asked, in that tone half-way between deference and disdain:

'Another cup, sir?'

Matthew looked up as if surprised out of far-away thoughts. 'No, I think not.'

The waiter took the empty cup, and placed it on a side bench. Then he bent, and with great deliberation and greater finality began to wipe down the table with a wet dish-clout.

Matthew had not received a clearer signal in all his sea service. As he made his retreat, he made his morbid calculation also, for the hundredth time: Lieutenant Lawe of the Royal Navy had never been higher, nor ever lower.

He walked slowly along East Cheap towards the river, and the Pool of London. It was better to watch ships than to brood over dry coffee-grounds . . . But whether walking, or sitting, or staring, he needed money and he needed a friend with influence. He had neither.

He had owed his great promotion to the favour of James Cook, who had made written recommendation of him in the last shipboard report he ever penned, and to 'Old Gore', who had brought the shattered expedition back to England. Promotion had seemed a triumph at the time, putting him more in debt to his dead captain than years of bitter mourning could express, and it had served him well—and then not at all.

In the spring of 1782, two years after that sad return, he had made one voyage on convoy duty to Quebec, as Third Lieutenant in the frigate *Albemarle*, and thence southwards in search of Yankee prizes. But the ship was a poor sailer, and poorly equipped, and they were stricken with scurvy before the mission was completed. This had forced a second wasted stay at Quebec, to clean the ship and put their sick ashore. On her return to England the

Albemarle was paid off. It was to prove, though he knew it not, the high peak of his new advancement.

Later he secured a berth in one of the guard-ships at Portsmouth: a dull task, scarcely touching the sea. Then she in her turn was put out of commission—this was dreary peacetime again, with the French calling it quits for a space—and for Matthew, like a thousand others, there was now nothing at all.

It had been 'Thank you kindly', and half-pay, and the beach—perhaps for ever.

This morning, as on every other morning, there were ships a-plenty in the Pool: fat and thin merchantmen loading and unloading, and all the attendants of their commerce—barges, fly-boats, hoys, ferrymen, store-boats, bum-boats, colliers, water-carriers, and those predatory skiffs which made their living by scouring the tide-way for corpses.

But there was nothing worthy of note for a competent Navy officer, and thus naught for him. In all his idle survey, he could only think of his friend David Nelson, the gardener who became a botanist, the botanist who became a mutineer's castaway, the castaway who endured all things for faithfulness, and died of exhaustion under far-off Pacific skies.

It was a death of honour, and a thousand times better than a life falling to rags and tatters at the muddy edge of the River Thames.

Matthew turned his eyes aside, and by chance sighted on the lip of the quay another such beachcomber: an officer of superior rank doing the very same as he—gazing nostalgically at anything which floated. But this was more than a chance stranger. In a moment of true astonishment he recognized his old commander from the unhappy *Albemarle*: the other Nelson in his life, Captain Nelson himself.

He could scarcely be called that friend in need whom Matthew was seeking. Indeed, this was the very last man in his thoughts, and never one from whom he could now expect help, for a particular reason concerned with that second visit to the City of Quebec.

At Bandon Lodge, in old Quebec, there was a fair maid dwelling: to wit, Miss Mary Simpson, daughter of Saunders Simpson, Provost-Marshal of the Quebec Garrison. She was but sixteen years old, but the sweetest sixteen ever likely to be met on a summer's day, anywhere on the continent of North America: a captivating beauty, tall and dark, with that strange blend of tenderness and aloofness which could entice a man to the most foolish of hopes—or to morbid despair.

The gossip-mongers of the *Quebec Gazette* had christened her 'Diana'; and it was as Diana, goddess of the moon, of the hunt, of chastity, of beauty, and Lord knows what besides, that she reigned that season in fashionable Quebec.

Since she was the belle of every ball, and since the lively city put forth its most noble efforts to entertain all visiting officers, she was much to be seen, and admired, and pursued, during the month's stay of the *Albemarle*. Captain Horatio Nelson's second and third Lieutenants, Bromwich and Lawe, who were often privileged to accompany him ashore, could not but notice that their commander had been sorely smitten, between wind and water, by this paragon of all the virtues and all the lures known to man.

Neither could they ignore common gossip which said that this passion was *not* returned: that Mary Simpson, as was her right, had other ideas beyond an impulsive young Post-Captain of twenty-three, without fortune, without looks, without a manly frame or a manly height, who had been pale before he met her and was now a positive death-mask of thwarted love.

Of course he would not do at all ... She was sixteen: there were endless years ahead of such warm attention, such vain pursuit, before Prince Charming, or perhaps Lord Charming, or even Monsieur le Comte de Charme, came sailing into Quebec and swept her off her delicious feet. Meanwhile, she enjoyed the gallant captain's ardent looks—and those of a dozen others besides. She was not heartless: she was sixteen, entrancingly pretty, and living and loving every moment of every God-given day.

Far too soon for Captain Nelson came the time when his ship the *Albemarle* must go down river and shepherd certain military transports on passage to New York. He had not yet declared himself, so it was rumoured by the free-talking scandaleers of the city; and certainly he would not be encouraged to do so by his experiences on their last day in harbour, when beautiful 'Diana' turned remote as the moon itself and scarcely gave him a glance or a word during an entire dinner-party of farewell.

Their host at this, when they sat down twenty-four promptly at two o'clock, was Mr Alexander Davison, a rich North-Country merchant and ship-owner with large interests in Quebec, and a spirit so hospitable that his house was the prime favourite among all visitors to the coast. Lawe, his captain's escort for the last time, and seated well down the table, had him under sufficient observation to note two matters.

These were that, although placed next to his *inamorata* by a kindly host, he was receiving but a pauper's share of her attention: and that he had become increasingly agitated and—it must be said—in liquor on that account.

Matthew thought, in the fashion of all junior officers: Squalls ahead! He will be in a cursed foul temper, our first few days at sea ... But the matter proved somewhat more weighty than could be measured by this simple balance of self-interest.

When the company rose from table, it was four o'clock and time to go. But Nelson lingered in the drawing-room, striving to keep his prize in sight, perhaps trying against all the odds for a moment alone. After ten minutes, Matthew Lawe knew that he must do his sad duty.

He came close to his captain, who was lurking—there could be no other word—in the lee of a curtain which gave him little else but a view of the unattainable, and said quietly:

'With respect, sir, time is up for us.'

Captain Nelson did not turn. He was holding on to the curtain, which was a bad sign, and his pale face was sweating even in the drawing-room's log-fire heat, which was a worse. The contrast between the gaiety round him—the laughter, the rustling silks, the gallantry of men and the soft compliance of women—and this lover's forlorn isolation was bitter indeed.

But his only answer was: 'We have time enough. Wait for me below.'

'Aye, sir.'

It was all that the bravest subordinate could answer. But after a further ten minutes of impatient loitering in the hall-way, Matthew felt obliged to return. There was a temptation to think: This is the captain: let him set his own timetable, and take the blame if it fails. But there was another law beyond this, which ruled all sailors, and ever would do so: Whether he be high or low, never leave a shipmate on shore when he should be afloat. Matthew obeyed it.

He found his captain in the same situation as before, save that Miss Mary Simpson was still in animated conversation with a scarlet-clad major of the Garrison regiment, and that Captain Nelson was now reduced to distraught misery. Aware of danger, and of many spiteful eyes enjoying this cruel masque, he said:

'Beg leave to remind you, sir. We must return on board, or the light will not serve us.'

This time his captain turned to face him. He was far gone, both in liquor and in despair. But there was something else in his pale face which Matthew had never seen before. It was the acknowledgment that, at this hour, on this day, in this place, there was no more battle to be fought.

'What time is it?'

'Near half past four, sir. We *must* pass Orleans Island to the convoy anchorage before twilight, or we lose a day.'

'I would gladly lose a day here.'

'But we have our orders.'

Nelson turned, looked at the fair Miss Simpson, sighed like a whole company of lovers, and turned back again. 'Aye,' he said. 'We have our damned Navy orders ... Attend me ... Where is my hat?'

In all his naval service Matthew had never heard more shocking words than 'damned Navy orders'. But he swallowed them in favour of victory. His captain, uncertainly, unwillingly, was at last on the move.

Back on board the *Albemarle*, which was creeping down-river in a somewhat hazardous dusk, Matthew reported the day's events to Second Lieutenant Joseph Bromwich. The latter, a wide-awake young fellow who owed all his advancement to his captain, and admired him, and no longer wondered at the surprises of the world, only at its continuous turning, gave his verdict:

'Well done. God will reward you, though you may have to wait awhile for our captain's approval ... I suppose it is the glorious Diana ... At least we have him safely back.'

But they did not have him safely back. Early next morning Matthew, off-watch and gratefully asleep, was awakened by some determined activity on the waist-deck over his head. Then the boatswain's pipe called: 'Boat's crew away!' and there was much running of feet in answer. Then there was Lieutenant Bromwich knocking on his door and entering without further ceremony.

'The captain returns to Quebec,' he began straightway. 'You have the duty, Matt. So rouse out!'

'What's this?' Matthew asked, bemused. 'Why does he return?'

'We shall be told . . .' Bromwich sounded his usual cynical self, though it would not affect his loyalty by a single whisker. 'Belike he forgot to say Thank'ee for his dinner . . . Be what it may, he leaves in a half-hour. The river current is slack and you have a good easterly breeze. All you will lack is your breakfast.'

'Cannot a midshipman take the boat?' Matthew grumbled, half-way into his breeches.

'The cook could take the boat, but it would not do, would it?' Bromwich, leaning back against the bulk-head, looked down at Matthew with an eye more serious than usual. 'Think well on it, Matt. We know what this sudden errand might be . . . He has been walking the deck half the night . . . The First Lieutenant must stay, to command. I have duties about the ship, with the master. So it must be you, *and no one less.*'

Matthew, with one arm into his jacket-sleeve, paused in doubt. 'But with what instructions? What am I? A guard-commander?'

'You are our delegate plenipotentiary. To see that the captain returns safely on board. We are due to sail tomorrow, and that cannot alter.'

'I have known better assignments, 'fore God!'

'So have I,' Bromwich answered. 'But let us all share this one. With you first!'

'Is it swords?' Matthew asked, busy with his belt.

'Grappling hooks, if you will!' Then Bromwich relaxed to a smile. 'Yes, Mr Lawe, it is swords. So buckle on! But pray consider yours as ceremonial.'

It was a strange voyage up-river, beautiful, tranquil, foreboding. By this, the middle day of October, fair Canada had shed her summer greenery with the main warmth of the sun; but now her autumn mantle was even more splendid—a great, continuing oriflamme of orange and scarlet and yellow, as a myriad maple trees signalled the onset of winter with a last blaze of colour against a pure cold sky.

The banks and heights on either side of the St Lawrence were alight with this fiery curtain, as the *Albemarle*'s boat, its oars comfortably stowed, wended its way up-stream under a billowing lugsail. Within the boat there was little answering fire. Captain

Nelson sat wordless and brooding in the stern-sheets. Matthew had the tiller, since he knew best the puzzles of their course. The captain's coxswain, thus usurped of his proud command, was as sullen as he dared.

Matthew had time to think, but scarcely wished to let his thoughts wander. They were sailing through the great Traverse again—but on what an errand! His captain, like General Wolfe before him, seemed set to try to scale a fortress. Would he succeed? —and could the two attempts be set side-by-side without futility and shame rendering this latter day a betrayal of every last duty?

Or was the covert guessing of himself and Joseph Bromwich a betrayal even more shameful?

When the boat, under easy oars again, came alongside the quay of Quebec's lower town, Captain Nelson seemed to be of two minds. Having stepped ashore, smoothed his coat, adjusted his sword-belt and his hat, he addressed Matthew:

'What duties have you on board, Mr Lawe?'

'Nothing of note, sir. Lieutenant Barstow has the deck in my absence. I attend you.'

'There is nothing for you here. I make a farewell call or two . . . Better return to the ship.'

Seeing that his captain was undecided—he had said 'Better return', for him an unusual phrase, rather than a straight 'Return' —Matthew made bold to persuade him otherwise:

'Shall I not wait, sir? You will wish to be back on board in two or three hours. There is the convoy muster, and the order of sailing. Since we lead the way down-river at first light tomorrow——'

'Oh, well enough!' Nelson said testily. He was looking down at his boots, as if they were the voices to guide him. 'Secure the boat alongside, and see me to the town.'

It was an advantage of sorts, and Matthew could be pleased with it. But in the event they never reached the town. A right royal dispute with an unexpected third party intervened, and changed, perhaps, many strands of many lives.

There was a stranger walking about the quay a little way off, taking the air on this crisp sunny day. He was, by his clothes and bearing, a man of substance. He now turned to stare at the two officers, then hurried towards them. But it was no stranger who greeted them: it was their generous host of yesterday and many other days, Mr Alexander Davison the merchant.

Seeing Captain Nelson ashore, he could hardly contain his astonishment, since he had four ships in his convoy and knew their orders. When the first salutations were over, Davison put his question directly:

'But what do you ashore, Captain Nelson? Is not the *Albemarle* sailing with the convoy?'

'In due course, yes.'

'Tomorrow at first light?'

'Even so. I have returned—' Nelson was already uncomfortable, 'to say farewell to—to other persons who have so kindly entertained us.'

Alexander Davison, who was some ten years older than his guest, and had not made his way in the world by bashful hesitation, looked at Nelson closely. He knew the gossip of the town; he had observed certain developments, in respect of a certain attachment, under his own roof; he was not the man to let a friend persist in error, for want of timely warning. He spoke his mind:

'I had thought all farewells were said . . . If you still have any hopes in a particular quarter, I must tell you that they are groundless.'

With Matthew an uneasy witness in the background, Captain Nelson spoke his mind also, from a bursting heart:

'I do not believe it . . . You are right in your supposition, though I must declare it to be none of your business . . . I have returned to offer marriage.'

'It will be useless,' Davison answered, gently enough. 'Pray listen to me. It *is* my business, because I wish you well. I can assure you, the answer will be No, and No again.'

'You have heard her say so?'

'No. Nor ever broached such a subject. One does not speak of these matters to young girls in her situation. Am I a village tattle-tale? . . . But I have enjoyed her father's confidence for many years, and I can warrant that I speak the exact truth.'

'Then I will bend the truth, and break it!' said Nelson in a sudden passion, quite careless of his company. 'I tell you, she will have me, if I persist! If my fortune is below her station, then I will seek another. I will leave the sea!'

'You cannot do that.'

'Who will deny me?'

'I mean, you should not. It would be madness. You have a fine future in the service——' even Davison, the urbane man of the world, was growing agitated. 'I beg you not to cast it aside.'

'There will be no *casting aside*, if she will take me as I am.'

'She will not do so.'

'That is for me to discover.'

With two such determined men, it could only end in a quarrel. Presently it reached a pitch so fierce that Alexander Davison had grown almost contemptuous of this dereliction of duty.

'You have your orders, sir,' he said. 'Not from myself, God knows, but from your own Commander-in-Chief! You stretch them to the limit already, by returning now. You are bound for New York, and then for the West Indies station—and do not ask me how I know, since *I know*! We are not fools, because we are landsmen!'

'You go too far, sir.'

'I go too far because I have your best interests at heart. Your duty is clear. It is to return on board your ship, without delay, and to forget this fantasy. I speak plainly, as a friend.' By chance he caught Matthew's eye at that moment, and, observing his demeanour, claimed him as an ally. 'I believe you have another friend here, captain. What say you, Mr Lawe? Have I not right on my side?'

It was a turn most uneasy. Matthew had been listening to their exchanges with great misgiving, not the least because he would rather have been a hundred miles away than have to see and hear Captain Nelson speak of leaving the sea. For one mere girl?—he must be a madman, or a lecher, or bewitched . . . But he could use none of these words.

Matthew answered with all possible care: 'I only attend the captain, Mr Davison. I am sure you have naught but goodwill for him . . . I can only say, as a junior in the same service, that we would feel it a great loss if——'

But even this humble essay in diplomacy was too much for his superior.

'Hold your tongue, sir!' Nelson cut him short, in a fury. 'Do you take sides in this matter? By God, we will have the boatswain's mate piping up next!'

'I beg your pardon, sir.'

'You will be begging for more than that, if you *say* more!'

The quarrel for two persisted: it grew hot, and their voices high. At last, when Davison had advanced onto most dangerous and delicate ground—the worth of one girl against the worth of future duty—Nelson in a fresh passion broke away, and tried to pass by him on the road into Lower Town.

Davison deliberately barred his way, and by mischance the captain, seeking to avoid, blundered into Matthew Lawe and was brought up all standing by the taller, stronger man.

Nelson, at bay, stared up at him with the most frightful venom. 'Would you use force against your captain?'

Matthew, appalled, scarcely dared any answer. Then he thought: Well, I am into it now, up to my neck and beyond. He cannot kill me. He can only ruin me, and he is in that humour already . . . He answered as bravely as he had ever done, whether to proud captain or boatswain with a rattan cane:

'I believe I would use anything, sir, *for* my captain.'

Five minutes later, in a reversal as swift and astonishing as the end of a summer thunderstorm, they were in the *Albemarle*'s boat and on their way down-river. Alexander Davison, now all compassion and care, travelled with them. 'I have some instructions for one of my captains,' he dissembled. It was the act of a true and understanding friend—and also, perhaps, of a prudent man who wished to see this matter to a safe end.

On board, with both ship and captain sailing according to their orders, Joseph Bromwich gave Matthew a heartfelt 'Well done', and for a space saw fit to address him, in private, as 'Anti-Cupid'. Where it was most dangerous, it was never spoken of again. But Captain Horatio Nelson's cold eye seemed scarcely to rest on his third lieutenant until the *Albemarle* was paid off in Portsmouth, half a year later.

Now, some eight years after, they met face to face, for Captain Nelson had sighted him—or sighted a familiar uniform and drawn closer to examine it. When he had established his man, his greeting was given without any air of surprise:

'Lawe, is it not? Good morning to you.'

'Good morning, sir,' said Matthew, returning a salute of the finest formality for many years. Not knowing whether he was to be hanged or stared out of countenance, he asked warily: 'I hope I see you well, sir?'

Matthew did not see his captain well. Horatio Nelson, never distinguished in person, seemed to have shrunk down to the very minimum of man since they had last met. He was drawn and thin: his keen eyes were the only lively feature in a face yellowish with fever, or exhaustion, or nagging care.

Though he never looked less than a captain—and it was the same now: though the air of command persisted, yet the frame of the picture seemed sadly distressed, as if he had taken more buffets and blows from the passage of the years than the sea had ever dealt him.

To Matthew's surprise, Nelson appeared to share this view. 'You see me *alive*!' he answered tartly. 'I have been wretchedly ill, since my cursed West Indies commission. It may be the station of honour, but for me it was the station of putrid fever! And half-pay idleness in England is *not* the best physic.'

He is the same as myself, Matthew thought, amazed that it should happen to captains and lieutenants alike. He is beached and dry . . . He felt the need to explore further, with a man—not a demi-god—who seemed ready to talk.

'Do you reside in London, sir?'

'Never!' Nelson said roundly. 'It is like Babylon for expense! I am no more than a farmer now; three years ashore, and many more to come, beyond doubt. It is bread and cheese and a humble cottage for me. *My* stone frigate is my own father's Norfolk rectory . . . Oh, I trudge up to London every quarter-day, so that they will not forget my face at the Admiralty. But to what purpose? I kick my heels for an hour in their Whitehall ante-room, and then it is ten minutes with some polite fellow—they are still excessively polite, like executioners who beg leave to strike off your head— and he says, "Well, well, Captain Nelson! You look in the pink, sir!" I do not look in the pink, damn it,' said Captain Nelson. 'I look in the yellow, like a drop't leaf, and a cursed ugly colour it is . . . Then this *fellow* tells me: "To our regret, there are no wars to fight. *Ergo*, there are no ships to be manned. *Ergo*, no officers are wanted, and there is nothing for you today. But do not fail to keep touch with us." And so on, and so forth, for three dry years.' Suddenly Nelson's face, up and down like magic mercury, cleared of all save comradeship. 'By God, Mr Lawe, I have not heard a longer speech since Prince William Henry—you recall Lord Hood's midshipman in the old *Barfleur*, who served under me later?—gave me, me his captain, his views on the royal prerogative as it concerned the precise number of saluting guns due to him on his ship entering Havana Harbour. *It was my ship!* Well, I have no ship now—and that, I fancy, is your own case.'

What a man this had become ... Rising and falling like a puppet, but he fingered his own strings ... The spark had shown in the *Albemarle*: now it was a working flame. Self-knowledge, energy constrained but readily on call, life abundant in the midst of dusty decay—they all shone out of him like the faithful candle-glow which was, in truth, a tall lighthouse twenty miles away.

But why did he speak so free *now*, to a man whom he had no reason either to trust or to honour in this fashion? Was he starved of talk? Was he perhaps not free to swear a trifle, in a Norfolk rectory with a father's reverend ear close by? Was he friendless? That could hardly be.

Beyond all else, Matthew knew that he must somehow cling to this man: not for advancement, but for the secrets of valour and endurance, without which life was no more than a weary, down-hill plod to the grave.

He answered: 'Certainly I lack employment, sir. I have had no berth for near four years.'

'What was your last ship?'

'The *Medway*, sir. Guard-ship at Portsmouth.'

'Prison service! I would as soon go for a soldier!'

Matthew became aware that Captain Nelson, as soon as he had dismissed his own troubles, had been examining him with some care. The sharp blue eyes must have already noted his woeful shortcomings in dress. Now they were busy on his face—and what he saw there only God and Nelson knew, and neither was likely to make a pronouncement ... He answered as boldly as he could:

'I had no better choice, sir, in those years. But I am ready for improvement now, within the hour!'

'So am I,' Captain Nelson replied, fretful again. 'It will come, it will come. Yet we wait on Fate as a farmer waits on rain. Or more likely, as the worm beneath that farmer's boots!' But he had done with self-pity and, it seemed, with his close inspection of his former lieutenant. Like any determined mariner, he straightway changed to another tack. 'I owe you a debt,' he announced, without any paltry sniffing of the breeze. 'Though I did not think so at the time.'

Matthew, morbidly aware of the past, was quick to say: 'Pray do not speak of it, sir.'

'I never shall!' Nelson answered forthrightly. 'You may be sure of that! But it is there none the less . . . Are you married, Mr Lawe?'

'No, sir.'

'I can recommend it,' the captain said, with a shade more formality than fire. 'I have been married these past three years. On the West Indies station—the only benefit that ever came out of it! His Royal Highness Prince William did me the honour of standing up to give the bride away.'

'I offer my congratulations, sir. Have you children?'

'No. Well, a stepson. My wife was a widow, niece of the President of Nevis Island. So you are free to move?'

'Aye, sir.'

'I need a handy man. A man of all trades—secretarial, agricultural, companionable, confidential. In short, a sailor. But it will not be sailor's work.'

'That will come again, sir.'

' 'Fore God, I hope so! Are you willing?'

'With all my heart, sir.'

'The place is Burnham Thorpe, thirty miles from Norwich. The time is not yet—we are painting the parlour, the kitchen, the larder, and the passage-ways, which is the same as saying, we are re-fitting a first-rate ship-of-the-line from Admiral's walk to the orlop deck! . . . Come at mid-summer. You may begin by helping me with the turnips. Agreed?'

'The twenty-first day of June, sir.'

'Even so. We talk of money matters later. Now I bid you good-morning.'

Thus, on a certain summer's day in 1790, Lieutenant Matthew Lawe took a spectral oar in his hand, laid it across his shoulders, and walked in-land to Norfolk.

II

It was, he found, no 'humble cottage' to which his footsteps had been guided. The Parsonage House of Burnham Thorpe was of two storeys, two red-tiled houses put together in the shape of an anchor with one fluke, set in thirty acres of mixed gardens, lawns, vegetable plots, and trees, with a running stream to divide them.

Beyond the trees the wind blew incessantly across bare flat fields, and the curling waves of a shallow North Sea for ever attacked the coast.

In this rural palace—for palace it was to Matthew Lawe after his crimped London lodgings—there dwelt a slender branch of the Nelson family: one man of God, one man of the sea, one wife, and one step-child not yet assigned to any duty.

The Reverend Edmund Nelson, now sixty-eight, was an ageing churchman in whom piety, humility, and a certain contented indolence had combined to fashion a classic country parson, half-way to heaven already and resigned to the remainder of the journey. His son Horatio, not at all resigned to any aspect of idle-ness, was yet in the same situation: waiting for *his* heaven, which was a different goal from his father's pious dream, but equally in the hands of the Almighty—which was, for him, the Board of Admiralty.

His daughter-in-law, Frances, born Woolward, widowed as Nisbet, now to be known as Fanny Nelson, was thirty-two, as was her sailor husband: a girl who had blossomed early in the hot-house of the Caribbean, and whose present transplant to frigid Norfolk had rendered plain, dull, and ill.

Her son Josiah, at the age of ten, was happily at boarding school for much of the year, and hopefully thought of for the Navy.

To this tranquil household Lieutenant Lawe, 'a valued sea-officer and friend' in Nelson's words of introduction, was presently admitted.

The warmth of his welcome varied, as did the household itself. It was kindly from Nelson, who seemed to need a trusted male confidant more than he needed any other member of the human species. It was courteous from the rector, who saw no evil nor threat from anyone, and reserved from Mrs Frances Nelson, whose preoccupation was a troublesome chilblain beneath a nose which, even at high summer, sought and found a climate inimical.

It was less than anything from the Parsonage House itself: in summer draughty, in autumn cold and damp, and in winter Arctic.

At this latter season the house was beset. The biting north-east wind from the sea, and thence from Siberia, seemed to have power to cut through the walls. It did nothing so cheerful as to whistle:

it sighed or it howled or it screamed, while all the time it tugged at a man's trees, or his house, or his animals, or his clothing if he ventured out-doors.

Out-doors was a land of tossing branches, of laden air coarse with sand and salt, and of whirling windmills which ground the corn which kept men alive to endure this malice.

Mrs Nelson would take to her bed for days, sustained by a log fire in the grate, a procession of warming-pans and hot stones wrapped in flannel, beef-tea spiced with cloves, and physicks to assuage her various ills. The father bore all with stoic Christianity, mufflers, and mittens. Nelson had rheumatic pinchings, and moved about like some relict from the Aged Sailors' Refuge. Young Josiah, home for the holidays, snuffled like a puppy with the distemper.

Matthew's attic room, with no refinements to warm it, might have been the mast-head in winter . . . There were times when he bound his feet with strips of buckram before retiring, and slept in his boat-cloak. Yet for him, nothing in this house could truly displease or discomfort. He had come to anchor in a secure berth, he had a chartable place in the world at last, and it was blissful ease to enjoy it and to earn it willingly.

His duties were what Captain Nelson had prophesied, on the quay above London Pool: a morsel of everything, from letter-writing to digging roots, from carrying prayer-books in church to drenching a sick calf. But they lay mostly in a realm never cited plainly by Nelson: which was, to serve as a companion for an achingly bored man.

First they had talked avidly, for nearly six months, of the past which was the sea: Nelson's lost love, his Holy Grail, his most pure lust. It was more than rank which dictated who was the talker and who the listener; of these two, the captain without a ship and the beached lieutenant, the captain had the main tale and a bursting tongue to tell it.

'The worst of idleness is that it breeds upon itself,' he said one day as they sat together under a beech tree in an angle of the house: one September day which, by some mistake of nature, was fair and sunny and warm. 'I am idle as a sailor, and thus the skill rusts. Soon I shall be idle in everything . . . I saw it in the West Indies, when I was captain of the *Boreas*. We did *nothing*, for months at a time: it was half-way between peace and war, so the order was: "Remain in harbour. Provoke no quarrels with anyone. But keep your ship trim." Keep your ship trim! We were nearly aground on our own beef-bones! You recall English Harbour in Antigua? I could wish it did not bear the name of England—a tideless pool, full of the filth from fifty ships at anchor. The mosquitoes swooped like sparrow-hawks. I had thirty of my good fellows ashore with the fever. And in this detestable place, I was ordered to keep an inactive hulk *trim*!'

His face under today's clean Norfolk sky had become so enraged that Matthew had to turn his eyes away, for fear of smiling. This was Horatio Nelson at his truest: seizing upon foolishness or imperfection as if the Almighty had conspired with certain of His more stupid creatures to place obstacles in his way . . . After a moment he said soothingly:

'At the least, sir, you provoked no quarrels.'

He had not been soothing. 'Fiddle-faddle, Mr Lawe!' Nelson answered him, as if Matthew were a part—and no mean part—of this same conspiracy. 'I never had more quarrels in my life! With my own Commander-in-Chief! With every planter and merchant in the Caribbean! After the revolution of '76, we were bound by the Navigation Act to forbid the former American colonists any commerce with our possessions in these waters. They were now foreigners, by their own rebellion and their own choice. My ship, with the rest, was there to enforce the embargo. But what happened?'

Nelson, his mind fixed firmly on an old quarrel, now paused so long that it seemed he might never return to the present. Matthew prompted him:

'Pray tell me, sir.'

'The Navigation Act was winked at!' Nelson answered, with instant indignation. 'By ourselves, the Navy! The Yankee trade was so long established that no American ship, and none of our islanders, would willingly give it up. So we were counselled to let them be. Monstrous cowardice! . . . I wrote eleven despatches to Admiral Hughes, commander in the Leewards, pressing for action. What did I have in reply? Soft soap bubbles! "Do nothing," I was told. "All will settle itself." When I insisted that we must act, I made such a name for myself that no one would speak to me. In some of the islands, I could not even leave my ship for fear of insult, or worse. But all I wished to do was to enforce our good English law, by Parliament established.'

The fierce tussle seemed a thousand miles away from a Norfolk garden in the peace and sunshine of late-relenting summer. But clearly it was most real to this man, still clinging to principles of law and conduct, still hazarding his career with senior officers and influential magnates alike, in the pursuit of honour untarnished. One must admire him, Matthew thought, even if one must also think: Perhaps a little caution, a pinch of the snuff of accommodation, would not have run counter to God's almighty plan for an ordered universe . . .

Matthew knew in his heart that this could never have been Captain Nelson's course-to-steer. But then, where was Captain Nelson today, five years later, and what might be his second thoughts?

They were, it seemed, the mirror of his first. He remained quietly content with his conduct.

'I was justified at the last,' he said, 'though a damned uncomfortable time I had, with lost friends and a deal of whispering within the service. As soon as the Act was enforced, though feebly, a few merchants suffered in their pockets, and since I had taken a dozen prizes they went so far as to sue me. A hazardous moment for a frigate captain, engaged to marry and without private means . . . But when I was sued for forty thousand pounds—they might as well have said, for the National Debt!—I was told that the Admiralty would support me and the Treasury Board would bear my costs.'

Matthew looked at his captain. There was no triumph in his voice, nor in his pale face. 'Sir, that must have been a fine moment.'

'Well, I did not feel so lonely . . . Until the next tug-of-war.'

'There were others?'

'Certainly!' Nelson said, and then smiled. 'There are more perils to captaincy, Mr Lawe, than a lee-shore or a main-mast lost to a broadside . . . The chiefest, at that same time, was a plague of Customs and store-keepers' frauds, by our own Crown Agents all across the islands. I intervened again . . . What could I do, seeing his Majesty robbed by these fat villains of landsmen? I uncovered huge peculations among the store-keepers in the dockyards. Two million pounds sticking to English fingers was too steep for me, and I said so. For that I was called an officious busy-body, and more. There were even those who said that I took this interest out of spite, because I had not received my share of the spoil . . . Well, I was proved right in the end, and some of the principal thieves went to gaol. But not before the *Boreas* was sent home, and paid off.' He sighed—the sigh of all just men who find the world lacking in gratitude. 'The moral was freely hinted at—one can be too zealous. *But I do not believe it, and never shall.*'

At Christmas-tide they were house-bound. The snow-scene, with the wind whining through bare trees before laying icy fingers on the eaves of their refuge, was mournful in the extreme. Norfolk might have been the northern wastes of Russia. Mrs Nelson became invisible, like a sick ghost not to be numbered among the living inmates. The rector, with two churches to serve, laboured mightily, with Matthew as his escort and occasional broad-backed steed, to make his Christian rounds.

Aloof in his study-room, Nelson read and wrote. He had formed an affection—and a respect—for another harassed sailor of an earlier day, Captain William Dampier, whose account of his voyage round the world some hundred years before had excited his interest. 'Now there's a brave fellow!' he would exclaim, to anyone who would listen—and this, in a meagre season, could only be Matthew.

Captain Dampier's journals set Nelson, prisoned in a dull brown room in a cold grey world, to dreaming once more of vanished glories. He read of adventures in mysterious Java seas, of buccaneers and honest sailors, of maroonings, rebellions, desperate fights, splendid voyages which ended in triumph: of Dampier's sailing-master Alexander Selkirk, the castaway already made famous in song and story as 'Robinson Crusoe'; of men challenging the mighty sea and conquering it by a greater strength and endurance.

He read of a sardonic footnote to a turbulent life: how William Dampier, returning from a three years' naval cruise to Australia and beyond, was charged with undue harshness to his crew and fined the whole of his pay for the voyage.

'Official robbery!' Nelson declared to Matthew, who was waiting to serve as secretary; and then, in brooding discontent: '*Nothing changes!*'

Nothing changed; and all seemed to give the Sneer Direct to his own futility.

He could be briefly happy with a smaller task—since he had no other: a correspondence on a matter dear to his heart, the saving of life at sea.

'If I cannot kill an enemy, I may at least rescue a friend,' he had said to Matthew, when he first received a letter from Mr William Wouldhave of South Shields in Durham, another coastal county under siege by the North Sea. Mr Wouldhave, a simple cottager who proclaimed himself an inventor, and his friend Mr Henry Greathead, a boat-builder, had conceived the idea of an 'unimmergible boat', which could put to sea in any weather and thus to the aid of ships in peril.

'I came to love their names,' the unaccountable Nelson once confessed, 'before I knew what they were at. Mr Wouldhave *will have* this life-saving boat. Mr Greathead has a *great heart* in the same endeavour. They seek my advice, which is flattering; I fancy that they also seek my money, which would be, alas, a triumph of hope over reality. I plan to give them plenty of the one and perhaps a little of the other. As to their plans——'

As to their plans, Nelson the sailor and Matthew Lawe his trusted bow-man could recognize immediately their value. They called for a small boat which could never be wrecked by the sea, only by the harsh rocks at its margin: a boat with a tough straight keel, and two high-peaked ends in which air-chambers and a filling of cork supplied buoyancy for ever; a boat which might roll over and still right itself.

'Pray record this letter,' Nelson now said, dismissing Captain Dampier's ill-treatment and the whole world of mutton-headed authority for an hour at the least. 'My dear Mr Wouldhave, Your new plans for the "Unimmergible" continue to claim my attention. I would that you could invent some Engine which might drive such a vessel better than oars or sail. But since that is impossible, we must ensure that our men survive to supply the muscle. If without loss of stability we might raise the height of the rowing-platform so that this will be self-draining as well as the boat self-righting, we would keep our oarsmen, lashed to their thwarts, steadily at work. I truly believe that this life-boat service, as I choose to call it——'

He truly believed. He always believed. Matthew, scribbling

furiously so as not to be outstripped, still found a moment for admiration. At the last, the boat would float, and the men would float, and this man would float, confounding more than the sea.

He would swim again, God bless his buoyant heart, and so would ships uncounted, and ships despised and rejected, and England herself, foundering in decay.

If only the tide would turn.

The turn of the tide was not yet. As soon as spring-time was advanced, Captain Nelson made yet another long and costly expedition to London, by post at twenty miles a stage and sixpence a mile for the journey, and returned as dejected and dispirited as he had ever been. Even though there were rumours of war and rearming, the answer to this man was the same as it had been for four degrading years: No ship for him.

Sitting with Matthew in the gloom of the study, with a log-fire to warm a weary body and mulled claret to cheer his spirits, Nelson recounted his day in London. He had been allowed a brief meeting with Lord Hood, a Naval Lord and the admiral best known to him from the old days on the West Indies station; but it was Lord Hood who gave him the official sentence, and Lord Hood who acknowledged his farewell bow before passing on to weightier matters.

'I expected nothing better,' Nelson finished sombrely. 'I know how the wind sits in that quarter ... I have heard the same words from his Lordship before—and more harshly too.' He hesitated, then decided to round out his story, as if the telling of it might somehow assuage defeat. 'It was some two years back—the summer of '89. I asked him straightly why, with my seniority, I could not have a ship, and he gave me a straight answer—the straightest of my life. He told me that the king had formed an opinion against me, and that this was the true reason, and that it would not change.'

'But how was this?' Matthew asked, astonished.

'I do not know. And could not ask.'

'But I thought his Majesty had commended you.'

'I thought he was my friend!' Nelson answered bitterly. It was clear that his spirit had been shattered, and that the deep wound remained. 'But I was wrong, and my nose was rubbed in it ... I can tell you, I was never more down-cast.'

A log flared, and showed a face down-cast indeed, at the mortal
end of hope.

'Was it perhaps the business of the dock-yard frauds?' Matthew
ventured. 'Or the troubles with the Navigation Act? You told me
once, sir—' he was unsure how to phrase this, 'you used the
words "too much zeal", or "too zealous".'

'It was all things,' said Nelson. 'I can even fancy that I was
blamed for Prince William Henry's scrapes and scandals. He
was put in my care on the station, and I was told privately to
keep a weather eye on him. One might as well keep a weather eye
on a charging bull!'

'Scrapes?' Matthew echoed the word, mystified. To the best of
his loyal belief, Princes of the Blood did not get themselves into
scrapes. Scrapes had been out of fashion, it was held, since that
merry monarch Charles the Second discharged his cargo of them.

'Among other indelicacies,' Nelson interrupted his train of
thought, with coarse deliberation, 'His Royal Highness fell somewhat foul of a girl in Antigua. Was that my fault? *My example?*
Was I to rouse him out of all the beds he strayed into? Should
I have stalked him about the islands with a tube of mercury?' His
humour seemed to be headed for a burst of fury, and then tiredness and disillusion put him down again. 'Oh, what matter now? A
prince suffers a passing blemish on the escutcheon, a post-captain
loses a ship. Put that to your philosophers, and see them scratch
their heads for the metaphysical answer.'

'Sir, it matters to me, and infinitely more to yourself.'

'You are a good fellow, Lawe.' Nelson was equable once more,
as the fire died and the mulled wine ebbed. 'The best of fellows,
but you lack the cynic's touch. Let me finish a miserable story.
After hearing Lord Hood's verdict, I did not touch Babylon for a
twelve-month. On half-pay of £120 a year, how could I afford
such idle waste? . . . When I told you that I travelled to London
each quarter-day, it was not the whole truth. I had always done so
in the past, full of hopes, full of dreams, but after that buffet I
could not bear it. On the day that you and I met, it was my first
visit for a year. The answer had been the same. Nelson? No ship
for Nelson. Wait until signalled.'

Slowly the seasons turned and turned again, crawling like cripples
with no hope but heaven. Early in 1792 the Reverend Edmund

Nelson, who was seventy and failing, left the household at Burnham Thorpe and removed to a nearby village, more convenient to his many tasks which he would fulfil to the end of his ministry. But at Christmas time, already making his farewells, he had talked to Matthew of his despairing son.

'I know Horace to be unhappy,' he said, when Matthew had spoken of the captain's discontent. It was 'Horace' within the family circle: 'Horatio' was for formal wear, in that great outside world which had forgotten him. 'Yet he has fared so well in the service. A post-captain at twenty-one—I was so proud of him on that day! Cannot he be content with it?'

'It would be asking much, sir, when he has so much to give.'

'But he has this farm to keep him occupied, and his dear wife who is so attentive, and Josiah growing up.'

'Even so.'

The rector sighed. 'If only he could see it as God's will . . .'

He went on to speak of resignation and humility, though Matthew could scarcely listen, so far from a sailor's world were these. They were the father's gifts and the father's solace, not the son's. But had the father never wished to be a bishop? . . . He watched the old man's grey face, which was puzzled and tranquil at the same time, and decided that he had not.

At the homily's end, Matthew said: 'With all respect, sir, Captain Nelson remains ambitious, which may also be God's will.'

'Yes, yes,' the old man answered, suddenly fretful. 'When one has tasted glory . . . But do you, who know him well, believe that he can never be happy here?'

'Not truly, no.'

'What then?'

'He lives in hope, as I do.'

It may have been that the rector, before he left the Parsonage House, spoke of this conversation with someone even more closely concerned. A few weeks afterwards, with the household shrunken and the daily round as quiet as a mouse with the dropsy, Mrs Nelson broached the subject of a husband in limbo, to a man who might know more of it than she did herself.

Matthew had talked little with his hostess during his two years' sojourn, and nothing of any intimacy. Now, as they sat in the faded parlour, she with her tapestry frame and he with idle hands, he expected nothing more than an hour's civil exchange before dinner which—it being a Tuesday—was boiled mutton, caper sauce, a side-dish of beef brisket, and cheese as pale and flat as the Norfolk landscape.

But to his surprise she had something of import to say to him, and embarked on it without any warning.

'Pray tell me, Mr Lawe,' she said, 'is Captain Nelson content with his life ashore?'

He was startled, and looked at her with a closer attention than he had ever dared—or wished. He saw what he had always seen: a woman withdrawn and silent, pale as her husband, suffering as ever from a winter cold: dressed in severest grey, as if she were widowed already: a woman apart who knew not why, and perhaps did not wish to know. Could he prescribe a cure for such ailments? Was this the first step in seeking him as an ally? He could do no more than feel his way.

'On my life, ma'am, I do not know. Of course he would wish to be at sea. He is convinced that he should be. So he waits, with patience.'

'And cannot be resigned?'

'Resigned' was the rector's word. Matthew understood it, and did not like it. Why should a man be 'resigned', if he still had the breath of life in him? It might serve for monks, and the feeble old, and prisoners who would never be let loose till their pauper's coffin was nailed down. But for men? For sailors? For *Nelson*?

'As to that, ma'am,' he answered warily, 'the captain must speak for himself.'

Mrs Nelson stitched for a space, and eyed him once across the tapestry frame, and stitched again, and sighed. It was clear that she had expected more, and was now herself 'resigned'—to disappointment. At length she said:

'We all wait with patience. I wait to return to the sunshine, but will not see it. So I do my best, loyally.'

The ground was extremely dangerous. It meant that her best was not good enough, and she was telling him so, with some resentment. But why no happy man at her side? Why no joy, no loving concord? *Why no children?* Not the bravest lieutenant in the world would have ventured his toes into such private strife.

By good fortune there was now an interruption, though not of the best quality. The child of another marriage, stepson Josiah, came noisily into the room, and without greeting them thumped down into an armchair. At twelve years old he was graceless and awkward, and stubborn in his conviction that the big world must revolve around his small person.

He was one of God's unlovable, and, Matthew was sure, would grow more unlovable still. The great peaks of imperfection still lay ahead.

'Mama,' he began, on a most determined whine, 'William says I may not have a piece of the fruit-cake.'

'That is so, dear.'

'He said, "Put that down", quite roughly! How can he forbid me? He is a servant.'

'He forbade you because it is near to dinner-time, and he knew I would not approve it.'

'But I am hungry.'

'Then you will eat a good dinner.'

Josiah wriggled and threw his legs about. 'He had no right to speak so,' he complained, and then grew spiteful. 'When I grow up, I will discharge him!'

If you grow up, Matthew thought, and wished he might say it aloud. But he could not—nor, to a loving mother, could he offer even a shadow of a hint that Master Josiah Nisbet might be no small element in that land-sickness which made sailors yearn for the sea.

The seasons turned once more. All that year they worked in the garden, and watched men digging and ditching and hedging and harvesting—a poor yield in a barren time. They dug themselves, and helped to pick fruit, and made a little water-course across a little desert. In the autumn they walked among the bare trees. 'Two thousand of those go to build a ship-of-the-line,' Nelson would say, eyeing a broad stand of oak. 'I *thirst* to see them cut down and put to use!'

There was no country sport for the men of the Parsonage House. 'I cannot shoot,' Nelson said—a strange confession indeed from a fighting sea-officer, 'and therefore have no licence.' It might have been his living epitaph, in this dead season.

Reading the political news from France in the tumultuous, murderous year of revolution, he went to London once more, as did a thousand other hopeful officers, to offer for service in the

fleet. The fleet remained deaf and blind. Nelson returned despondent, and shivering with more than the miserable cold, and
took to his bed for three days.

'Malaria, or the remains of it,' Matthew was informed, when
at last the pale yellow ghost of a naval officer emerged from his
sick-room. 'I have suffered from it since I was a midshipman of
seventeen. Trincomalee, on the Far East station—another cursed
trap for young sailors. When it strikes, I lie down and sweat awhile.
Then it leaves me alone for a year or two.'

'I did not know, sir.'

'Oh, I have every disease in the Pharmacopoeia,' Nelson
answered—and indeed he looked as if this might be true. 'But
they seem to forget me when I am afloat . . .' He was standing at
the study window, gazing out across a garden in the bleak grip of
winter towards a farming flat-land waiting for the next spring.
There was something in the slight drooping shoulders which said:
That spring is six months away. 'After these wretched bouts, I
have a kind of curtain over both my eyes, a pale film which comes
and goes. I cannot see the distant land—even now, I cannot see
the far ten-acre.' He turned about suddenly. 'I do not wish to! I
hate the land! Miserable grudging stuff that will not *give*! What
am I doing here?'

'The same as ever, sir,' Matthew answered. He knew from long
past that there was nothing he could do to rally Captain Nelson
from these desperate moods: when the tide turned, it must turn
from within. 'You wait for better days.'

'Suppose that there are none. What are we? Navy officers? I
have not been a true Navy officer since the *Boreas* paid off in 1787.
Five full years. Five empty years! . . . It drives me near to madness! We are sea-animals, you and I. We came from it, and we
must return to it. Either we swim, or we die.'

'There are lives to be lived on shore,' said Matthew, who was
beginning to believe it.

'Worthless lives. There is no *life* for me at Burnham. My dog
does better. I tell you, I must get back, if I have to take service with
the Russians!'

He had never been lower in his waste-land of despair. After a
silent, brooding supper, he went to bed that night shivering again,
in pain from his eyes, in greater desolation of heart. The sea-animal,

trapped above the tide-mark, had begun to gasp for that element which alone would keep it alive.

Then all changed.

III

It changed with the ultimate in gory signals: the severing of a man's head from his body, a great gout of royal blood which drenched the executioner's feet, and a falling trophy caught neatly in a wicker basket as the mob howled its triumph. King Louis XVI of France, guillotined, had more power to move hearts and minds than he ever enjoyed as a whole man.

The story was heard with horror in England, which had not beheaded a king for 144 years and thought that such barbarities were vanished with the evil past. But it was only the last chapter in a sequence which now seemed to threaten their own health.

The fever in France, reaching its first climax with the storming of the Bastille on the fourteenth day of July, some three years earlier, and now flaming up again with the execution of their monarch, had been too violent for France to contain. It seemed likely that she did not wish to.

The fever spread, not only throughout the continent of Europe but across the Channel to her nearest neighbour, England; and most men of consequence and sobriety did not welcome the infection. Aping the fiery Gallic mode, revolutionary clubs were formed even in peaceable Norfolk, with tap-room meetings for 'Friends of the People', 'Resolution Men', and such-like.

Soon the mobs, the riots, the rick-burning, the insolence of common men over-blown with the promise of power, threatened to outstrip the magistrates. Presently another king gave another signal.

George III of England, 'Farmer George' to the majority of his friendly subjects, delivered himself of a speech to the faithful Commons. Carefully phrased by more experienced hands, it stated in cold terms that England could no longer be neutral in continental affairs, since foreign persons were fomenting troubles within his own happy realm. It was therefore the royal will that his military and naval forces must be increased, to preserve internal stability and external strength.

The vulgar translation of this was more direct: the 'French disease' was *not* to spread to these purer parts. If it needed an Army to quell it at home, and a Navy to ward it off, then an Army and a Navy were the twin answers. If it needed a declaration of war . . .

The hint was certainly enough for the Navy. Within a week their ships were being floated out of dock: their sailors were summoned, and came running.

Among these was Captain Horatio Nelson, who did not wait for a sign which might never reach him. At the first lifting of a January winter storm which had made the flooded roads impassable, he posted to London; and on an impulse, a strange prejudgment which was to prove exactly true, he took his lieutenant with him.

Lieutenant Matthew Lawe, whose three-year employment had enabled him to purchase at least one dress-uniform worthy of the name, and who now wore it with pride, waited in the Park of St James for the return of Captain Nelson from the Admiralty building nearby. This had been so thronged when they first presented themselves, and the office flunkeys still so proud and peremptory in their dealings with the rush of naval place-seekers, that Nelson had decided he would make better progress alone.

Matthew divided his watch between the Admiralty and the royal ducks which had made the park their domain. Though the former remained austere and aloof, with no indication of fortune good or ill, the omens from the ducks were excellent. Never had they swum about so briskly: never had they seemed happier or more prosperous in their chosen element.

They swam in convoy, in line ahead, in line abreast: they executed turns and circlings and bursts of speed which would have left a squadron of frigates limping in disarray. If the royal ducks were in such jaunty commission, then surely the royal ships could not be far behind.

He would know soon—and he would know *now*! Hasting towards him, almost running, was a figure easily to be recognized and yet almost unknown in its present guise: one Captain Horatio Nelson, with a light in his eye, a smile on his lips, and colour in his cheeks which had not been seen by mortal man or woman for more than five years.

He drew near: the smile became almost laughter, and the light in his eye a gleam bright enough to serve the mast-head lantern at midnight. There could be no mistaking: Nelson was his buoyant self again, and bearing perhaps the greatest tidings of his life.

'I have a ship!' he said as soon as he was within earshot. His meagre figure seemed to have grown tall and proud as a main-mast. 'A ship-of-the-line! My first! What say you to that, Mr Lawe?'

'I say, God be praised.' The bustle and stir of their meeting was such that the nearest flight of ducks took wing, as if heaven-bound. Another true omen ... Matthew, overjoyed, stepped forward. 'That is wonderful news, sir. I beg leave to shake you by the hand.'

The hand-clasp was warm and fervent: at that moment, Nelson would have shaken hands with a blind beggar—or one of the ducks. But there were people round them, smiling or staring, and he recollected himself. By agreement they turned aside, and began to stroll across the green of the park.

'A ship-of-the-line, you said, sir?'

'Aye. I am promised a sixty-four. *Now* they smile upon me!' But Captain Nelson was too happy to dwell on past neglect. 'A seventy-four was offered, if I wished to wait some months, but I closed for the one most ready. Five years is long enough ... So I am off at last.'

'What ship is it?'

'That is not yet decided. One of three, I believe. But as soon as I know it, you must return to Burnham.'

'Burnham?' Matthew said, taken aback.

'Burnham.' Nelson glanced sideways at him. 'Be not downcast, Mr Lawe. You are with me, named for a berth already. And Josiah too, as a youngster—he will benefit, I hope, from mid-shipman's service under my direct eye ... But *your* first task is to raise men in Norfolk. I fancy there are many who will come forward, and the fewer pressed men and gaol-birds in my ship, the better we shall be.'

'I thank you most warmly, sir.'

'You have fully earned it, these past foul-weather years. And so, by God, have I!' Nelson let his eye wander round the pleasant green of the royal park, and then to the stern Admiralty building which had proved such a palace of patronage. 'Well, we must part for a space. I seek an appointment with Davison.' He looked at Matthew with a hint of mischief in his eye: on this happy day which had turned grey to blue, a stagnant pond to a free-flowing sea, there was no aspect of the tormenting past which could not be treated as a jest. 'You may recall Mr Alexander Davison, of the city of Quebec?'

'Why—yes, sir.'

'He now resides in London. He is still my man of affairs, having acted as my prize agent in the old days. Another true friend. I hope to repay him for the lean times.' Then he nodded abruptly, becoming a post-captain again with more than half-pay, half-penny pinching on his mind. 'Meet me at our lodging at two o'clock. We will, for a change, *banquet!*'

They waited nearly three weeks, in pleasurable industry and ease, for the name of Captain Nelson's new command. Then it was given: the *Agamemnon*, the promised 64-gun battle-ship, lying at Chatham in the last stage of dockyard care. She needed a full crew, and that was all.

Nelson, once more, was pleased beyond measure. 'She is one of the finest afloat,' he declared. 'Built at Buckler's Hard in Hampshire. Down there, they can still tell a ship from a plough-share! I would not exchange now, if I were offered the *Victory* . . . I must attend her at Chatham, and you are needed back in Norfolk. Spread the good news. Find me men, willing men. Post bills in the taverns, speak to the mayors and the councillors. Promise nothing but hard work, fair treatment, the chance of action and prizes, and the utmost care as long as I command. If that is not enough, then the breed of English sailors is dead!'

The breed was not dead, however neglectful had been their treatment. In the time allowed him Matthew Lawe raised 200 volunteers, ready to serve (as he was not ashamed to declare) 'a leader worthy of love'. When Nelson returned to Burnham to collect his sea-gear and bid his family farewell, he had two things to say to his lieutenant.

The first was 'Well done,' and the second: 'Our mutual friend Mr Davison sends his warmest regards to you.' After that, it was all packing, and a brief round of visits, and then the coach to London with Lieutenant Lawe and Midshipman Josiah Nisbet in attendance. The tie with the land, so long and wearisome, was severed at least as readily as a French king's head.

Thus, in the sixth year of his exile, leaving behind a proud father now past seventy and an obstinately childless wife of thirty-five, Nelson went joyfully back to sea. He was to join Lord Hood's squadron in the Mediterranean: Lord Hood, now a friend liberated from royal petulance, whose flag would be hoisted in the *Victory*.

Of *Agamemnon* it was said that she could out-sail the fleet. Thus she did, and so did he.

By the time that Lord Hood and his squadron of fourteen line-ships had spent a month cruising off the great French port of Toulon, in the hope of enticing their fleet to come out, it had grown clear that the enemy was not planning to join such a hazardous game. The reasons might be guessed at, and were presently confirmed by British agents.

Fleet discipline, under the heady wine of revolution, had broken down, and more than half the officers, unrepentant Royalists, had been led ashore to be guillotined at Marseilles. Thus the French ships chose to stay where they were, for as long as they might. Toulon was strong, well-fortified, not to be breached from the sea. Let the British fleet waste their time on it, if that was what they wished.

There was only one answer, one alternative to assault: a taut blockade and the pinching hunger which must go with it, since a poor countryside in turmoil could not sustain this crowded port. Hood put this in train, and achieved his aim sooner than he had expected. Toulon, with its ships and docks, its reputation as the stronghold of Europe, had first been invested in June; at the end of August, broken by starvation and internal strife, the city hung out its white flags and surrendered.

There was no time wasted in rejoicing, though the extra gills of rum went round and the men lined their bulwarks, cheering and counter-cheering whenever they passed another British ship on blockade duty. Lord Hood knew that Toulon could only be held by soldiers, who must enter and take firm control before an effort was made to rescue it by a French land force. The nearest troops, Neapolitans who were allies by treaty, were at Naples. They must be summoned to join.

For this commission, which might require a cool and persuasive man as well as a fast ship, he detached Captain Nelson and *Agamemnon*, which had in a short space already made a name for herself in the twin realms of speed and competence.

Nelson's orders were clear and yet liberal, as suited a Com-mander-in-Chief who was learning to trust him: to deliver urgent despatches to the British Ambassador, Sir William Hamilton, who

enjoyed the more sonorous title of Envoy and Minister Plenipotentiary to the Court of the Two Sicilies at Naples: to supplement
these writings with any argument, explanation, or detail which
Sir William Hamilton (who had negotiated the treaty of friendship) might need in his representations to the King of Naples; and
to remain on that station for as long as was necessary before rejoining.

He was not to neglect any opportunity of harassing such enemy
ships as might be within his reach, once his dispatches were
delivered.

It was blissful freedom and important service wrapped in the
same enterprise, and Nelson had never been happier in his life.
With some five hundred sea-miles to run, coasting past the sublime
heights of Corsica on the way, his *Agamemnon* was, for the term of
her detachment, a private ship on the direct business of her
sovereign. As soon as he set his prow eastwards into the gentle
Mediterranean blue, he entered a sailor's heaven not glimpsed for
six years, yet no more than a prelude to the noble music of his
time.

On a certain September morning when the *Agamemnon* had one
more day of her voyage to run, and all the world was fair, and
all the sky light blue and all the sea dark, and all their billowing
canvas rose tier by tier in triple towers of white strength, Captain
Horatio Nelson made his customary appearance on deck.

He looked what he had speedily become in recent months: a
small resplendent lion of energy on whose narrow shoulders the
Agamemnon was borne. As soon as he was sighted, all his officers,
and all sailors not immediately at work, withdrew to the lee side,
leaving their captain with half the breadth of the quarter-deck
for his private use.

The withdrawal was a mark of respect, and a mark also of
the loneliness of command, even more apparent in a big ship-of-
the-line than in anything else afloat. There was a vast gulf separating this single man from the six hundred others on board, and no
human being in sight would dare to cross it unless beckoned,
nor address a word to his captain until commanded to answer.

Nelson waited his own good time before acknowledging the
presence of his ship's company. Having glanced aloft at their taut
sails, and astern at the helmsman, whose eyes were suddenly

linked to the compass-card as if by marine glue, he took a turn
or two along the length of the quarter-deck.

He bent to examine a coiled rope which was the finished end
of one of the weather braces, while John Wilson the master watched
him anxiously. That coil had been fairly stowed, not five minutes
earlier . . . Then he turned and walked aft again, and stood still
under the shadow of the wheel, his feet braced against the gentle
roll which was the *Agamemnon*'s eternal dance when running proud
before the wind.

Matthew, standing alert among a group which included Wilson
the master, Martin Hinton the first lieutenant, and young Josiah
Nisbet, now marvellously transformed into a midshipman as
smart as their paint-work—Matthew watched his captain as
keenly as any. He continued to be astonished at another, greater
transformation, the change which had been wrought in the man
now enthroned.

Less than a year previous, they had both been dull clodhoppers
on a Norfolk farm, thirsting for that sea-employment which never
came. Now the one was captain—captain next to God—of a fine
fighting ship under his most exact command, where not a sail
would furl nor a gun be silenced before duty was done, according
to a rule not to be questioned, nor neglected in the smallest
degree.

Matthew Lawe had prospered well in the shade of this eminence,
and, giving of his best, had found the best reward, which was trust.

Captain Nelson's horizon was now wonderfully wide. He had
been seen challenging his midshipmen to a race aloft to the top-
mast tree, and had arrived first, to welcome certain nervous young
gentlemen who now knew that the great ascent must be easy, if a
decrepit old fellow of more than thirty years could groan and grind
his way to this pinnacle. He had also (so far as a post-captain
could) encouraged Lord Hood, his commander-in-chief, to per-
severe in an arduous blockade, and led the way in its pursuit.

Work indeed was the balm for all the wretched boredom of the
past. Work was the magic wand of advancement. Work was—
work was now!

The instant fact was signalled by a public greeting, customary
yet still as new and as welcome as the dawn of each day. Captain
Nelson, fashioning his own dawn, stood a trifle taller as he turned
to beckon his first lieutenant. It would have been unthinkable to

summon any other man before this one. If the rank of captain was sacrosanct, so was every step on the ladder which might lead to it.

Martin Hinton came forward, uncovered, and waited in silence. He was tall, and strong, and confident. But he would not have spoken first, any more than an oyster would have opened its maw to cry 'I am a native of Whitstable! Do me honour!'

Nelson returned his salute, and opened a new day with new speech.

'Good morning, Mr Hinton.'

'Good morning, sir.'

'Anything worth the reporting?'

'No, sir. Course, east-south-east. Distance made good since yesterday noon, one hundred thirty miles. No land in sight. No ships spoken or sighted. No men sick. One man to be reported to you.'

'For why?'

'Sir—' and now Lieutenant Hinton changed to the disciplinary speech of the master-at-arms, 'I report top-man Adam Ratcliff, being slow to turn out, slow aloft, and slow off the yard. Second offence.'

Captain Nelson looked at him carefully. 'In your opinion, Mr Hinton, a flogging matter?'

'No, sir.' Hinton, aware as was everyone else on board of a strict yet humane captain, tailored his advice in the same direction. 'Better he were logged as incurable sea-sick, and put among the waisters.'

'Sea-sick?' Nelson, after half a life-time afloat, was still subject to this same sailor's curse, as all his officers knew, and yet he still survived every wretched morning of every rough day at sea, in order to command and direct. He might have been fiercely contemptuous of another's weakness, but he was not. He was Nelson. 'Sea-sick? Forgive the poor cripple . . . No flogging in the world will ease a putrid stomach . . . Tell Ratcliff his only cure—go sit under a tree!'

'Dismiss the charge, sir?'

'Aye, dismiss. And see him taken off the top-yards. Better a live cook's mate than a dead seaman.' Done with domesticity, Captain Nelson now returned to the broader business of sea-command, and he had it set down in his head, and read it out as if from the first page of a daily journal. 'Bring me Lieutenant Lawe, Doctor Roxburgh, Mr Nisbet, the master, the master-gunner, and the gunner's mate. All at one hearing. We must not waste the day.'

'Aye, sir . . . Is it gun-drill?'

His captain smiled. Hinton had earned the right to a question or two. 'You read my mind, Mr Hinton, like a gipsy maid. Yes, it is gun-drill of a most particular kind.'

When they were all assembled, and salutes exchanged, Nelson made known his orders without delay.

'It is broadsides today, and broadsides until I am satisfied. We use blank cartridge, else I shall be in trouble with their Lordships of the Admiralty. Broadsides, first to starboard, then to larboard. I wish to hear a true ripple-firing, from stem to stern— the shots as close together as may be, but *not* one great thunder-bolt to shake the ship to match-wood. Is that understood?'

There was a murmur of assent, and Nelson proceeded to his closer directions.

'Dr Roxburgh, set up your tables in the midshipmen's berth, with instruments ready as if for action. I will send down wounded men for your attention. Dispose of them well. Do not amputate . . . Master-gunner, leave us now, and prepare the gun-crews . . . Mr Wilson, when we are finished with the starboard firing, come smartly about and lay a course opposite for the larboard guns to bear. The first lieutenant will give the word . . . Mr Nisbet—' he stared at his stepson as if at a small stranger who might prove to be human, 'you will attend me with your note-book. Go fetch it now, and do not lag behind.'

They began to disperse at a nod from their captain, who was left with Matthew and Martin Hinton. To the latter Nelson said:

'I will observe this myself from the gun-decks. So let us make a small game of it. I shall be wounded, and put below. You will have the command. You will give the order to fire, then fetch round on the other tack and repeat it.'

Nelson turned to Matthew. Already, at the prospect of action— even action in pantomime—he was flowing over with energy and spirit. 'Take station on the lower tier, Mr Lawe. You have heard what I have in mind. Ripple-fire broadsides, from both decks. It will go upper, lower, upper, lower, stem to stern; as swift as possible, but no two guns together. If any gun is early or late, I will have its number. If it is number seven of starboard again, I will have its captain's shortest hairs plucked out one by one! . . . Let the gunners listen out sharply, and they will learn the tune.'

Back he went to his first lieutenant. 'Now, in five minutes, beat to quarters. Let me hear old *Heart of Oak* thunder out! Report to me when all is ready.'

Within a short space, all *was* ready, and all was activity and the bustle of simulated war, as if Nelson's overflow of spirit had flamed outwards and set them to a willing task. When the drummers beat out their rhythm of 'Ta-ta-TON, Ta-ta-TON', on that foreboding note which could fire one man to courage and another to deathly fear, there came from below the rumble of guns running out, and the sharp slap of gun-ports raised and secured.

The marine sharp-shooters clambered up the rigging—like one-legged spiders, said the sailors, with affection and derision mixed—and took up their favoured posts on the yards: while John Wilson the master mustered his shrunken force of top-men and yard-men and pull-haul men for the braces, since a ship for ever remained a ship, which must be steered into action before the mere gunners could bring their weapons to bear.

Now *Agamemnon*'s crew stood ready for whatever fate might bring, and that, by Nelson's pocket-watch, in short order. 'By God,' he said, in a private word to Matthew Lawe, 'they are willing enough!'

They were willing for him.

As he spoke, Lieutenant Hinton returned to his side, with the report: 'Crew at quarters, sir, and ready.'

'Hoist signal for close action,' Nelson commanded, and the signal lieutenant ran to the halyards. Then the captain's head sank upon his breast. 'I am wounded, Mr Hinton! Not mortally, but I will go below to bleed awhile. My ship is yours.'

Though he could make a 'small game' of it, the game was deadly serious. One day, such disaster might befall, in the bloody heat of battle, and then Hinton must prove himself, like all the rest.

To Lawe he said: 'Below with you,' and Matthew went swiftly down, and into another world.

On his way to the lower gun-deck he passed through the after cockpit, the midshipmen's rough quarters which now became the surgeon's lair. Here all was in grim order, though it was a neatness which would not inspire the most house-proud wife. After the clean air of God which blessed the deck, this seemed the entrance to the Infernals.

The makeshift hospital was ill-lit: the large candles burning in their tin sconces which had been set upon the tables and chest already gave off more evil-smelling smoke than light. This furniture had been ranged in rows, and covered with sailcloth, to serve as the surgeon's chopping-blocks. Below them on the deck were buckets full of sponges and dressings, and larger empty mess-kids for amputated limbs.

At one side stood a rack of saws sturdy enough to delight a master-butcher: another of knives and probes; and a barrel of rum for sufferers to quaff before they came to this encounter.

Such was the *Agamemnon*'s haven of mercy, to which, in action, a procession of groaning, screaming, silent, and always bleeding men would be carried down, within moments of the first enemy fire, to take their turn and their chance of life: a hell below, to match the hell above.

Presiding over all were two necessary men, Dr Roxburgh and the chaplain. While the latter looked grave, as befitted his cloth, the surgeon, whose Scottish humour was gaunt as his native crags, gave Matthew a smile as he passed by.

'Fare thee well, Mr Lawe,' he called out. 'But never forget—we are At Home to you, at any hour!'

With not the merriest of smiles, Matthew found the head of the ladder and stepped down to the fiery heart of the ship.

There was order here also, but of a more martial kind. The guns were already run out, prepared for the first broadside, and their questing snouts seemed to sniff the air beyond the open ports, alert for the view, the chase, and the kill.

On their blood-red carriages—a colour repeated all round the gun-deck, to mask the morbid stains of battle—the lashings had been doubled, against that fearful recoil which could turn a gun into a mad bull, charging red-hot across the deck through men and weapons till it finished up in gory chaos on the opposite side or, since a 32-pounder weighed two tons, tore a hole in the ship's hull and disappeared into the sea.

The men who might have to survive this thrashing fury in battle stood to their guns, their tarry pigtails safely coiled, their head-scarves tightly bound about their ears. Round them were ranged all the paraphernalia of this cauldron of war: the shot-racks, the priming-tubes, the copper powder-shovels, the sheep-skin sponges for swabbing out the barrels, the rammers to force home the cannon-shot, the handspikes for training their weapons towards a mortal wound.

Gun-captains nursed the lanyards which served to trigger their flintlocks—the great new adaption which had now replaced the old match-and-powder train, and would flash the priming powder within a second of the signal.

All was as ready as human wit could devise. Success still lay, and always would, in the lap of chance. A flint might fail to spark, and the gun misfire, destroying the rhythm. A gun-captain might be deafened by the monstrous noise bursting out upon him from a few feet away, and miss his turn.

In true action a heavy-shotted gun might explode, or break its bonds and begin its rampage; or an enemy hit might fill the deck with huge flying splinters, which like hellish hornets inflicted the most terrible of all the wounds of war.

Nicholls, the master-gunner, the man at the centre of this iron tangle, greeted Matthew as if they were both guests at some sedate family breakfast. He had a true gunner's face: grey like his guns, weathered as the timbers surrounding them, disillusioned like a midwife's in her third decade of delivery, rock-like in its refusal to admit that a broadside firing was any different from a dawn-chorus of sweetly singing birds.

This was his own small empire, and the emperor had it in his grasp as a shepherd his gentle flock.

When Matthew asked if he were ready, he replied: 'As ready as ever we will be.'

Matthew had expected something more forceful and eager. 'It is an important trial of our strength,' he cautioned.

'Aye.' The master-gunner turned aside to shout to a powder-boy: 'Cover that cartridge with your jacket when you leave the magazine! You know the orders, damn your hide! Do you wish to blow us all to kingdom come?' Then he said to Lawe: 'It is always important.'

'Captain Nelson is looking for perfection.'

'So am I, sir, so am I. And *my* perfection is that I am still alive, after thirty-four years of service.'

At least one knew where one stood, with such a man . . . Then a stentorian voice from above relayed the order: '*Broadsides!*' and on the instant the brazen tongues began to roar.

On this misfortunate morning, they did not at the first test roar so well; the ferocious lion proved to be a hyena, laughing at men and their designs, mocking even the law of the jungle. It was as if the proud *Agamemnon* had suffered an attack of the hiccups, at the very moment when they could least be endured.

Guns misfired; other guns went unheard; when their ship turned on the larboard tack, a single shot was followed by a ghastly silence, since the next gun-captain below had not the wit to fill the gap. If Captain Nelson was looking for perfection, then he was still looking for it as 'Cease fire!' was called.

Their captain was in no two minds when, at mid-morning, a sulphurous inquest was held.

'I do not know,' he told his assembled officers, who avoided his frosty gaze as if it might shrivel them, 'what effect our gunnery will have upon the enemy, but by God it terrifies me! Our next exercise, beginning in ten minutes, will be *broadsides*, and after the noon break it will be *broadsides* again, until I am satisfied. We are fortunate,' he observed, with extreme distaste, 'that we still have a ship from which guns may be fired . . . Pray convey my message, by deaf-and-dumb signs if need be, to every block-head on board. Dismiss!'

Even on an easy, innocent passage, Nelson did not waste his sovereign's time.

But towards sunset there came, at last, a break in these lowering clouds. The mast-head look-out hailed the deck, and reported land ahead; and it was Matthew Lawe, climbing to the topmast tree, who was able to give it a name. Far to the south-east, perhaps seventy miles away, there was a gilded crown of land, dark at its base, pearly pink at its peak, cropped off short against a fading yellow sky. It was the distant crater of Vesuvius.

'Shorten sail!' Captain Nelson ordered his sailing-master as soon as he had the news. 'We will dawdle the night, and present our compliments at the Godly hour of nine.' He added, for the benefit of any within earshot who might profit from the message '*With exact gun salutes!*'

At the Godly hour of nine, Master-gunner Nicholls was striding to and fro between his two lines of saluting cannon set upon the fore-deck, intoning the hallowed, mournful dirge of all the men of his trade at this testing moment—the rhythm of a naval salute, which began with a secret murmur and ended in crisp command:

'If I were not a gunner, I would not be here—FIRE ONE! If I were not a gunner, I would not be here—FIRE TWO! If I were not a gunner, I would not be here——'

The *Agamemnon*, gliding forward into the matchless beauty of the Bay of Naples, was paying the respects due to a foreign potentate: twenty-one guns for the King of the Two Sicilies, with a reserve of nineteen more for the sole British presence on this scene: the Ambassador to His Majesty's Court at Naples, if he should come on board.

The arm of the bay, enclosing a shining bowl of blue-green water, a small armada of painted fishing-boats, and terraces of sun-warmed buildings with noble hills beyond, was one of the most fair in all the world. Poets had hymned it, travellers had awarded it the raptures of fresh discovery; lovers had languished in its embrace, painters had mourned the day when they found that their skill could not match its glory—and had stayed to enjoy, without envy, its pure perfection. Sailors, simple sailors, could note it down in their sparse log-book prose: Fine harbour, good bottom, dependable sunshine, *most* dependable girls.

The *Agamemnon* ran on, and the saluting guns kept her company, across a sparkling sea which slowly shelved to a trusted anchorage. Nelson, his telescope busy, felt all the satisfaction of a fine landfall. Who would not be in this charmed corner of the Mediterranean Sea, on a morning of such brilliance and warmth? Who would not be proud of a handsome 64-gun ship of war, now dousing and furling her greater sails as their way fell off and the catted anchors were lowered to the waterline? Who would choose to be in Norfolk?

Then his eye was caught by a sudden splash of colour at the entrance to the inner harbour. First it was a small vessel decked with flags; then it was a gilded barge headed in their direction, impelled by lusty Neapolitan oarsmen. Then it could only be a visitor of great consequence.

Nelson, standing on the quarter-deck with his immediate officers about him, spoke his thoughts aloud. 'If that is not the Doge of Venice, come out to marry the sea, it is the British Ambassador. We are honoured . . . Mr Nisbet! Pass the word to pipe the side! Mr Nicholls—nineteen more when he boards!'

IV

They were honoured indeed. Not only was this the British Ambassador in person, rowing out to greet a mere Navy captain;
but he bore a message of welcome from the King, who would be
waiting on the quayside to receive his visitor.

It was difficult to know who was the most flattered by this
singular condescension: Captain Horatio Nelson, whose acquaintance with the great had not yet prospered beyond an unsatisfactory
British prince and a sea-lord or two, or Sir William Hamilton, who
now saw his long-continued, weighty efforts to cement this Sicilian
alliance crowned with such brilliant success.

Captain Nelson had stepped ashore in Naples Bay with, as his
escorts, First Lieutenant Martin Hinton, Matthew Lawe, and
Midshipman Nisbet. Now, with the next turn of fortune, Hinton
returned on board alone, as duty dictated.

'It would please me greatly,' Sir William had told Nelson, as
soon as the King had ridden back to his palace and they were
alone, 'if you would make your home ashore with me for as long as
you choose. The Palazzo Sessa is tolerably comfortable, Lady
Hamilton would be delighted, and you and I have much to discuss
before we say good-bye. I must tell you that we dine with his
Majesty tomorrow.'

Though Sir William could be cool enough when he chose, he
was invariably a mirror of courtesy and charm with those whose
acquaintance he valued—or might value in the future. A veteran
envoy of sixty-three, whose interests were finely balanced between
archaeology, diplomacy, and elegant self-indulgence, he had held
his present post for nearly thirty years.

No one had cared to disturb this long tenure, since no one could
perform his office with better grace or competence. Who in
authority would dream of replacing an antiquarian so enamoured
of Etruscan pottery, an ambassador on such excellent terms with a
pliable King, and a man who, despite his spindle-shanks and
scholar's pale face, had ascended Mount Vesuvius no less than
twenty-two times?

At this moment, Sir William was making Nelson especially
welcome because he wished to see a British ship-of-the-line at
anchor in Naples harbour, as tangible evidence of British goodwill, and also in order to buttress those treaty rights which he
had spent half a lifetime in fostering.

It might also be that the newcomer, strange-looking little fellow as he was, might amuse his wife, who at twenty-eight was some thirty-five years his junior, and was, even after a bare two years of marriage, sometimes in need of distraction.

At the Palazzo Sessa, a town-dwelling set half-way up the hillside so that one might obtain matchless views of the bay, yet close enough for a fatigued ambassador to enjoy a daily 'grampus-puff' in its benign waters, the naval party found sumptuous comfort such as none of them had ever known before.

Noble corridors of marble opened upon room after room of splendid proportion. A vast pillared portico led to a garden—indeed, a succession of gardens—where humble men had toiled throughout a century, and were toiling still, so that the great might enjoy the ordered prodigality of nature.

Nelson himself was accorded the royal suite, and was lost in its magnificence—and then emboldened. After this, it seemed that he need never be daunted again: not by display, not by hand-and-foot service, not by the touch of aristocratic luxury. The contrast with the harsh cramped life on board ship, where a sailor's lot was 'honour and salt beef', might have been laughable. To him, it was only a step ashore. This deck, more friendly than his own, stood firm, and would ever be remembered as the due reward of a tired mariner.

They dined privately, *en famille*, with Sir William presiding like a lean Bacchus at one end of the oval walnut table, and his wife like Bacchante at the other. Emma Hamilton had long been a surprise to all who met her in such surroundings, and this day was no exception. She was perhaps at the peak of her beauty: a flame-haired charmer, perfect of skin, opulent of figure, natural and amiable in manner, in conversation, or in silence.

She seemed, as the wine flowed and the sun slanting through the louvered shutters warmed the bones of northern visitors, to become all things to all men. To her husband she was attentive and dutiful: to Nelson full of subtle flattery, and to Matthew Lawe (in whom she must have divined a lesser quality) companionable and reassuring. For the gangling young Josiah Nisbet she was half mother and half partner in the turbulent discontents of youth.

At the dinner's end, when the last glass of Marsala was mingling agreeably with the sun-blushed peaches which had furnished their dessert, Captain Nelson was bold enough to embark on a compliment.

'Your Excellency,' he said formally, raising his glass towards his host, 'after this splendid repast, I hope I may be permitted to propose most grateful thanks to our hostess—and to her cook.'

There was a moment's pause, scarcely perceptible save perhaps to a musician who knew, and understood, what the small separation of one note from another might truly mean. Then Sir William, with a courtly bow, answered:

'Thank you, Captain Nelson. I am sure they will be equally grateful for such a kind expression.'

He smiled: save for a modestly grave Lady Hamilton, they all drank; and then, as the liveried servants withdrew, peace and good humour reigned supreme.

It was five o'clock. While Lady Hamilton, seemingly with the utmost willingness, entertained Josiah, Sir William conducted his two other guests on a tour of the Palazzo Sessa. It proved a treasure-house of antiquity, assembled over thirty years by a most determined and most knowledgeable collector.

Sir William must have known that he was showing his 'lumber', as he called it, to two simpler men who were in the very cellar of ignorance, as far as such jewels were concerned. Thus he sketched lightly his picture of magnificence. This was a fireplace of Robert Adam—some might call it modern Etruscan: these were cameos of a certain respectable age—the age being eight centuries: these scrawls were preliminary outlines, by Michelangelo, of the muscles of the forearm of God, now to be viewed in all its splendour in the Sistine Chapel.

This was a gold chalice attributed to Benvenuto Cellini; and that—ah that!—an amusing *graffito* from Pompeii, showing what the ardent citizens were up to, or wished to be up to, before a fiery cascade of molten lava snuffed out all spark of desire and all chance of its performance.

The easy, agreeable parade continued. They might have been wanderers in Eden, with a trusted guide, now past his best, who had once been its young, ambitious gardener. Only once did the ambassador allow a private sentiment to intrude. He had stopped before an empty space, a carved niche in a wall, handsome in itself but now, for him, desolate as a robbed grave.

'I wish,' he said, pointing to this cavity, 'that I might show you what once stood *there*. The greatest treasure of my life . . . It was a vase, a blue glass vase, the most beautiful I have ever seen. It had an outer casing of milk paste, carved in the form of a goddess under a tree. Even its handles were perfection . . . It was as old as Anno Domini itself—some eighteen hundred years.'

The deep feeling in his voice demanded a sympathetic answer from his guests. Captain Nelson supplied it:

'But what is the story, sir? Where is it? Did it belong to the Sessa family?'

'No, no. It was mine. I bought it, many years ago, from the Barberini Palace, which had it from Frascati where it was discovered. It is no secret that I paid one thousand guineas for this privilege . . . Then I was *prevailed upon*—' the diplomatic voice, astonishingly, now conveyed all kind of intimate pressures and persuasions, 'to sell it to the Duchess of Portland. I am told that it is now to be known as the Portland Vase. For my part, I see no reason why I should not for ever think of it as the Hamilton Vase.' Suddenly he was smiling again. 'But what I *think*, my dear Nelson, and what I say, are two different matters. I am a diplomat in all things . . . Now—shall we retrace our steps, from Anno Domini the first to this urgent year of seventeen-ninety-three?'

They retired early, on that first night, after the dispatches which Nelson had brought were closely discussed, and a candle-light supper had been served, and Lady Hamilton, gowned in shimmering blue, had sung to the piano, and—at the urging of her husband —had mysteriously promised that she would perform her 'Attitudes' for their amusement, if not tomorrow, then on the day after.

They retired early: there had been almost too much to enjoy, for mortal men transformed into pagan, bemused gods. Sailors who had not set foot on land for nineteen weeks found in the soft *palazzo* beds a foretaste of paradise. Whom they might have wished to share these with, in the drowsy sensuous twilight of consciousness, was for each a secret, and for each as transparent as a gossamer gown.

Matthew wakened from a lascivious dream, which was not new in a sailor's life, to the bounteous comfort of a canopied feather-bed, which was. But he did not start into wakefulness, like a maid from a fainting-fit, with a piteous cry of 'Where am I?' He knew where he was, like all watchful sailors who, having curled up in a corner for the night, found themselves next morning in the same accustomed place.

Last night he had 'curled up' in a tapestried guest room of the Palazzo Sessa, with a glowing fire in the open hearth to warm his toes and a bed as big as a bomb-ketch to comfort the rest of him. Now he awoke without surprise to the same, save that the fire, secretly replenished while he slept, was flaming afresh, and the bomb-ketch bed seemed no more than suitable for a man of leisure facing the fatigues of the morning.

Sailors had a phrase for such chance, unearned good fortune. They had found themselves, they would tell their envious ship-mates, 'like a pig with its arse in butter'. Matthew Lawe, if he were not a commissioned lieutenant instead of some common sea-faring fellow, might have used the same uncouth words. Now he was content to say, or to think: 'This is what I was born to—and will *retain*.'

There was a discreet, almost apologetic knock at the door, and Matthew sat up against his pillows as the door was opened. Last night a footman had taken away his clothes after assisting him into bed. Now he saw that they were being returned, brushed and pressed, by a much more imposing figure.

This was a man whom his host had indicated on the previous evening: Gaetano Spedilo, the ambassador's personal valet, whom Sir William had called 'the true mainstay of this household'. Tall, swarthy, wizened and bewhiskered, with the gait of an actor and the self-importance of a man who knew himself indispensable, Gaetano might have been its presiding magician.

He advanced, bowed, and murmured: '*Buon giorno, signor tenente.* His Excellency wished to know if I might be of service.' Since he was not in the least doubt of this, he turned, and beckoned in a footman bearing an ambrosial dish of hot chocolate. Then he drew back the curtains, fussed with the returning clothes, helped Matthew into a silk robe, poured his chocolate, and then, without dissembling, stayed to gossip.

The weather, he said, was as always beautiful. The lieutenant's ship, he added with a winning smile, still lay safely at anchor in the bay. He hoped the lieutenant had slept well. A man needed sleep after such a fearful voyage. Was there anything—anything in the *world*—he might do for the lieutenant before preparing his bath?

The ambassador would not leave his room before nine o'clock.

Her ladyship might even remain invisible until ten—ah, but women had more cares in the world than anyone knew, and they needed their sleep as birds needed their warm nests.

The beauty which resulted, he said finally, was worth all the anxiety with which one awaited its appearance.

For a servant, Gaetano Spedilo was unusually free: there had been moments when an awakening Matthew, listening to the flowery phrases, aware of the smiles, nods, grimaces, and sudden heavings of the shoulder, had thought the man demented. Then he began to suspect that the valet, as snobbish as all servants, had seen in him a twilight figure hovering between the drawing-room and the servants' hall, and was suiting his song to an over-familiar tune.

Then, sipping his chocolate, enjoying every last element in this unaccustomed world, Matthew decided: No. The extravaganza sprang only from the simple fact that Gaetano was a foreigner, and could know no better. He might be all the more entertaining on that account. For idle enjoyment, Matthew decided to explore an alluring coast.

He put his own over-familiar question. Lady Hamilton had spoken, last night, of performing her 'Attitudes' for the enjoyment of her guests. What, he asked Gaetano, were these 'Attitudes'?

The valet came to immediate attention. His fingers began to weave the air, and his face to assume an almost alarming degree of fervour.

Ah, the Attitudes . . . *Bellissimo!* The *tenente* must understand that *la signora* had been of—of the theatrical persuasion, before she married the ambassador. It was said that she had the whole of London at her feet. The Attitudes were—well, the French would call them *tableaux*.

She would dress in a certain way, and pose herself in a certain way, and the ambassador would hold up a candelabra so, to light the scene, and *Ecco!* there arrived an Attitude! It might portray the goddess of love, or a warrior queen, or a girl of the streets, or— *scusi*—the Queen of Heaven herself. Such were the Attitudes. All Naples society was in raptures with them. Well, perhaps not all the *women*.

'I see,' said Matthew, and, as an early-rising man in a warm bed, he saw only too clearly. 'I look forward to this entertainment . . . How long have they been married?'

'Two years, *tenente.*' Gaetano's face, never composed, never wholly normal for the English taste, now became positively bizarre. 'But she was residing here for five years, before the matrimonials.'

The ground seemed to be growing delicate, and none the worse for that.

'She was, perhaps, a friend of the family?'

'*Exactly!*' Gaetano exclaimed, nearly beside himself with the lust to paint a true picture. 'It is said that the ambassador's nephew— Mr Greville, a difficult name—relinquished her to please his uncle. Thus she came here to improve her Italian. And here she stayed.'

Gaetano now busied himself with the disposal of the pot of chocolate and the cup and the tray, and then with making straight, for the third time, Matthew's uniform clothes and linen. The air grew calmer, but only, Matthew realized, because Gaetano, after such rank indiscretion, wished it so. Finally the valet, turning round towards the bed again, became confidential rather than theatrical.

'Sir, is it true that your *commandante*, last night at dinner, drank a toast to the cook?'

'He said that he wished to thank her for an excellent meal.'

'*Prodigioso!*' Gaetano was delighted. 'Such a thing has never happened before! The *commandante* is so brave, so direct!'

Matthew was mystified. 'It is no more than a custom.'

Gaetano's smile had progressed beyond delight to the purest mischief. 'But not at the Palazzo Sessa. Here, the cook is *la madre!*'

'The mother?' Matthew asked, astonished. 'The mother of whom?'

'*Della signora!* Her ladyship's own mother.'

'But how can that be?'

'Sir, it is the course of life. And thank God for it. I could wish that I had such a daughter.' Suddenly Gaetano, like a puppet-master who could close one scene and open another at his own sweet will, became formal again. 'You dine today with the King, *tenente.* A fine man, though a little free with the common people. We call him *Il Re Nasone.* You would say, King Nose.'

'Why is that?'

'You will see, beyond a doubt.'

The valet, with a bow, turned to go. The audience seemed to

be at an end. Matthew, aware that he was not controlling these domestic tides in the smallest degree, said:

'Thank you, Gaetano. You have been most helpful.'

Gaetano turned again, holding open the door as if he were obeying an actor's call before the curtain. His arms spread out, miming a warm embrace.

'There is only one reason for that, *tenente*. We love the English!'

'We love the English!' King Ferdinand declared, with such emphasis that his jowls were still trembling after the last echo of his voice had died. 'They are the saviours of our nation! That is why,' he told the man seated on his right hand, 'I was so happy to see your brave ship enter my harbour. May it long remain with us.'

The guest who had been set on the King's right hand, against all rules of protocol but to the general satisfaction of most observers, was Captain Horatio Nelson of the *Agamemnon*. Why he had been singled out for such distinction, displacing his own ambassador in the table of precedence, was unclear, though deeply satisfying.

Privately he might have wished that the King of the Two Sicilies were not such a boorish peasant, huge and fiery of nose, scarlet of complexion, raucous of voice, and gross of appetite. But one must, Nelson thought, take kings as they came. He had not met so many of them that he could afford to be dainty.

It was a vast and glittering throng which had been assembled at the Palazzo Reale. From the outset it had been clear that this was a dinner-party to honour the naval guests. All of the *Agamemnon*'s officers who could be spared had come ashore for it: even Midshipman Nisbet had been found a place: the meal was prodigious, the red wine flowed in rivers, the noise was enough to shake the chandeliers.

Sir William Hamilton, in confidential discussion earlier, had told Nelson that this would be so. He had also told him why, in certain phrases which paid his guest the compliment of being shorn of all diplomatic nicety.

'The fellow needs us,' the ambassador had said. 'If not, it would be bread and a coil of macaroni for us, and the dregs of Naples fire-water to wash it down. You will see his quality: a *popular* monarch whom the people love as long as their stomachs are full, but without any external strength except through our alliance. 'Tis a pity his wife will not be there: she has all the brains, all the spark, and all the cunning too. But she is not too cunning to escape pregnancy, year after year. God knows how many children that right royal ram has planted . . . There is one thing to remember: Queen Maria Carolina is sister to the late lamented Marie Antoinette, and thus a little concerned for her neck, in these mutinous times . . . At dinner he will flatter you. You need do nothing much but smile, and agree on all matters, and stiffen him if he raises hands to heaven and tells you he cannot spare any troops for Toulon. He can spare them well enough—they will be glad to be paid promptly . . . I am sorry for his nose,' he concluded, with a return to suavity. 'I would rather ascend Vesuvius. At least that volcano is dead.'

God bless our ambassadors, Nelson had thought, as he listened to this frank appraisal. They can be as free as sailors . . . And Sir William might have been gifted with the devil's own second sight, as far as this dinner party was concerned. Even to the raising of the royal hands to heaven. But in the course of the meal, these had been lowered again.

'Yesterday I thought it impossible for me to send the troops you ask,' King Ferdinand was now declaring. He beckoned to his major-domo, who signalled to a wine-butler, who refilled their glasses from a great beaker of best Sicilian. 'But I am ready to do so,' the King said, after a monstrous swallow, 'to help an ally and a friend in need . . . Tell me, what is the least that would satisfy your government?'

It was something new for Nelson to speak for his government, but on this evening, and at such a brave moment of his life, everything was new. Having also had the benefit of advice from Lord Hood and Sir William Hamilton, he could not be said to be a lonely gladiator. 'Sire, I believe, six thousand.'

'And what is the most?'

'Sire, I believe, six thousand.'

Their eyes met, and suddenly the King burst out with a great bellow of laughter. '*Gran Dio*, I like your spirit, captain!' he said, spluttering, as all eyes turned towards them, and far down the table even Sir William's lips pursed to a careful smile. 'Very well—six thousand it shall be, and that is a promise. Are you content?'

'Indeed, your Majesty. And thank you kindly.'

'There is one condition.'

Foreign weasels! Nelson thought. Royal weasels! . . . What does

he want in return? A million pounds? Six ships-of-the-line? Lady
Hamilton in an Attitude? But he was to be surprised.

'I wish to see your ship,' the King said. 'May I come on board
tomorrow, or the next day?'

'Your Majesty, I should be deeply honoured.' Then he thought
of the *Agamemnon*, recovering from the long grind of the blockade,
with a hundred men sick and scarce enough cups and plates for a
tea-party in Norfolk. 'You will find us rough, I fear.'

'I will find what I seek,' the King answered grandly. 'A British
fighting ship with a friend on board to greet me. Tomorrow, then?
Or Sunday?'

'Sunday,' Captain Nelson answered. 'At one o'clock, sire—
if that will not interfere with your devotions.'

'What can you know of my *devotions*?' the King asked, and
unloosed such a monstrous roar of laughter that the whole vast
room was silenced, and certain of Nelson's officers, peering across a
sea of heads, could only think enviously: This is our captain
make him laugh? Could he not save a jest or two for us?

The sailors quit Sir William's hospitable roof early next morning.
There was much to be done on board, even though the ambassador,
after expressing grateful thanks to Nelson both for his conduct and
his lively perception, had undertaken to provide plates, cutlery,
food, and wine for the royal visit, from the resources of the Palazzo
Sessa.

There was already, between the two men, a most sympathetic
understanding, with admiration on both sides. Yet the sea-captain
could not help but feel that their lives would never truly mingle.
The one had an embassy to command, the other a ship; though
these were two arms of the same royal service, they were two
different arms, as far apart as the outstretched finger-tips of
both.

If one strayed into the other's territory, even with the best
intentions, a certain floundering might ensue, or a tournament of
influence, or a rage of jealousy, or simple chaos ... It was a
thoughtful party, affected by excesses of indulgence, second
thoughts on duty and pleasure, and early-morning regrets, which
sat in the stern-sheets of the captain's barge and headed towards
the *Agamemnon*.

Young Josiah had the first queasy headache of his life: Nelson was musing loftily on certain battle-plans if his single ship were cornered by the piratical French in Naples harbour, and, at a lower level, on the number of plates and spoons, even chairs and privy-stools, necessary for a royal visit; and Matthew, with fewer cares and longer hind-sight, was wishing that he might be marooned in Naples forever.

The oars thudded against the thole-pins, the bright bay unfolded, the sun began to gain strength, and the *Agamemnon* to emerge, in all her towering majesty, from a drifting surface fog. Matthew, who had become a stranger to silence during the last two days, felt it his duty to bridge an unusual gap. He chose what was certainly uppermost in his mind.

'Sir, what did you think of our entertainment last night?'

Nelson was preoccupied and grumpy. 'What entertainment? At the royal palace?'

'I was referring to the Palazzo Sessa, sir. The later entertainment —the Attitudes.'

'I thought,' his captain answered austerely, 'what we should all think: that they were tasteful and enjoyable.'

'I agree heartily.'

Nelson, though cold and grey in the early sunlight, seemed to be awakening. 'I have never seen so much done with a shawl and a tambourine.'

'Nor I, sir.'

'What did you think, Josiah?' his stepfather asked. 'Did you enjoy Lady Hamilton's play-acting?'

Josiah, who could have matched Captain Nelson for colour, answered with unaccustomed fervour: 'I thought it the most beautiful thing I ever saw, sir.'

'Did you so? Then we may be all in agreement.'

'Her ladyship,' Matthew said, 'must have had a great vocation before her marriage.'

Nelson seemed to see some slur in this. 'And why not? It is *now* that we should think of. I believe,' he added, somewhat portentous, 'that she is an amiable young woman, who does honour to the station to which she has been raised.'

'Yes indeed, sir.'

'That should be enough for all . . .' He then altered course in a most inconsequential manner, uncovering for a moment a whole mine of private thought. 'I have said before, that once past

Gibraltar, every man is a bachelor. By which I mean, Mr Lawe, that he is married only to his ship!'

He was to prove this, beyond a doubt, within the space of a day.

On a red-letter Sunday morning when a Navy captain, with some assistance from his friends, was to play host to a king, all was bustle on board the *Agamemnon*. It was the sort of bustle which Nelson hated: the bustle domestic, the bustle social, the bustle unseamanlike.

It demanded that his first lieutenant must supervise the twining of coloured cords and ribbons round the honest rope-rail of their ladder; that his sailing-master should make sure that only sailors with unpatched breeches and trim tarred pigtails would assemble on the starboard side of the quarter-deck: that Lieutenant Lawe was mustering wine-glasses, and inspecting them for fingermarks: that the master-gunner was fashioning fireworks, the marine trumpeters practising courtly flourishes, and Midshipman Nisbet tying up nosegays for the ladies.

It was *war*!—with macaroni for cutlasses, Catherine-wheels for broadsides, sun-awnings for topsails. It was a damned imposition. Though this was what his career dictated, his manhood fiercely rebelled. It was no more than Nelson's own Attitudes—and be damned to the whole false tragi-comedy!

The company, which must assemble promptly and discreetly before the royal barge left the inner harbour, came on board in droves, like silk-clad cattle. There were bishops, noblemen, men of politics, men of distinction, men of impudence and guile. There was the *élite* of Sicilian society, and of the British colony, and the naval, and the military, and the merely well-dressed. Led by Sir William Hamilton and his lady, they boarded and thronged the ship like passengers—and that was the best and the worst that could be said of them.

Then happily all was changed. The small pinnace which presently put out from Naples fish-quay might have carried some late-arriving notables. Instead, it bore an urgent message from a certain person ashore for the captain of the *Agamemnon*. It informed him that a French convoy of one man-of-war and three merchant sail was at anchor off Sardinia, and could be surprised, if his ship were ready for action.

First Lieutenant Hinton, who had read the message, stood

waiting by his side. After a suitable pause he asked: 'Your orders, sir?'

'Are we fit for sea, Mr Hinton?'

'We can be, sir,' said Hinton, whose disposition at this moment was not at all sociable. 'Right promptly!'

'Then we sail,' Nelson said; and to Sir William Hamilton, some minutes later, he repeated: 'Sir, with great regret, I must forgo this occasion, and leave harbour instantly.'

His resolve was crystal-clear, his motives strangely mixed. He wished to show this fashionable world that a British man-of-war never forgot her prime purpose. He must demonstrate this curtly, even to a king. He must remind himself that sailors were not groomed for the hot-house of idleness, nor born to indulge themselves ashore, when duty beckoned.

He must exorcize certain improprieties of conduct, or intention, or sensual ambition. There was only one way.

There was also a tempting prospect of prize-money.

Brusquely cutting the tie with the land, he shed the burden of his guests, weighed anchor, and sailed within two hours; not to entertain a king, nor embrace the softer shores of life, nor see Sir William or his wife for five fateful years.

V

The dreadful contagion which was the naval mutiny of 1797 spread, like a stain upon the sea, as far as the West Indies station and the far-away Cape of Good Hope, and could not fail to infect the Mediterranean, bringing to that southern lake dishonour, shock, and death. But it began its foul journey in English waters, in those sailors' heartlands of Spithead and the Nore. It sprang, some said, from continental dreams of liberty and equality; but also, undeniably, from the inhuman treatment which disgraced such a multitude of ships in the King's Navy.

Its leader and principal villain—if villains must be provided for the comfort of right-thinking citizens—was one Richard Parker, whose journey from the cradle to the noose occupied him barely thirty years. He had been, by turns, midshipman, prize-money spendthrift, maker of Scottish golf-balls, schoolmaster, midshipman again, disrated and discharged able seaman, prison debtor, and volunteer able seaman once more.

Within six weeks of joining his last ship, the 90-gun first-rate

Sandwich, he was leader of the Nore mutiny and 'President of the Floating Republic'.

There was no need, despite the suspicions of certain men of politics, to see some hideous continental, or Irish, or radical plot in this swift advancement. Parker was a good seaman, a born trouble-maker (even as midshipman, he had challenged his captain to a duel), a man of education, a man of permanent discontent, and a lowly sailor with little to lose.

He was thus a natural leader and, above all else, an orator; and he need not look far for facts of crushing substance to inflame his speeches—and his hearers. He had two ready themes: the conditions within the naval service, and the atrocious cruelty which kept men in its bondage, under daily threat of death.

A slave's wage, a dog's food, and a tyrant's discipline furnished the score for this anthem of the damned.

The payment of England's most valiant fighting men was indeed so villainous that it demanded no mutineers to proclaim the fact—though it needed a mutiny to bring, at last, a Parliamentary cure. The scale of pay had seen no increase since the reign of King Charles II, nearly one hundred and forty years earlier: though the value of money had now shrunk by one third, a seaman—married or not, father of a family or not—remained trapped, as if by holy command, at a pittance of nineteen shillings a lunar month, which was £12 a year.

By the custom of the service, these wages were never paid until they were overdue six months; and if a voyage or a ship's commission lasted long enough, they might lag behind three or even four years. The silver-tongued slop-sellers who came on board would always advance credit against wages due, but only at a monstrous discount. The wives of these sailors, deserted for so long, might choose their fate: they could either go a-whoring, or, lacking any protection, retreat in misery to the workhouse.

Any seaman who was sick or wounded, and must stay in his hammock, was ruled to be not at work, and docked of his pay for every hour he spent in idleness. Of prize money, won at such cost of blood and sweat, missing limbs and torn bodies, an officer might receive £200, and a seaman sharing the same peril a paltry fourteen shillings. It was small wonder that sullen sailors expressed the hope that the cannon-shot received in action might be distributed like the prize-money—the lion's share to the officers.

It was no wonder either that Richard Parker, in the full flight of oratory, could make his hearers weep as he described destitute sailors pulling behind them little painted ships on wheels, as they begged their bread on the streets of Portsmouth.

The bread they begged ashore was a hundredfold better than any likely to be offered them afloat. Here it spawned maggots and mould, just as the crack-jaw ship's biscuits bred weevils, and flour sprouted mealworms. The water-casks, filled and refilled, were never cleaned; river water from the sour-running Thames first turned green, then incubated unnamable wriggling creatures, then 'settled', and then was drunk.

But it was the meat—salt beef and salt pork—which was the mainstay of the sailor's fare, and his principal curse. Much of it might well have been supplied in the reign of that Merry Monarch of long ago. Since the victualling yards insisted that the old meat must be eaten first, each new voyage began on a disgusting note; for this 'old meat' was meat returned by other ships at the end of a cruise, routed out from odd corners of the store-sheds, and might have stayed for years in the brine of the harness-casks before being steeped for twenty-four hours on board, then cooked and served.

It could hardly be eaten. Old sailors, resigned to making a jest of it, talked of horse-shoes found in these casks, of the braying of donkeys suddenly silenced, of negroes who mysteriously disappeared after straying near the dock-yards. They could only pride themselves on the fine carvings to be made out of this stalwart substance, which took a neat polish, like the very best mahogany, and might furnish an heirloom, if a man ever saw his family again.

There was another hallowed custom of the service which, as with the payment of wages in arrears, had been fixed like an ancient barnacle on the sailor's back. Though the victualling yards ashore supplied the correct weights of these daily rations, those served to the crew were reckoned at fourteen ounces to the pound, and seven pints of liquor to the gallon. The missing two ounces of meat, the vanished pint of beer, stuck to the Purser's fingers, or the cook's—anywhere but the sailors'.

Last, looking beyond the cramped, stinking cage of misery which was a crowded ship at sea—last was the treatment the men had from their officers. An officer's uniform was a licence to scourge: a boatswain's rank the liberty to beat with canes and knotted rope-ends any man whom they saw as a laggard.

In some ships, the last man down off the yards after furling or reefing sail was invariably flogged; there were cases when a skilled top-man, after half a voyage of such torture, would cast himself into the sea rather than bare his back again to the cat-of-nine-tails, and hear behind him the whistling of this whip before it began to strip the flesh from his body.

Some captains, and some lieutenants, were clearly mad. Two dozen lashes for failing to sew a number on a hammock: the same for spitting: constant floggings for slowness, for murmuring, for discontent: a beating over the head with a speaking-trumpet—no mean instrument of punishment, being brass-ribbed round its mouth—for anything or nothing: such was the lot of sailors for whom protest or resistance might mean hanging.

Even little midshipmen, secure in their uniforms, could goad an experienced sailor twice their size and age into anger, then thrash him with a knotted rope until their puny arms were tired.

A brave or hardened man who had the spirit to smile after one such ceremonial flogging was ordered re-rigged to the grating, and given three dozen more. His captain, beside himself, was dancing with rage. 'He bares his teeth at me? I'll see his backbone!' He did, as it emerged through the bloody ruin of raw veal which had been his back.

A new flogger was put on after each two dozen lashes; and it was this captain's delight to find a left-handed boatswain's mate who could cross the cuts scored by a right-handed man.

So on and on, up this fearful ladder of suffering, until the last penalty became due. For striking an officer, for attempting mutiny, or for desertion, the sentence was either a hanging from the yard-arm or a flogging through the Fleet. This latter was intended to kill: as many as 800 lashes might be ordered, and no man had ever survived, even to the crippled wreckage of manhood, more than 350.

The wretch was borne through the Fleet in a ship's boat, to the drum-beats of the Rogues' March; at each ladder, a sturdy boatswain's mate descended, to administer his share. A surgeon went with the culprit, perhaps for decoration, since the punishment must still go on, to the point where a dead man was transported from ship to ship, and received his unexpended ration of stripes when already beyond the grave.

When sailors at last reached harbour, even before they set foot on shore, they might be press-ganged again, to serve for another voyage. If they ran, they landed penniless, and would be hunted for ever. There were some who spent five or six years afloat, without an hour of sweet rest ashore; and then, returning to cash their pay tickets, were trapped again, and pressed again, and condemned to this barbarous treadmill.

Such were the slaves, brutalized, cheated, and caught in the foulest web ever spun by man, who at last rebelled.

It was done with the utmost decorum, and marvellous patience, seeing what goadings they had had to bear. After careful consideration by certain 'fraternal delegates', the red flag of liberty was hoisted at Spithead, off Portsmouth harbour, and the Channel Fleet refused to put to sea. Instead, the crews submitted in orderly fashion a list of their grievances, under headings as familiar as the caustic tyranny of the lash: their pay, their food, their short weight of rations, their liberty to go ashore, and the harshness of their officers.

With dutiful respect they affirmed their undying loyalty to the person of the King, swore that they would sail out instantly if any French ships attacked, and sat back on their haunches to wait.

The moment was well chosen: England *was* at war and in peril, and their Lordships of the Admiralty, after some blow-hard blustering maintained for as long as the national safety would permit, took fright. The grievances, submitted with such temperate skill, were allowed to be justified. Immediate redress was promised. Not a man was court martialled. Certain officers of notable disrepute were sent ashore. The Channel Fleet sailed out to do its duty, under the King's own pardon.

This was the foolish hour when Richard Parker, on the Nore station between Thames-Mouth and the River Medway, determined to spring into the same action. It was his misfortune that he had risen to full prominence and power just as the Spithead demands had all been granted. It was his misfortune—but he had chosen his path, he would not retrace it, and the path was fatal.

He was already overblown with self-conceit: he had swiftly talked his way to a peak of influence: on one intoxicating day, this President of the Floating Republic had thirteen ships-of-the-line, together with frigates, under his command. As admiral of a formidable squadron he refused all orders from ashore, turned his guns on certain ships which had not yet hoisted the red banner, and sat back, like the Spitheads, to await the surrender of the realm.

It was his further misfortune that his flag-ship the *Sandwich* was a wretched old hulk so rotten and foul that she was condemned and sent to the breakers as soon as his reign was over. Seated on this malodorous throne, inactive and brooding, he found it, day by day, more difficult to rule.

After five weeks of a hot summer, Richard Parker's support began to ebb. Ships ran out of provisions as the tie with the land was cut. Raging arguments divided the fraternal delegates. When reminded that all their demands had been granted at Spithead, Parker retorted that this could be changed in a single morning by those rogues of Parliament.

But what, he was asked, of the King's own word? 'The same!' Parker sneered. 'Does a crown make him a man of honour?'

This treason did not sit well with scandalized shipmates who were, incredibly, still loyal to their country, and ready to fight like tigers in its defence. 'You are alone, out on a yard-arm!' he was told. The foreboding words were not lost on any hearer, nor on Richard Parker himself.

Presently a far stranger mutiny began, and he found himself deserted. Neither soft words nor dire threats could stem the outgoing tide. Ship after ship ran the gauntlet and broke free for the open sea. His proud fleet dwindled, and disappeared. At the end he sat alone in a usurped captain's cabin. He was taken, and sentenced to be hanged, along with twenty-four other leading malcontents.

As so many brave but misguided men before him, he met his death with fortitude. A last glass of white wine: a last speech in which he asked that he should be the only one punished: a handshake from his captain before his arms were bound—and he dangled from the yard-arm until sunset, as did all his brood of comrade-mutineers: all the soiled bunting of treason, which might have earned honour, just as, too late, it earned liberal and lasting reform.

The tideless Mediterranean did not escape this swirl of foul water from afar. Certain ships which had been concerned at Spithead

had been sent southwards for their health—or the health of the
Navy; they had carried their poison with them, for poison it had
now become, in a climate so much more benign.

Once again, ships refused duty, hoisted the bloody pennant of
revolt, and waited for 'redress' which had already been granted.
But now the current ran firmly against them, and, lacking any
massive urge, their end was always the same: surrender, court
martial, and death for leadership.

One such sentence, on board His Majesty's Ship *Marlborough*,
was ordered to be carried out by the mutineer's own shipmates,
while the guns of the Fleet were trained upon the offending vessel,
and ship's boats crammed with marine sharp-shooters circled
about, alert for any delay and prepared to put an end to it at the
beat of a kettle-drum.

The fashion had changed, and luckless men, both leaders and
dupes, lived only long enough to learn this hard truth.

While certain pious officers remonstrated to Lord St Vincent,
their Commander-in-Chief, that even such men as these should not
be hanged on a Sunday, another admiral whose new flag-ship was
the *Theseus*, 74, admitted no such soft feelings. 'On a Sunday?' he
snapped, when he heard of this objection. 'I would have hanged
them had it been Christmas morning!'

This had occurred some two months after the Spithead business
was settled, and the distinguished officer was angered by these
laggard dregs of mutiny. But he was no crusty fire-eater. Earlier,
he had written to a friend:

'I am entirely with the seamen in their first complaint. We are a
neglected set.'

The word 'we', and the comradely term 'set', were still at the
very core of his belief. For him, the service was never less than a
true brotherhood. But when the erring sheep of this flock strayed,
Rear-Admiral Sir Horatio Nelson, Knight Commander of the
Order of the Bath, could only transform to ferocious wolf.

Thus there was both bark and bite in this swiftly-rising man,
and they did not always match. Aboard his own *Theseus*, for
example, a moving incident confirmed it.

During one moonlit middle watch, with the ship on easy
passage, Matthew Lawe's attention was caught by the gleam of a
piece of paper which seemed to have been dropped on the quarter-
deck. It could hardly be heaven-sent . . . Matthew picked it up,

conscious that all around him were men of the duty watch, staring at him out of the shadows while pretending to be at work.

He smoothed out a crumpled note and, imagining the heavy breathing and manfully squared shoulders which had governed its composition, read by lantern-light:

> Success attend Admiral Nelson! We are happy and comfortable, and will shed every drop of blood in our veins to support our officers.
>
> SHIP'S COMPANY.

No decoration within the Sovereign's gift could have matched this accolade.

The battle-honours had come thick and fast, like gilded laurel leaves when the turn of the seasons commanded them to fall. In the old *Agamemnon*, his first triumphant love among his line-ships, he had seen continuous service on inshore work and blockade, and after two full battle actions against the French was promoted commodore and given command of the larger *Captain*.

The *Captain* inspired him to his greatest exploit so far, at the battle of Cape St Vincent, the crown of 1797. This 'Valentine'—it was fought on the fourteenth day of February—brought him to two acts of singular daring: the first of which was to leave the English line-of-battle on his own impulse, which was a court martial offence, one minute before his Commander-in-Chief signalled him to do so, and to cut the Spanish squadron into two confused halves.

The second was to lay alongside the nearest enemy, then to lead a boarding-party which stormed upon this, one of the largest of the Spaniards, the 80-gun *San Nicolas*. She was quickly taken, but in the confusion of the fight she in turn fell foul of a consort even larger, the towering *San Josef* of 112 guns.

Commodore Nelson did not delay for an instant. Seizing the chance, using the *San Nicolas* as a convenient gangway, he led his men up and up to the deck of the *San Josef*, made a determined assault at hand-to-hand—and a second Spanish ship surrendered to him, within the same hour.

It was a victory which resounded throughout the Fleet, and 'Nelson's Patent Bridge for Boarding First-rates' swiftly became a legend, like the man himself. So did his reply, when asked by an eager inquirer what was the principal hazard of this capture. 'To find a man strong enough to carry their surrendered swords,' Nelson answered. 'Fortunately, my best bargeman William Fearney has two stout arms.'

In all this he had been forced to surrender a little of his own private armament, here and there . . . At the assault on Calvi, in Corsica, he had been struck by a splinter on the brow, and lost the sight of his right eye; ever afterwards he wore a green shade over his left, that sole survivor which must be cherished. Later, when leading the boats from his new ship *Theseus* to cut out and capture a Spanish treasure-ship in Santa Cruz, Teneriffe, he had forfeited his right arm.

This had seemed the worst blow of all, the crippledom of body and of hope. Sent home to England as an invalid, in constant and monstrous pain from a nerve imprisoned within the ligatured stump, he had resigned himself once again to bread-and-cheese and a labourer's cottage—all that a one-armed admiral could now command, or expect.

It might have been the end, but it was no more than another beginning.

Within half a year, with the nerve subdued and the stump miraculously healed, he was appointed to the *Vanguard*, of 74 guns, and sent to join Admiral St Vincent off the Portuguese coast. There he was given a detached squadron of ships with a roving commission in the Mediterranean; and there this sailor with one eye, one arm, and one ambition began to forge a steel circle of friends, a band of brothers, fellow captains who were never to falter in their loyalty until he, the fixed star in this firmament, fell from the sky.

Their names were all to become famous: Hardy, Fremantle, Alexander Ball, Blackwood, Foley, Hallowell, Berry, Troubridge, Collingwood, Saumarez, Hood. These were the men whom a small battered Admiral led to glory. They were men he could trust, men not envious, men who did not need the Nelson touch to make them brave, only to bind them together into a single sword.

They were his, and he was theirs; and the harshest word, and the greatest tribute which ever came out of this brotherhood, was from one of them who growled, some few minutes before the last battle of all:

'How I wish that fellow Nelson would not crowd us with signals. *We all know what to do!*'

VI

Men to be trusted, men not envious . . . We all know what to do . . . It was displayed to noble perfection on one particular night, under the glare of lightning, amid cruel curling waves, in a raging gale which seemed intent on murdering all sailors, when Nelson's ship the *Vanguard*, and perhaps his own life, was saved by one of this band of brothers, Captain Alexander Ball of the namesake 74-gun *Alexander*.

Admiral Nelson's squadron, of *Vanguard*, *Alexander*, and *Orion*, with four frigates and a sloop, were intent once more on the eternal chase of the French Fleet, now loose in the Mediterranean but as hard to find as a mouse in a hayrick. Off Sardinia a storm blew up, and increased to a gale as fierce and relentless as any could remember. On a certain Sunday night in May, with the threat of a lee shore to add to the violence of nature, the *Vanguard* suddenly fell into disaster.

At two o'clock in the pitch-black chaos of the middle watch, the main topmast cracked and went over the side, with a sail full of men never to be seen again. Then the mizzen topmast followed it, and one hour later the foremast itself. It had needed no enemy save the raving wind and sea to reduce the flag-ship to a sullen lump of timber, drifting down-wind towards a hostile shore.

The shattered spars thumping against the ship's side sounded their knell as the *Vanguard*, rolling terribly, shipping water by the ton, was brought to mortal impotence.

Nelson, with his Flag-Captain Edward Berry, First Lieutenant Galwey, Matthew Lawe, and a dozen officers, were never to forget that night, nor the two nights after it. They were alone in a wilderness of storm; never a ship's light was to be seen, and as dawn broke, not a ship either. Their consorts had their own troubles to endure. They had not deserted. They had been struck by the same malicious fury, and it had scattered them.

For two days the *Vanguard* drifted helpless, with no sight of the sun nor break in the foul weather to show them their position. They only knew that enemy Corsica must be down-wind from them, and if not Corsica, then the jagged rocks of Sardinia's west coast. Though they had hacked their wreckage free, retrieved a few serviceable sails, mourned their lost men, and slaved at the pumps, they were still no more than a wreck themselves, with stumps of masts, like the Admiral's own arm, showing gaunt above a hulk swept by the sea; with a raging scud all around them, and nothing to give them hope.

Tuesday evening brought, at last, a distant glimpse of a topsail as the *Alexander* came racing down-wind to their rescue. 'God bless Ball!' said Nelson, who had not always warmed to an officer so different from himself. Alexander Ball was less like a sailor than a sage among schoolmasters: a studious man, tall and grave, whose only pastime was reading—with an austere embargo on works of amusement.

But he was still a valiant seaman. Handling his ship, of a size equal to the *Vanguard*, like an obedient yacht, he floated down a grass rope, then a heavy hempen towing-hawser which, after killing work by exhausted men, was presently made fast to the victim. Then the *Alexander* took the strain, and the flag-ship, though she could make no headway out of danger, was at least brought up with her bows towards the roaring waves, and could enjoy a little peace.

The *Alexander* held them in this grip all night, until, on the morning of Wednesday, the wind relented a little. But a heavy swell was still driving them down into danger, and now the danger could be heard.

It was the growl and thunder of surf under his lee which brought Nelson, who had been dozing with his back against the rail like any sheet-man of the duty watch, to full wakefulness. After two days and nights of howling wind, cold, care, and doubt—and a vilely throbbing stump of an arm—he looked ghastly: a tattered waif of an admiral who might lose his ship.

He stood up, and stretched his aching limbs, and stared about him. Ahead was the hazy outline of the faithful *Alexander*; between them was the tow-line, now taut, now lost in the scudding sea; under his feet was a rolling, useless hulk; and at his back the sound of waves breaking on a tormented shore and, a few moments after dawn, the sight of forbidding cliffs, jagged little islands of rock, death and destruction for all.

The two linked ships were not gaining. They were losing.

Nelson turned to Matthew at his side. They were both chilled to the marrow, drenched with spray, scarcely human. They had enjoyed a mouthful of soup six hours ago. Then the galley fire had been doused by a rush of water, and if they wanted something to pass their lips, it could only be the sea.

He said: 'We lose ground. They must cast us off.'

Matthew could not answer. He could hardly listen. For the first time in years not to be measured, he was beginning to be afraid. This was like following old Drake's star again. Then, his Captain-General had soared to the skies, only to lead them too far out. Now the crew of the *Vanguard* was in the same desperate trap: a bunch of wind-tossed scarecrows, to be seen all around him: officers and men standing, lying, crouching in the lee of the rail, their occupation gone.

They manned a ship which had lost the name, and was now, in that not too distant surf, to earn a new one, drift-wood.

He awoke to hear Nelson say, above the sighing wind: 'Signal them to cast off. Better lose one ship than two.'

Unwillingly, fearfully, Matthew beckoned to the signal-lieutenant, who at least had attended to his narrow duty, and had rigged a signal halyard on the half-ruined mainmast. Presently his two flags fluttered upwards. The wind tugged at them, but they stood out clear and cold in the dawn light.

The splendid *Alexander* was watchful, even in this chilly hell of mountainous waves, mournful gusts of a dying breeze, and hungry breakers drawing ever nearer.

'They answer, sir.'

'What answer?'

The lieutenant had no need to consult his code book. 'A negative, sir.'

Nelson the admiral roared: '*What?*' Then Nelson the man smiled, through thin blue lips. 'I intended to say, Mr Lawe, "What a friend." ' Then it was the turn for anger again. An order was an order, never to be questioned, never to be disobeyed, and he had given it. He crooked his finger to the signal-lieutenant.

'Is there a flag signal for "Obey me instantly"?'

'No, sir.'

'Why not?'

'I do not think it is ever necessary, sir.'

'I can believe *that* . . . Well, what else have you in your magic **book**?'

The lieutenant held it in his magic head. 'We have "Conform to my signal," sir.'

It was not the most ferocious outburst in the world, yet it must do. 'Hoist it.'

There was a pause, and then an answer. But it was not a hoist of flags, it was a voice.

On the quarter-deck rail of the *Alexander*, a tall figure appeared. In the light of sunrise, it could be recognized. Then its voice through a speaking-trumpet came down the wind, a voice scholarly, slow, and serious. Captain Alexander Ball had a message, even a lesson, to impart.

'I feel confident that I can bring you in safe . . . I will not slip my tow-line, and I will not leave you now.'

'By God!' Nelson began, and then fell silent, for so long that his signal-lieutenant was presently forced to ask: 'Have you an answer, sir?'

'Yes. Acknowledge.'

Though the pace was grindingly slow, it was done at last, and one man's confidence, and another's trust in him, became living fact. The *Alexander*, by superb sailing, by pinching the wind as a housewife makes cold meats from a joint of roast beef, a ragout from cold meats, cat's meat from the ragout, and Thursday's broth from the bones—the *Alexander* clawed her way off the hungry coast, at a rate of half a sea-mile in a daylight hour, towing a sodden lobster-pot of a ship in her wake.

Then, guided by Captain Saumarez, Nelson's second-in-command, in the *Orion*, *Vanguard* was nursed gently to her anchorage under the lee of San Pietro island, off southern Sardinia, and there found blessed sanctuary.

The first human animal to make a move, as all fell quiet, was Admiral Nelson, who left his wreck at anchor to pay a call upon Captain Ball.

Whatever Ball might have been expecting, it could hardly have been the warm handshake, the professional and private compliments, and the ignoring of everything in the immediate past save Nelson's gratitude towards a friend in need. He gave his rescuer unstinting thanks, for a service never to be forgotten. Then this sleepless commander came to the moment of *now*. What was to be done, to recover from disaster, and how to do it?

Alexander Ball, who had had time a-plenty to survey a shattered consort, gave a man-of-reason's answer:

'It must be the Gibraltar dockyard for you, sir.'

'Never!'

For Nelson, the dockyard it must not be, since his reverse could not have come at a more perilous moment, and no single ship—let alone the flag-ship—could be spared. The great French force which had been building up at Toulon and other Mediterranean ports was now at sea, and vanished from English eyes. Its destination was a matter of wild rumour, and no certain knowledge.

Naples had been guessed at, or Portugal; and Ireland, the West Indies, Egypt, England itself . . . The best intelligence confirmed that it carried unnumbered troops, 12,000 cavalry, and Napoleon himself; and the best conjecture was that Napoleon, after looting Malta, was bound for Egypt, and then for the conquering of India.

With the best part of Europe already prostrate under this tyrant's rule, his whirling dreams must be brought to a halt. To look for him, not a ship could be missing from the squadron; and if he were found, and brought to battle, not a ship could be spared either.

Nelson, with all the honours of gratitude done, looked round the sparse comfort of San Pietro, and decreed:

'I must be made good, here and now.'

There was only one answer, in this Fleet and in this service:

'Aye, sir.'

Despite an early word-of-mouth verdict which said: 'It cannot be done,' his friends set to work with their customary goodwill. Captain Saumarez had a choice of spars to spare, and sent them on board immediately. Captain Ball dispatched a man somewhat like himself—a dour Scot of few words and many talents, one James Morrison who was shipwright to the *Alexander* and a silent maker of miracles.

Flag-Captain Berry of the *Vanguard*, determined to show that his ship which carried the Admiral was not unworthy of her burden, whatever misfortunes might have struck her in the past, laboured like a Hercules afloat to retrieve his reputation.

For four days the bay of San Pietro echoed to the sound of hammering, sawing, planing, hoisting, and hammering once more. New sails from old cloth sprouted like clouds in the upper air.

Vanguard indeed performed a sailor's miracle; those four days were sufficient to make her fit for sea again. It was a patched-up jury rig, but good enough: the flag-ship was ready to lead the Fleet, and suddenly the Fleet itself sprouted like the sails.

With reinforcements pressing in from England, which brought his strength to fourteen ships-of-the-line, among them the finest to be found afloat, Nelson took up the chase. Though hot on the trail, he missed the enemy once, and once again; and then, after a last forlorn cast to the eastwards, he came upon his quarry, on the first day of August, in Aboukir Bay at the mouth of the Nile.

The Admiral, and the roster of officers old and young who had been invited to dine with him on that first day of August, had just seated themselves at his snowy, shining table—the appointments here were always so much more appetizing than the food—when, at two o'clock of the afternoon, a sudden burst of cheering sounded from the deck above them, startling them all.

No sound could have given more astonishment. There had been no such cheering, anywhere in the Fleet, since time out of mind; the wearisome chase had brought them nothing but disappointment, and the fear that the eyes of England, fixed upon them, greedy for good news, must by now be growing critical, even derisive . . . Nelson looked towards his Flag-Captain, Berry, at the other end of the table, then at Matthew, his natural messenger on this or any other occasion of doubt.

'Mr Lawe——' he began, and was forestalled. A clatter of boots on the ladder heralded the entrance of the Officer-of-the-Watch, so excited that he could hardly get his words past his lips.

'Signal from *Zealous*, sir! Enemy in sight!'

There was no need to hear more, since all understood. The two ships leading the Fleet on its last cast eastwards were Captain Hood's *Zealous* and Captain Foley's *Goliath*, two 74s whose fine sailing qualities had earned them a long advantage over the rest. There was no need for anything, save a dash on deck.

Nelson said: 'Gentlemen, dinner will be delayed,' and sped upwards, followed by what might now be called a running rabble of guests. All around them as they reached the open air were cheering sailors—and officers—gazing towards the east. All that could be seen there was the distant hazy shore-line of Egypt and, far ahead, the twin topsails of the *Zealous* and the *Goliath*. But these twins, having given them one jewel of news, swiftly followed it with another.

The *Vanguard*'s signal-midshipman, perched high on the royal main-yard, hailed the deck. It was his proudest moment, and he could scarcely fulfil it for excitement.

'From *Goliath*, sir.' The voice through the speaking trumpet was piping high, both with youth and with wild anticipation. 'Enemy in sight. Moored in line of battle.'

'*Moored!*' Nelson repeated. 'By God, we might have them!' But he delayed no more for marvels. A shower of signals, like lights flickering within his brain, came cascading out, and were translated into hoists as fast as hard-worked men could send them soaring upwards.

'Repeat *Goliath*'s signal to all astern.'

'Make to *Goliath* and *Zealous*: "Rejoin me and take station ahead".'

'Make to all others: "Form up in line astern as planned".'

'Make to all: "Prepare for battle and for anchoring by the stern".'

Then, with all these signals acknowledged by a Fleet suddenly in ferment, the Admiral took a turn about the deck. The preparation of *Vanguard*, the sound of guns running out, the bustle and the shouting, were all as they should be. There was nothing new here, but this time, this time, it might turn at last to gold and glory. Rejoining his Flag-Captain, he said:

'Press on, Berry. We have two hours or more. But we must make the most of daylight. When those two—' he pointed eastwards towards *Zealous* and *Goliath*, 'are in company, hoist "Signal Book, Plan Four" to all.'

Then, a diminished company, they dined. Nelson grew calm again as the meal progressed, though he sometimes stroked or pressed an itching armless shoulder; and Matthew marvelled afresh at this masterful control, which could shower sparks or glow gently as his steely spirit dictated. At the end, when they must break up, Nelson led them in a wordless toast to victory, and then, never at a loss for a phrase, said:

'It is either a peerage, or Westminster Abbey.'

A glorious victory, or noble death ... That was *his* estimate, *his* private portion on earth or in heaven ... But what of others? Was this man beginning, Matthew wondered as he followed his

Admiral up the ladder, to believe himself set apart from the company of mortal men? Mortal men remained.

What mortal men had sighted, two hours earlier, and what the Fleet now viewed in all its promise or peril, were thirteen ships of the French line, anchored within Aboukir Bay in a curving column, a phantom of that old demi-lune of the Armada, some two hundred years before. As the clock, and the sun, moved towards half past five, and one fleet on swift passage closed upon another which lay still, the view enlarged like a dream unfolding, so that both sides could gauge the trial of strength which lay ahead.

In the centre of the French line was the towering bulk of their flag-ship, *L'Orient*, a monster of one hundred and twenty guns, against the seventy-four which was the best to be mounted by any English ship in company. Their line had been anchored as near as they dared to the shallow water of the bay behind them.

There looked to be no room enough, nor depth enough to pass inside them. They had made themselves an embayed fortress, and at their back was a forest of French tricolor flags, on every house, on every hill, on every beached boat.

From this bold display, or this subservience, it might be deduced that Napoleon the conqueror had already landed, and placed his giant footprint on this land; and, as became known by the end of that day, had already won his vaunted Battle of the Pyramids and made Egypt his own.

But he had made it his own as a land animal. The sailors and the ships remained, to dispute the claim in their own fashion.

By now, Admiral Nelson had concluded all his signals. They had ranged from the courtly 'Open fire as convenient' to a rebuke, less courtly, for a luckless midshipman who had dropped his ceremonial dagger from the mizzen yard-arm to a sacred spot within two feet of his Admiral. He was greeted by a fearsome: 'Young man—are you *French*?' as he came shamefaced to retrieve it. Now this Admiral was in the mood to talk, and Matthew Lawe was the beneficiary.

The mood had become familiar, and as precious as words whispered on a pillow, or at the communion rail. Sometimes these thoughts were kept close, like secrets in a vault; sometimes Nelson must share them, or burst from solitude. This evening was a sharing time, when the insane spirit of battle, the nearness of death, began to possess them all. It was the time when Nelson was the man to love, and nothing else.

Standing braced with his back against the quarter-deck rail, confident that Captain Berry had the *Vanguard* in his grasp, just as his own arm embraced the Fleet, Nelson looked about him. The long column of French ships was coming into closer view; behind them was the evidence of their command ashore—the flags, the field artillery which had begun to waste cannon fire on ranging shots destined for a shallow sea-bed, or the smallest fish that swam.

Others might see the strength of this battle array; Nelson saw its weakness. Already he was strangely elevated, like a lover, like a confident thief, like a jolly giant about to spring.

'What happens on land is for soldiers,' he told Matthew. 'Let them flaunt their little flags . . . But here we have a line of their ships. A serpent of ships. The serpent of the Nile.' He glanced at Matthew, and put a question which seemed idle. 'Who was the serpent of the Nile?'

'Cleopatra, sir.'

'Well done, boy!' Matthew had never heard him so free. 'What do you do with Cleopatra?'

'Put another serpent in her bosom?'

'Such ingratitude! But you are right! Here is Cleopatra the sea-serpent, lying still, but watchful. You may tread on its head, and risk its poison, or you may aim lower, and break its back. I tell you, we will break this back, by the time the moon is up!'

'No doubt of that, sir.'

There was every doubt, for certain anxious minds, though this was not the moment to voice them. Matthew needed no reminder that this 'breaking of the back' was the core of 'Signal Book, Plan Four' for which Nelson's band of brother captains had been brilliantly schooled. Its purpose was to concentrate their fire on one point, whether it was one ship out of two, or one prime target out of twenty: to destroy *that*, and then to enlarge the scene of destruction.

For this, and for a score of other enterprises which might go this way, or that way, or come to nothing and need instant repair, they had been constantly summoned on board the *Vanguard* for conference, for questions to and fro, for the sharpening of memory. All they need do now was to remember their lessons, never to lag behind, and never to fail a friend.

The English Fleet ran on, backed by a sun already lowering in the west, and its leaders would soon begin to overlap the French line of anchor. On board the *Vanguard* telescopes were busy, just as the enemy's signal flags now started to chatter like a flock of starlings at dusk. It was likely that these were recall-signals for their men ashore. But Nelson still had time—carefully calculated time—to finish what he had to say.

He pointed towards the great bulk of the French flag-ship *L'Orient*, far ahead and seventh of the line. 'In that ship,' he said, 'is one Admiral François Paul de Brueys.' The Norfolk pronunciation might not have passed muster in Paris. 'He thinks himself safe, I have no doubt. He thinks he can out-gun us, ship for ship, and he is right. He thinks that no ship of ours can pass inshore, and engage his fleet from behind. *I believe he is wrong!* Capel!'

Sometimes it was hard to maintain a link with this jumping-jack of a man. Only when Lieutenant Capel sped across the deck to his Admiral's side did Matthew wake to the fact that the last word had not been some outlandish oath, but the name of his signal-lieutenant.

Swiftly the flags were hoisted, taken from a well-thumbed battle-plan of two months earlier: they were to remind certain captains that, according to their own judgement, they had leave to quit the line and pass inshore of any anchored enemy, if the risk were acceptable. The risk was surely there; entering a strange bay cursed by rocks and shoals, with dubious charts, no pilots, and darkness coming on, was a most formidable test of seamanship, if nothing else.

It was a measure of the Fleet's perfect understanding that six ships and men elected for it: the two leaders, Hood's *Zealous* and Foley's *Goliath*, followed by *Orion*, *Audacious*, *Theseus*, and *Culloden*. The last-named ran aground on an outlying spit of clay and, though serving as a useful buoy for some late arrivals, gave Captain Troubridge the most desolate evening and night which could befall any sea-captain.

All the rest passed safely by, pressing themselves between the French ships and the shoal water, firing broadsides as they went, then taking station opposite their agreed adversary, anchoring by the stern, and letting loose, one by one, a most bloody hammering.

On their way they had all noted that the enemy had been caught unawares: their landward guns were useless, the decks and gun-ports being cluttered with stores, furniture, bags, boxes, and lumber of all kinds. 'Good splinter stuff!' observed a happy English gunner. Some had been half-shorn of their crews, who were trapped ashore, storing, watering, even digging wells. The blind side was a gift from the gods.

The *Vanguard* led the seaward squadron which anchored in its turn, so that most French ships had a foe menacing them on either side. After that, there was nothing under the rising moon but darkness, blood, and death.

Caught in a deadly cross-cannonade, ships began to lose their masts and take fire, ships began to sink, ships began to cut their cables and run for the beach. There was no lack of sublime French courage, but there was also misfortune, foolish negligence, and a murderous disadvantage in fire power. The great three-decked *L'Orient*, having beaten one English ship, the *Bellerophon*, to such a gory pulp that she fell out of the battle altogether, was assailed by two more, Hallowell's *Swiftsure* and Ball's *Alexander*, which steadily reduced her.

Ships began to haul down their colours and surrender. The thunder and the glare of battle was so awesome, so tremendous, that weaker men died of shock. Then, after two hours of this pitiless grappling, one small strong man was directly hit and, as he thought and so declared to his Flag-Captain, brought to his death. It was Admiral Nelson.

Standing on the quarter-deck with Matthew Lawe and Berry, directing the *Vanguard*'s fire onto the crippled *Spartiate*, Nelson was struck on the brow by a piece of langrel shot, that evil charge of metal scraps, jagged iron, bolts, bars, and links of chain so beloved of the French. It fell upon a spot already tormented enough, cutting to the bone the old wound above his sightless right eye, and carving out a flap of flesh which, bleeding in finest pig-style, blinded the other.

He fell to the deck, and gasped to Berry: 'I am killed. Remember me to my wife.'

Matthew Lawe, whom all this monstrous uproar of torn ships, crashing spars, and men shambling down to their death had sickened and terrified, was now witness to a scene so strange that he wondered if his wits had at last fled and mad fantasy had taken their place.

Together with Captain Berry, he assisted Nelson below; the

ghastly figure, pale as the moon and drenched with blood, was laid in a cot and the surgeon summoned as soon as he could be spared from all the other wounds which the *Vanguard*'s men had suffered.

Despite the gloom of the great cabin, Nelson seemed determined to die in the glitter of play-house history. He refused medical treatment until his turn came; he sent constant orders to his Flag-Captain, and other signals to other ships which had supported him so bravely, and made wandering statements which seemed to illuminate a vision of noble death, to be nobly ornamented before it was too late.

He commended his soul to God, comforted his chaplain, and once more sent fond remembrances to his wife.

Matthew nursed his sinking Admiral as best he could. But Surgeon Jefferson, freed at last, declared the injury to be a flesh wound only; stitched it up, bandaged it, ordered his patient to remain quiet, noted prudently that a wound to the head must always be dangerous, and went on to other, grosser mishaps.

Presently Nelson quit his cot, and was carried down to the bread-room, a cavernous space where he could escape some of the noise of battle. From this murky vault, lit by one wavering candle, he sent Matthew to summon his secretary who, on seeing this fearful vision of mortality, promptly fainted. Flag-Captain Berry returned many times, to report French ships sinking, or burned, or surrendered. The total was hardly to be believed. Soon Nelson feebly beckoned Matthew to his side.

'*Write!*' he commanded. He might have been a dying King Harry, or a living Moses with certain precious tablets under his arm which must be transcribed. 'To the Commander-in-Chief. "My Dear Lord, Almighty God has blessed His Majesty's Arms ..."' He went on to report a brilliant victory in modest, concise terms. Then: '*Write!*' he said again, with the same stubborn determination. It was an *aide mémoire* to the effect that Mr Alexander Davison, late of the City of Quebec and now of St James's Square, London, was appointed as the sole prize-agent for all ships captured in the recent battle.

Then Captain Berry reappeared, fresh as his paint-work, full of good news. The French flag-ship *L'Orient* was afire, and like to sink, or explode, or otherwise to die. Few of her consorts in the enemy Fleet would survive.

'Help me on deck,' Nelson said. 'This is a last moment which I *must* see.'

He saw it, as did twenty thousand other sailors, fearful or triumphant or sullenly indifferent on that hell-brew of a night. *L'Orient*, tigerish-brave, unfortunate, and doomed since the first day of August dawned, had been painting ship for many previous hours. A First Lieutenant who should have been crucified on a cross of wooden legs had allowed a vast collection of paint buckets, oil jars, and wiping-rags to accumulate on the upper deck. These had now caught fire, and, as ten o'clock approached, so had the whole vast ship.

The flames spread greedily from spar to spar, sail to sail, cabin to cabin, deck to lower deck. The gunners on both sides, as if by agreement, grew silent and slack-handed as they watched the tongues of fire, the whirling sparks, reach up past new-tarred rigging to a black, smoke-laden sky. A blood-red glow illuminated the whole bay: not a ship, not a flag was hidden as the vast torch flamed brighter and brighter.

None had ever seen a mightier, more desolate funeral pyre. But there must be a last fearsome rage to come. As the fire crept and leaped about the ship, and ate downwards into her vitals, it would also devour her powder-magazine.

Matthew, supporting Nelson until he found his feet, watched in horror as the perilous moments raced towards their climax and the flames turned merciful night into unnatural, brutal day. He had never been more afraid of what he must soon see. He could not purge this quailing spirit, the sickness of his doom. As one man's courage rose, so another's faded. For him, fire had once more proved the greatest of all terrors. Once again, the moment proclaimed the malediction of Francis Drake: anyone who stood near to a valiant man stood in danger of death.

At ten o'clock, *L'Orient* blew up, with a frightful gust of flame and heat like a lid lifted from hell. The shattering roar of it might have been heard ten miles away. After the heavens split, there was a pause, like a giant intake of breath, and then the bloody rain began.

Torn spars soared upwards, then plunged back into the sea. Flaming timbers hissed and glowed as they reached the fouled water. Burning scraps of sail and rope, like fiery feathers, fell on other ships. Dead men dropped from the sky, as neglected, mistreated dolls which had lost the once-loved outline of life, and been abandoned.

Then the fire was quenched, and the flag-ship, with a gallant Admiral, hundreds of his men, and the principal treasures of Malta, was engulfed forever.

On Aboukir Bay, silence came down like a pall.

Nelson himself was the first of their company to speak. Compassionate in victory, he said—as he thought—to Matthew: 'If we have a floatable boat left, send off Lieutenant Galwey to search for survivors.'

But there was no answer from Matthew, and when Nelson turned towards him it was to find his trusted lieutenant lifeless on the deck.

A guilty and ashamed Matthew Lawe, who had woken in the cavern of the bread-room among wounded men who were not expected to live, came furtively on deck as sunrise signalled the second day of August. He could not guess what his welcome might be; even the feeblest of sailors should never faint from fear as he had done; the prospect of meeting his Admiral's eye again seemed a worse ordeal than the fire-bomb of *L'Orient*.

He might invent excuses; men *had* been clubbed senseless by the rage of battle; but they were not seasoned officers who were expected to set an example and to maintain the standard of the Fleet. If only he had been struck, as Nelson had been struck; if only he had shed honest blood, suffered demonstrable pain, gained one puckered, livid scar to advertise his valour. If only he had played the man, in place of the mouse.

Astonishingly, he reached the *Vanguard*'s tangled deck to find himself innocent. The few weary men working on the mizzen-mast, which was a useless wreck, gave him a smile or a nod as he appeared; a boatswain's mate came forward to lend a hand up the last step of his ascent; and it was First Lieutenant Galwey who greeted him, with nothing in his voice save welcome.

Galwey, who looked as haggard as a crow, was a large and cheerful man who did not stand on ceremony, nor needed it to preserve his authority. When he said: 'Good morning, Matthew! Are you well? Are you alive?' it was as one triumphant survivor to another. Since Death the Leveller had not felled either of them, the ladder of rank might take a holiday also.

'Well enough,' Matthew answered, feeling his way back to a world of repute. 'I know not what put me down . . . All I remember is the French flag-ship bursting apart . . . I am sorry to have been of so little service.'

'Think nothing of that. 'Tis the luck of the lottery. There was a man at the foremast who was killed by the same shock. Not a mark on his body, but overboard he must go . . . I may tell you, you were mourned for already. The Admiral thought you were dead.' He winked; he was a man who could even wink at a junior, and still command him. 'He near forgot his own wound!'

'But how does he do?'

'Well enough, like yourself. It was a fair deep butcher's slice, but they heal, thank God. He is resting below. So is Captain Berry, who drove himself more than the men . . .' Galwey stretched his weary arms, and then opened one of them wide, encompassing the bay, the sunrise, the gory fruits of battle. 'Look about you, Matthew. Victory! You will never see a finer scene, if you live to be a hundred!'

The scene was not 'finer' than anything, save for a painter requiring the shores of hell for his canvas; but victory it was. Though there were still bursts of cannon-fire at the rear of the line, where the tireless Alexander Ball remained at work, scaring some Arab pirates away from a stranded prize which they hoped to loot before its rightful owners could take possession, the rest of the bay was an utter, mournful, silent desolation.

Aboukir was empty of live enemies. Most ships had surrendered: two had been destroyed or burned: some were to be spied ashore, as smoking hulks with no colours showing, and no life either. When the sun began to lift, dead fish could be seen gleaming among the scum of victory. Already a waft of death mingled with the burnt smell of Africa. There was an inlet full of wind-driven corpses, with certain scavengers busy among them.

The rest of the bay was a soiled pond of charred wood, sodden canvas, spars which would never serve upright again, split barrels, split humans—the refuse, the very excrement of defeat.

For a long moment the two men surveyed the waste around them. Sombre thoughts warred with the triumph which lay at the heart of the morning, and could not banish it. Turning, Matthew noted once again the grey, almost spectral colour of Galwey's face. He must be near to dropping with exhaustion.

'You have done enough, sir,' he said.

'Well . . .' Galwey was not ashamed to admit it. 'We were out all night, fishing for swimmers. With other boats, of course. But we lifted a poor crop.'

'How many?'

'Perhaps eighty.' Yet there were other, greater totals to compare with a paltry sum. 'But think of it. *All* the French Fleet is captured or burned! *All!* Well, two line-ships made their escape, and two frigates. But they will not make their fortunes by themselves . . . The battle-tally is two ships sunk and nine surrendered. That is more than victory. It is *conquest!*'

'Wait till they hear in England.'

'Wait till they hear in France!' But Galwey, after a brief burst of spirit, was drooping again. 'Hades, I am tired! . . . Are you fit to take the deck?'

'Aye, sir. And gladly.'

'Then I to my cot. I wish it were for ever!'

'Orders, sir?'

'None but common sense, which you have. We stay at anchor till we are all pieced together. Burials at sunset. There will be captains coming aboard with great tales of cock and bull, and by God, for once we may believe them! But I pray that they sleep late. As I must, or fall to pieces.'

First Lieutenant Galwey, who now left him, was not alone in his exhaustion, his lust for oblivion. All over the Fleet, after twelve hours of fighting and slaving and enduring the thunder-crack of war, men had dropped asleep at their guns, at the capstan-bars, at the tail-end of a rope. They could not be driven more, nor was there need to do so. The Battle of the Nile was won, and peace after carnage, blessed peace, came swiftly upon them.

It came also to certain women of the Fleet. Some of these had carried powder to the guns, some had tended the wounded and given, to dying men, their last glimpse of life in the form of a mother's care and gentleness. One, at the height of the turmoil, had borne a son.

But another mother's son, a bright French lad of ten, had quit the world in great grief and agony, as a weeping survivor, naked and gasping on board the *Vanguard*, related to any who would listen, in the strangest, most moving story of the night.

Some breeds of men caught fire of their own accord. Some others took marvellous heart from this example. A few might pass on the spark of valour to their children, before it was too late. The French flag-ship *L'Orient*, in her last moments, had given to mankind a magic, shining pattern of this chain of courage.

Admiral François Paul de Brueys, commander of the French, had watched the slow, inexorable destruction of his flag-ship and his Fleet with disbelieving eyes. He could not credit the disaster and thus would not accept it. Though he had been wounded early, he directed his men in fighting off the English ships, and then in battling the flames which had taken hold of his own, with the most stoic determination.

Then he was wounded again, so cruelly that he must die. With a courage which might have been called noble or animal, and was certainly divine, he decided to make a fitting end. His body nearly severed in two, legless, with tourniquets on both his stumps, he chose to die facing his foes in an arm-chair on his own quarter-deck.

This he did, as the fiery ruin of what had once been the greatest ship of France prepared its own death.

His Flag-Captain, Comte Louis de Casabianca, a fighting sailor set in the same mould of bravery, elected to follow his Admiral without the smallest moment of doubt. Now in supreme and lonely command, he fought his ship, eaten to the heart by flames, until her very end. Then, within minutes of an explosion which must be fatal, he was hit, and fell, and on his deck he died.

He was not, in death, so lonely. He was loved. He had with him his son Jacques, just old enough to be *aspirant*, the strange French term for midshipman. Ten years was never a great age, but great enough for love and valour mixed. Jacques de Casabianca cradled his father's body until the last, and then, drenched with the dearest blood of his life, stood up and looked about him.

He found himself alone, within a roaring curtain of fire; alone save for the terrorized men now leaving their guns, and every other duty, and leaping into the sea.

He did not leap, he stayed. Thus, till the end of time, the boy stood on the burning deck, whence all but he had fled—including the poor shattered ghost of his father; and was ripped to mortality as the ship went up in a last volcano of fire.

On her slow and would-be contented journey back to Naples, the

Vanguard, already sufficiently battered in action, was again dismasted in a gale. The jury-rig of San Pietro Island could endure much, but not the Battle of the Nile followed by a foul *sirocco* wind. In what Nelson wryly called 'Our customary *pas de doo*'—he might have mastered the French, but never their language—she was taken in tow by the frigate *Thalia,* and thus made harbour as a limping cripple.

Matthew observed Nelson to be in a wretched state, in spite of all achievement. 'Captain Hallowell has promised me a coffin made from the mainmast of *L'Orient,*' the Admiral confided to him once. 'By God, I could use it now!' It looked almost to be true. The raw wound above his useless eye was giving violent pain to his head. The *Vanguard,* rolling day and night in true pork-barrel fashion, made him miserably ill. Now he must enter harbour, as the Victor of the Nile, like a beaten dog on a leash.

The Bay of Naples transformed all things to happy and glorious triumph.

News of the greatest victory of British naval arms for a hundred years or more had preceded them. The whole city was *en fête,* and determined to remain so. There were flags and fireworks and crowds everywhere. The bay was filled with hundreds of pleasure-craft, some with bands sawing and trumpeting out a very fair version of 'Rule Britannia!'

Though doves and rose-petals were held in reserve, the saluting guns ashore did not wait for the *Vanguard*'s courtesy; they began to blaze away as soon as the ship was sighted, to make clear to the world, on this day of all days, where the greatest honour lay.

There could be no mistaking the message. The French had been rendered ship-less in the Mediterranean, at one glorious stroke, and Naples was to be the first city to celebrate it, in wild rejoicing all the better for being a tax-free, unearned gift from Britain.

The anchor plunged down: the weary ship rounded up and settled to her cable; and, as in a new command performance of an old play, the barge of His Britannic Majesty's Ambassador Plenipotentiary left the quay and ploughed manfully towards them.

It was Ladies First up the ladder, and the first lady, followed by Sir William Hamilton in his best black knee-breeches and the golden emblems of diplomatic rank, was a sight for the weariest of eyes. When Nelson **saw** Emma Hamilton, after a lapse of five years, he could be in no doubt that a warm welcome was his for the asking.

At thirty-three, she had grown a little stouter—or, in accordance with taste, to a most noble armful. She wore a billowing blue gown, a scarf embroidered with gold anchors wherever gold anchors might find a resting-place, and a spangled head-band inscribed 'Nelson and Victory'.

What she saw so suddenly was what Matthew and all others who sailed with him saw, and had grown accustomed to for many years: a small ruin of a man, pale as a shrimp, battle-weary, with one empty sleeve, one shaded eye, and now a fresh bandage to mask yet another wound in the service of his country.

She might have been overcome by his appearance, or by the great occasion, or the blast of saluting gun-fire, or by real and deep feeling. She might have rehearsed a truly dramatic Attitude, and brought it to full flower for this, the first and only performance. It mattered nothing to either of them.

Emma Hamilton gasped: 'Oh God, is it possible?' and tottered towards him. Then, as Sir William looked on with appropriate if controlled concern, she was taken by what seemed to be a fainting fit, fell upon Nelson's breast, and there remained for ever.

VII

If blessed old Sam Pepys had been casting 'The Accounts of Lord Nelson' for the previous two years, Matthew Lawe's quick-writing fingers would, in July of 1800, have noted as follows:

ITEM: He is Rear-Admiral of the Red, Baron Nelson of the Nile, Duke of Bronte in Sicily: Viscount Pyramid and Baron Crocodile to his mistress; Freeman of the City of London; possessor of a vast collection of tasselled swords of honour, gold medals, jewelled stars, diamond-studded boxes, gifts from Eastern potentates and Western men-of-business; recipient of a grant of £10,000 from the East India Company, and a Parliamentary pension of £2,000 a year.

ITEM: His *amour* is the accepted scandal of the Fleet, the grim reproach of all established society, and the delight of the populace.

ITEM: He has frittered away much time at an indolent and degenerate Court of the Two Sicilies, and disobeyed a Commander-in-Chief's order so that he may remain there.

ITEM: He has directed the blockade of Malta to such good

purpose that the French have been expelled from it, and his brave and loyal friend, Captain Alexander Ball (soon, one hears, to be Baronet) has been welcomed with the most open arms as the first Civil Commissioner of the island.

ITEM: He is the object of such scurrilous comment in England that a print-seller of Seething Lane may be bold enough to publish the following: 'Coat of Arms for a Certain Sea Officer newly ennobled: A Pair of Globes Rampant upon a Sail-Cloth Spread, with Admiral Attached, and the Motto: Stand Firm, England!'

ITEM: He has lost arm and eye, health and spirits, in the most gallant public service; knit together a Mediterranean Fleet which is paramount over six thousand square miles of blue water; earned the love of every sailor who comes near to him, and a grudging respect from disapproving masters at home.

ITEM: He is now recalled to London, with Sir William Hamilton, both out of favour where it hurts, yet rapturously admired where it warms.

BALANCE: I swim against the stream with the Bard of Avon, and must declare: Take him for all in all, he is a Man. We shall not look upon his like again.

The trio of Nelsonians, as they were now named by shocked, long-suffering, or envious society, took their own sweet time on their journey back. Starting from Malta in the line-ship *Foudroyant* which, as a saucy young midshipman remarked to Matthew, 'had not seen such close action since she left the slip-way' (Penalty: a mast-heading from noon till sunset), Lord Nelson, Lady Hamilton, and a contented Sir William completed, in the *Alexander*, a halcyon summer voyage to Leghorn.

On certain evenings, Emma Hamilton played the harp on the quarter-deck, while discipline among enraptured sailors gave evidence of lagging. On certain others, their course was trimmed so that the ship would not roll—or, according to a lewd shipboard version, so that no one of rank would fall out of bed.

Then, by easy stages, skirting the enemy France, they made a European progress almost royal to Florence, Trieste, Vienna, Prague, Dresden, and Hamburg before embarking for England.

They spent four months on their journey. At every significant pause, British ambassadors and ministers, some of them willing, gave hospitality to a party which could never escape notice and often invited it. The tremendous travelling throng included, as well as its official principals, Queen Maria Carolina, five of her princesses, and Mrs Cadogan, the hard-working mother of Lady Hamilton.

The 'principals' were of course a subject of scandal; Matthew sometimes wondered if the scandal was required to be such a noisy one. The phrase 'Lady Hamilton *and all that*' grew commonplace among those who must entertain them; and 'all that', as well as acknowledging their notoriety, paid mournful tribute to their behaviour.

Her ladyship had grown larger in every particular; her voice was louder, her laughter more piercing, her appetite for liquor noticeable, and her figure opulent as a dream of Venus. She loved company, and could scarcely endure without it. Among many other pastimes, she had formed a passion for gambling. The game of Faro was now the rage, and Faro became her own: it might be translated to Pharaoh, the Lord of the Pyramids . . . Thus, everywhere they travelled, everywhere they lodged, there must be concerts, dinner parties, balls, and card-parties stretching to the small hours of the morning.

Lord Nelson, it was remarked, shared no such taste for gaming, and would sometimes fall asleep at the table. But he seemed to dote upon her in all else. For him, Emma Hamilton could do nothing wrong.

Sometimes, it was further remarked, she was taken with sudden sickness, and had to leave the company. Sir William, with whom, by a natural equation, Matthew now spent more time than with anyone, dismissed it as a matter of small concern whenever anxiety was expressed.

'It proceeds from a foul stomach,' he would declare, and go on to talk of other things. Recalled after thirty-six years of service in a country he had grown to love, uprooted with a positive ship-load of antiquarian treasures, he preferred the golden warmth of the past to anything which might await him in the future.

He made wonderful company, urbane, knowledgeable, courteous, ready equally to listen and to speak. He was a man of the world, the truly civilized world of serene accomplishment. Yet Matthew could not quite credit him, in the realm of medicine.

Certain events on board the *Foudroyant*, even before they left Malta: certain whispers, certain compliant glances between two people who maintained, on the surface, no more than an admiring friendship, had alerted a man very close to his Admiral that something, which might mean disaster for all, was now woven fatally into the tapestry of their lives.

But Emma Hamilton would always recover briskly, and regain her spirits, and with them her taste for feasting, loud laughter, outrageous play, and the general uproar of celebration. In Dresden she performed her Attitudes . . . It was small wonder that a certain ambassador, on the evening after her departure from his distressed and bursting roof, begged his immediate family on sitting down to supper:

'Tonight, let us all be very, very quiet.'

They landed at Yarmouth, in the common mail-packet, no frigate (though requested) having been sent by a censorious Admiralty to accommodate the Hero of the Nile. But the good citizens of Norfolk did not hesitate to make amends for this neglect of their own returning son.

Knowing well enough both the common gossip and the disapproval of authority, they had made an English decision that they liked a naughty Admiral, with victory tucked under his belt, better than any stiff London Sea-lord who thought his pen to be mightier than the sword.

As the church bells pealed, the fireworks pierced a gloomy November sky, and lantern-lights proclaimed a now universal motto, 'Nelson and Victory', the Nelsonian coach was pulled upwards from the harbour by loyal and loving townfolk. The Mayor and Corporation gave their own hero the freedom of the town, and set him down to an honest English dinner.

Ale was flowing freely as, in night-time turned to day by illuminations of every kind, a distinguished party advanced to show themselves on the balcony of their hotel. Lord Nelson delivered a modest speech of thanks: Emma Hamilton made a brief appearance to the roars of a lustful mob; Sir William bowed his thanks; and Nelson again, his glittering medals positively aflame under the lamplight, with Matthew Lawe holding his scarlet-lined boat-cloak, bowed and waved before retiring to God-knew-what voluptuous delights.

The hero was home again.

He was home to the greatest storm of all his forty-two years.

What was sweet triumph in Norfolk was poison in London: poison to the Court, to their Lordships of the Admiralty, to society in general. It was poison, of a particular kind, to a blameless, innocent, injured Lady Nelson.

She had waited so long, in the dullness of a workaday world, in dry virtue, for the rewards of constancy and the fruits of a glittering career: hearing nothing meanwhile save tales of sunshine far away, splendid palaces, royal favours, victories, honours, and the ineffable goodness of Lady Hamilton. Now, for more than a year, she had been left to listen to the whispers, the whiff of cruel gossip, and to read in coarse detail the scurrilous prints which kind friends did not neglect to send her.

Now the man was returned to her. But though they shared the same roof at last, she waited still for reassurance, for love, for anything.

Time and chance had stolen it away. He had been too long absent. In the great wave-pattern of life, this was an amorous peak for him, a down-trough for his friends and admirers, a piece of sport for the idle, and desolation for his wife. It did not need an hour to discover that she was out of favour. Just as Emma Hamilton could do nothing wrong, so Fanny Nelson could now do nothing right.

An infection seemed to have taken hold, feeding on itself and all about it. Apart from his fascination with a beautiful woman, when set against his duty to share bed and board with a dull and sallow mate so far out-distanced, he had a deep source of irritation, pricked by his own guilt, in their only remaining bond, Josiah Nisbet.

Josiah, the whining child, awkward youth, and young man spoiled by great connections, had gone down-hill like a shambling bullock. His step-father had always advanced his career, pressed for favours, begged pardon for his shortcomings, admonished him, and stood him up again. But there seemed no means whereby an uncouth lout could be turned into a good officer, or an officer of any kind.

As a lieutenant, Josiah had disgraced his rank at a Palazzo Sessa *soirée* by drinking himself into a mumbling stupor, making unmistakable advances to his hostess, then complaining, for all Naples to hear, that his lack of success was due to the fact that his step-father was performing the office of Sir William Hamilton.

Captain Troubridge and other friends had silenced him with discreet if muscular skill, and shipped him out of the gathering. He was forgiven: tearfully he promised reform: soon, beyond common prudence, he was promoted and, after representations to Lord St Vincent which Nelson would never have made for any other human being, given command of the frigate *Thalia* at the age of nineteen.

St Vincent, whose private reservations were outweighed by his admiration for a brilliant fellow-admiral and friend, had written: 'Captain Nisbet is on his own bottom now.' He spoke most truly. Captain Nisbet, with hardly any delay, landed firmly upon it. He was now in dire disgrace, for drunkenness, incompetence, and insufferable conduct towards his officers.

His immediate commander, Admiral Duckworth, with all respect, confessed to Nelson that he was unable to allow this useless fellow to remain on his station. He added a suggestion, a curative proposal, that Josiah should spend some months with Lady Nelson.

In all the circumstances, there could have been no prescription more unfortunate in its undertones. Unfairly, with hatred and frustration mixed, it was something else against Fanny.

London was a disaster and, for the compassionate of the world, a tragedy. Lord Nelson was the beloved target of the mob wherever he went. Though he was snubbed by a King, and almost hissed by society, he was adored by all lesser people. Flaunting themselves in a theatre box, he and Emma drew all eyes. Infuriated tenors were halted in mid-song as the late-arriving pair stole the scene— and the applause.

But they were never a pair. Behind Nelson in his gleaming orders of chivalry and valour, and Emma in seductive ocean blue, were two others: a courteous, withdrawn Sir William, and a grey shadow who was the hero's wife. Fanny was lost, and Emma was everywhere. A frightful scene at an evening supper party, in surroundings of great consequence, finally brought down the curtain on a nuptial dance of the dead.

When Matthew Lawe presented himself at Lord Nelson's new house in Dover Street, alone and with a hired carriage to convey his Admiral's wife to an evening party, he knew the awkwardness of his arrival. The arrangements made for their appearance at the home of a relenting Lord of the Admiralty had been, until that very morning, entirely different. Now he must explain away an uncaring, last moment alteration.

'Good evening, your ladyship,' he said as he was ushered into the drawing-room, still at that uneasy stage between being fully furnished and half ready for human occupation. 'I hope I see you well.'

She did not look well at all. She was, as usual, pale and drawn; though her *toilette* was carefully done, the purple satin and the strange jewelled turban which was now the foolish rage of London looked no more than hesitant play-acting. She could not match the most modest rival . . . When she glanced beyond him in search of another figure who should have been there, Matthew knew, not for the first time in this horrid London visit, that all excuses would seem lame as a one-legged man.

'Good evening, Mr Lawe.' Fanny Nelson never treated him with more than cold formality; she now counted him as one of the enemy. 'Is his Lordship below?'

'No, my lady. He sent his apologies for a change of plans, and asked me to escort you to Admiralty House. He will join you there.'

Lady Nelson considered this. He saw the look in her face which he had seen so often: the look of disappointment, of resignation, of defeat. There had been so many slights in the past month that there were no longer any graduations in this expression: it remained constant—constant in despair. She said:

'Where is he detained?'

'At Grosvenor Square, my lady. He had some business with Sir William.'

For the first time, she returned a bitterly revealing comment: 'Perhaps he may one day have more than he expects.'

Matthew, taken aback, could only answer: 'I know nothing of that.'

'Nor I.' She rose, a small sad figure in her absurd finery. 'But I expect that I shall . . . Well, I am ready.'

Below, gathered round the front steps, there was the usual group of idlers waiting for their hero. They raised a stir as Matthew's dress uniform was sighted, and then fell silent when he was recognized to be a lesser man. For Fanny Nelson there was nothing at all.

It was the same story as they arrived before the great house in Whitehall. There the crowds were thicker, drawn by private intelligence or the sight of carriages rolling up to a noble front door. The link-boys with their flaring torches stood ready to light the way. But Lieutenant Lawe and Lady Nelson stepped down, and passed within the portals, with scarcely a whisper to greet them.

The company which presently assembled upstairs had to wait almost an hour for the last three guests to make their appearance. Yet, without question, they were forgiven their tardiness. Sir William Hamilton was his usual quiet self: Emma came in under full sail, like a swan among sparrows; Lord Nelson, haggard, lean as a drum-stick, was gold-braided from head to foot and be-medalled from arm to empty sleeve.

They were applauded, led to their chairs, and then to the shining supper table as if Drury Lane had come to Admiralty House.

An hour later, all harmony was shattered as a cruel farce was played to its very end.

The distinguished supper party, some of them still reserved, a few censorious, all agog, were no match for the Nelsonians. Nelson himself, seated between his hostess and Lady Hamilton, was for all his resplendent presence only one half of a marvellous *duo* such as had not been seen in London since Punch and Judy first arrived from Italy, and never at this august table.

Emma Hamilton was at her very best—or, according to guests ever ready to inform the gossip-mongers, at her worst: blooming like the most over-blown rose in the Garden of Eden, noisy, screeching with laughter at the smallest display of wit, demonstrative to anyone who would intercept her gaze, flushed with wine, and above all fondly attentive to the hero of the hour.

She cut up his meat as he directed, and might sometimes be seen passing choice morsels to his opened lips, her eyes liquid with adoration. He received all this display with the satisfaction of a boy who had fallen among shepherdesses in some pastoral romp, and found one in particular who would satisfy, with good fortune, every appetite.

It was all absurd, and scandalous, and touching, and beyond belief. This was the man, Matthew Lawe thought, from his safe station far down the table, who had boarded the *San Josef*, and led the storming parties at Calvi and Teneriffe, and routed the French Fleet at the Nile, and fought like a tiger wherever he was engaged.

This was the man, the hero, the Admiral?—he was more like Cupid, with table-forks for arrows and a folded napkin for a bow.

It was all absurd. For Lady Nelson his wife, seated across the table with a full view of this *charade*, it was worse than anything she had seen, or heard of, or imagined. It was worse than rejection, it was public insult. But brave in her desolation, she must have decided to make a small fight of it.

When, at last, dessert was served, she took a small handful of walnuts, cracked them, peeled them, and leant forward to offer this small tribute to her husband. It was irritably brushed aside, almost struck away, by a man who could not bear to have his attention disturbed. Guests and servants alike looked on aghast as the violence of the gesture sent a broken wine-glass reeling across the table.

In an atmosphere electrical, Lady Nelson was seen to be in tears. She could only be an object of pity. But even now her moment was stolen from her. In the silence, Emma Hamilton with her usual freedom announced that she must retire. She rose, and suddenly it was no *charade*: she was trembling, deathly pale, and near to fainting.

Nelson, glaring across the table, exerting a monstrous will, seemed to be directing his wife to support her; and, after a moment of breathless confusion, it was Fanny Nelson who helped her stricken rival from the room.

Time passed: conversation languished, save for the low voices of scandalized guests: the supper party seemed in suspense, and likely to wilt away. There could hardly be another such tempestuous scene . . . Lord Nelson, with an empty place by his side, had turned glum and irritable: one or two knowledgeable officers among the company observed that he was 'working his fin'—lifting and then pressing down the stump of his right arm.

Presently, after an oppressive silence, he drank off his wine, and turned until he could see Matthew Lawe. Then he signalled, with a covert yet imperious turn of his head. Beyond doubt, he was telling his lieutenant to go scouting ahead and see what was in the wind.

Followed by the stares of the company, Matthew rose, and bowed to his hostess, who had no difficulty in conveying, by a single look, that this was not the customary style of her entertainment. Swiftly he passed through the doorway, and out into the passage.

He had not long to search before he came unexpectedly upon a dark scene. Emma Hamilton had not been able to reach the safety of a retiring-room before she succumbed to violent sickness. She was, with Lady Nelson, in the partial seclusion of a side-closet, hung with the coats and hats of the male guests.

Fortunately it also held a sofa, and a wash-hand-stand with ewer and basin. One small lady was still holding the head of one large one, to assist her in some last spasms of vomiting.

Aware of a shadow, Fanny Nelson looked up, met Matthew's eyes, nodded, and bent down again. Presently, in a sort of pale alliance, they assisted Emma to the couch, and Fanny held a vial of *sal volatile* under her nose. Recumbent, her clothing in disarray, Emma now revealed more than anyone in England had yet been privileged to see.

Her recent stoutness was unmistakable in its origin. Even under the voluminous fashions of that season, her pregnancy could be seen to be far advanced: perhaps as much as seven months.

Matthew in his common-sense heart had known of it; Fanny Nelson had not, and the shock and astonishment pierced her cruelly. At a stroke she perceived the degrading truth, and all its implications, and its insult to her own barren state. She ceased her ministrations as if her patient were already dead of the plague, and rose to her feet. Then she looked down, as Emma opened wan eyes, and spoke:

'Your ladyship should really take more care.'

It was the second and the last barbed comment of the evening. With it she quit the room, and the house, and a world of treachery, and saw her lord no more.

In the ensuing turmoil, Lieutenant Lawe had the task of acting as go-between for two people who could never meet again, but had certain mundane arrangements to make. Nelson, with many other preoccupations, always questioned Matthew keenly on his return. How was her ladyship? Was she comfortable? Was she in fair spirits? Was she in need of anything? What was her mood?

Once Matthew was forced to tell his Admiral: 'My lord, she has formed an opinion that she has failed in marriage.'

The response was instant, and memorable. 'Upon my life,' his

Lordship declared, 'I can swear that I have never found anything wanting in her conduct!'

Great news at last.

Then, all too soon for such domestic cares, Lord Nelson was ordered away by a Board of Admiralty which could trust no other man, to prepare a new Fleet for a new battle, and another matchless victory in far-off Copenhagen, where the slaughter and the lion-hearted courage of both sides, Dane and Briton, prolonged a fight to an agony, and then to exhausted silence.

Matthew, a man for all occasions, was left behind in London to settle his Lordship's papers—in fact, to aid in the delivery of twins, the burial of one and the spiriting away of another. This was a live girl child, carried off in a muff to the arms of a trusted wet-nurse.

Sir William, after expressing hopes that his wife, who had kept to her room for three days, had recovered from her recent indisposition, retired to peaceful sleep: the most temperate of men, and surely the wisest human being of all.

VIII

The man at the edge of the tide-mark stood a little apart from the crew-men and the marines as the Admiral's barge from the *Victory* waited at the foot of Southsea Steps, close by Portsmouth Harbour. He had his watch to keep, and they their legs to stretch and their tongues to wag: the Admiral's proud coxswain would keep order enough, without any aid from Lieutenant Lawe. Thus Matthew waited alone—though no more than a few yards from the greatest Portsmouth throng he had ever seen.

Above him the crowds on the harbour parapet pressed against the scarlet coats holding them back, so that Lord Nelson when he appeared might have a safe passage down. Behind these, the ramparts of the town were crowned with a huge press of people, come to bid farewell to the *Victory*, which they regarded as their own, and to her Admiral, who was the darling of all England, and ever would be.

It was to avoid these Portsmouth crowds, which had besieged his inn all the morning, that he had decided to embark from the humble boating-steps of Southsea. But the townsfolk had got wind of his change of plan and, not to be denied, had run for their lives from the inner harbour to the outer beach. Since it was a Saturday—the fourteenth day of September, 1805—the holiday throng was immense. Thus he was delayed.

Lord Nelson was always delayed whenever he stepped ashore, the patient Matthew Lawe thought, as he watched a woman above him threatening the sergeant of marines with all sorts of indignities if he did not let her three children pass to wave farewell to 'a greater man than any tyrant of a lobster'. She had her own husband on board the *Victory*!

He had been delayed in the same fashion all over London, where he had spent his brief leave after two years of ceaseless shipboard life, ceaseless watch—'Darning the ocean,' he had called it ruefully—upon an enemy who would not come out to fight, ceaseless chasing of the French from Toulon to the far-off West Indies. People could not bear to let him pass without saying what they had to say, whether with words, or staring silence, or outthrust arms.

Their message was as confused as Babel, or clear as that magic amber which had the property of attracting sparks from one element and passing them to another. They said, or yearned to say: 'You are England. You are us afloat. Go out to fight—but we beg you to *live*!'

All those who thus delayed him, loved him, and must see him, and cheer him, and touch him if they could. It would always be so, in admiration and honour, until the very end, which might have come at Cape St Vincent, or Calvi, or Santa Cruz, or the Nile, or Malta, or Copenhagen: an end which must fall due one day— and by now, that 'one day' was already late, in the envious calendar of chance.

By now, he owed God a death, and he who paid today was quit for tomorrow. It would not come *today*, unless the Admiral's bargemen from the *Victory* lost their senses, or their skill, or their love. But tomorrow? The next week? The next month?

Every story in the world must have its stop, and every myth of greatness its reduction to the bare bones of truth.

The myths abounded: some were true, some were false echoes, all were vital to the man, and to those who placed their eternal trust in him. They had started long ago with Nelson's Patent Bridge for boarding first-rates, and continued with the lost eye, and the lost arm, and the triumph of the Nile: gone downwards with the scandal of Naples, which seemed to worry great folks almost to a fainting-fit, and simple citizens not at all; then upwards again, to the heights of achievement, with the victory of Copenhagen.

Here Nelson had been second-in-command to a hesitant Admiral, Sir Hyde Parker, who in the heat of battle had hoisted a signal of recall. It was disobeyed. Nelson first pretended that it had not been reported to him; then, reminded by an anxious signal-midshipman who was bold enough to accost him as he walked to and fro on a blood-slippery deck, he said: 'Acknowledge,' and did nothing about it save to keep his own signal Number 16, 'Close Action', firmly hoisted.

Then he put his telescope, supported on the shoulder of a grinning seaman, to his blind eye, searched the Admiral's flag-ship, and remarked to his Flag-Captain, for all the world to hear:

'I have only one eye, Foley. I have a right to be blind sometimes.'

He had gone on to a victory complete and undeniable. His superior officer had not liked the fashion of it, and said so; the Board of Admiralty had temporized not at all before replacing Hyde Parker as commander-in-chief—with Lord Nelson himself. The common people, who had loved every moment of it, spoke out loud and clear; and another true myth had gone to swell a packet of such delights.

There was, as always, some swinish radical comment, from those who, if they could not bring down a Government, were ready to attack, wound, and destroy if they could its silent servants. A cartoon of the day made a merry meal of one battle-item, the blind eye.

Ignoring all aspects of valour and victory, it showed a picture with three characters portrayed: a huge Emma Hamilton surrounded by empty wine-bottles and unpaid bills of account, a tiny lecherous eye-patched Nelson leering in her direction, and a dotard labelled 'Old Antiquity' dozing in an armchair. The caption to this masterpiece was, in all the circumstances, singularly vile:

'Which has the Copenhagen Eye?'

Old Antiquity himself had died at his house in Piccadilly some two years before, having moved there for this very purpose from the estate called Merton Place in Surrey which the three had shared since Copenhagen.

He died in Emma's arms, and holding Nelson's hand. He died in perfect peace, though not before he had expressed himself, in the only forcible outburst of his life, as irritated by the constant bustle at Merton—'Never less than fourteen to dinner'—and the consequent household expenses, which sometimes approached a monstrous £200 a week.

He died proclaiming to the end the honourable conduct, beyond any reproach, of a distinguished naval officer, and named him, in a last codicil to his will, as 'my dearest friend, Lord Nelson, the most virtuous, loyal, and truly brave character I ever met with; God bless him, and shame fall on those who do not say Amen.'

It was at Merton, extravagant Merton, blessed, peaceful, rustic and private Merton, that Lord Nelson had spent his leave of twenty-five days, enjoying for as long as possible the company of his 'wife in the eyes of God', and playing with a daughter, baptized Horatia, now nearing five years old.

This was the bliss which, with urgent news that the French Fleet was at sea off Cadiz, he must now exchange for battle—and now, here he was to do so!

Far away along the ramparts of Portsmouth and the lesser beach-wall of Southsea, a distant roar of cheering swelled, heralding the man. Matthew, climbing half a dozen steps upwards for a better view, saw what he had been waiting for, and knew, and loved, and often feared for its close acquaintance with death. A small figure, but gleaming, bright as the buttons on God's creation robe, became Horatio Nelson, walking along the shore-line, the fatal edge of the tide-mark, before he took to the sea.

Matthew was quick to step down again, and send the barge's crew back to their stations—oars raised upright, caps off, legs crossed, backs straight—and to acquaint another man of rank with the news. This was Flag-Captain Hardy who, as duty dictated and pleasure inclined, had been sitting in the stern-sheets of his Admiral's barge, ready to welcome him back to the *Victory*.

Matthew announced: 'He is sighted, sir,' and Hardy, huge Hardy who could not walk upright in the *Victory*'s great cabin for fear of breaking his skull on the massive beams above, also stepped ashore to greet his master.

Lord Nelson came into view. Huzza! was the general cry as the greatest sailor in the world passed through the crowds. Some pressed forward, avid for a close look, a greeting which might be returned, perhaps even a touch of this greatness. But many others knelt as he passed, and kept silence, or wept. This greatest sailor was off to battle again. Battles wounded, battles killed. It was a time for prayer, not picnic jollity.

The Right Honourable the Viscount Nelson, Vice-Admiral of the White, shook a few urgent hands, and began to descend his last sixteen steps towards the sea.

He was pale, infinitely composed, yet jaunty as a magpie on a bough—something which Matthew Lawe had never ceased to marvel at. This was either the greatest Admiral afloat, or the finest actor of his age; or both at the same moment. Matthew had private reasons for knowing, beyond doubt, that Nelson had quit a damned warm bed at half past ten o'clock last evening, climbed into a post-chaise, changed horses at Guildford, travelled seventy miles to Portsmouth, dozing on the way: arrived at six o'clock this morning, listened to speeches and farewells from civic worthies, sent signals to Captain Hardy of the *Victory* and Captain Blackwood of the frigate *Euryalus*, the only other ships in near company; breakfasted like a lion at the George Inn, and now walked half a mile from his coach to the Southsea Steps, ready for immediate sea-service.

Such men were always the same. Old Samuel Pepys had been the same. Evil Henry Morgan had been the same. It was said that Napoleon, man of boundless power and intellect, did not stray from this pattern. Who could tell?—perhaps God in His creation had been the same. The act of intercourse, the discharge of love, did not dull them, nor tire them, nor send them to sleep.

Sleep? Sleep was for the grunting, snoring, base piglets of the world! Sleep was for half-men . . . In Nelson, such urgent activity had always released a second torrent of energy: an immediate electrical charge, as if his privates could spark out their own alchemy; clearing his mind, cleansing him to his finger-tips, fitting him out for keen thoughts, stern resolve, true action, *now*.

Matthew had once heard, in argument, a drunken yet perceptive man of politics deliver his verdict upon Nelson, dismissed as an arrant fornicator: 'If I were King of England, I would wish that *all* my Admirals could jump from a triumphant bed, and put straight to sea, and do the same service to the French!'

Now Lord Nelson, taking his last step downwards to the beach, acknowledged the salutes of his Flag-Captain, his lieutenant, and his colonel of marines. He was newly warmed, fresh from loving human contact, as he said:

'I wish I had two hands, to shake all those on the way ... I wish I had ten ...' He smiled up at Hardy, with whom he had a rare bond of friendship and trust. 'Are we ready for sea?'

'Aye, my lord.'

'And *Euryalus*?'

'The same.'

There had never been any other answer.

'God bless Blackwood,' his Lordship said. And to Hardy: 'God bless you.' And to Matthew, with another intimate smile: 'God bless us all ... Now—on board, and off to work!'

Five weeks later, and some twelve hundred sea-miles from Portsmouth, at noon on October the twenty-first, 1805, while the world checked its breath, the battle of Cape Trafalgar began. At that moment, as the two Fleets faced each other—27 English ships-of-the-line, 33 fine Frenchmen and Spaniards—Lord Nelson held the whole of Europe in his hand.

One mistake, one hesitation, one change of wind, one false signal or wrong helm-order, might deliver up the freedom of the seas, and a whole continent, to the strangling will of Napoleon. The invasion of England would only precede a universal tyranny.

He knew it all in his head and heart, and with his head he had made his dispositions. After skilful night-time manoeuvres, with plenty of false signals from rockets, guns, and Bengal lights for the enemy, and none but the truth for his Fleet, dawn found him with his two lines of battle poised to strike, ten miles to windward of the enemy—a sailor's masterpiece, planned a full year before. Now only the heart, with its courage, was needed.

At noon—the great scale of time was shrinking down to half a day—that plan was unfolding with a majestic slowness, the wind having fallen light, but with no lack of determination. The two columns of the English Fleet, one led by the *Victory* with twelve battle-ships, the other by Admiral Collingwood in the *Royal Sovereign* with fifteen, were moving to cut through the enemy line, a massive bent bow five miles long. Once there, they would lay about them till the fight was finished.

It was a plan unique and perilous, since before the general action was joined the two leading ships would be unsupported, and must take their punishment. It was the swift use of the knife, rather than a dull battering with the bludgeon. But all Nelson's captains had long been in accord with it, and all knew their duty, which was forever governed by their Admiral's final words:

'If signals cannot be seen, no captain can do very wrong if he places his ship alongside that of an enemy.'

Certainly he had noble ships in company: the fighting *Téméraire*, the beloved *Agamemnon* under Edward Berry's command: *Bellerophon* and *Spartiate*, old Crocodiles of the Nile battle, *Polyphemus* and *Defiance* of Copenhagen, *Euryalus* his most faithful frigate, even the *Orion* which long ago had helped a dismasted *Vanguard* to make good her wounds.

To Nelson, it was all perfection. Old ships, old friends . . . The wind fell lighter still as the two British lines in their livery of black and yellow, copied from the admired *Victory*, came slowly within range, and the thunder of the guns began.

Captain Blackwood, who as commander of the frigate squadron had remained on board the flag-ship to receive his last instructions, was bidden farewell when the first French shots were fired. He had spent some time in trying to persuade his Admiral to transfer to the *Euryalus*, and direct the battle from this safer vantage point.

But Nelson would have none of it. 'For the sake of example' he would not quit the *Victory*, where men were stripped bare-chested for action and gun-crews itching to blast off, where bands were playing 'Rule Britannia!' on the poop-deck, and sailors in fantastic spirits were dancing their last horn-pipe.

Blackwood, taking his leave, was left to remember a foreboding good-bye, delivered in unmistakable words as he stepped down the ladder:

'God bless you, Blackwood. I shall never speak to you again.'

At half-past noon—the time scale of this sailor's world was now reduced to hours—the pell-mell battle was in full roar, with frightful carnage on both sides. It was close work, ship to ship, marksman to marksman, and presently hull to hull. Shots were seen to meet in mid-air as the gun-duels became point-blank. The

Victory, which had taken with patience a merciless battering from three of the enemy before she could bring her own guns to bear, at last ranged up to engage the *Bucentaure*, the flag-ship of Admiral Villeneuve.

Villeneuve, one of the two French captains to escape from the Nile and live to fight another day, had led Nelson a fine dance to the West Indies and back again, in a feint to withdraw him from the Channel and leave England open to invasion. With him, there was a score to settle and a valiant man to beat. In the event, it was quickly done.

The *Victory*, a three-decker, had one hundred guns at her disposal, and her own favourite and fearsome piece of ordnance on the fore-deck to back them up: a 68-pound carronade which fired a round shot preceded by a keg of five hundred red-hot musket balls.

One single broadside, double-shotted, with the help of this fore-deck monster, reduced the *Bucentaure* to disabled silence. Then, on the other side, the French *Redoutable*, Captain Lucas, was seen to be approaching, and raking them with langrel shot. A moment later Mr Scott, the Admiral's secretary, standing close to Lord Nelson and not five feet from the huge-framed Hardy, whose shoe-buckle had already been torn off, was cut in two by her second broadside.

It was Matthew Lawe, searched for diligently all over the ship, who must now take his place.

Since the first thudding, splintering shots had begun to assail the *Victory*, Matthew had been skulking—he could not, in his own soul, find any other word for it—in the dark vicinity of the cockpit, where the first wounded of the battle were beginning to pour and tumble down, or be carried in such a fashion that, between one gory deck and another, they became cold torn flesh fit for nothing but a butcher's funeral over the side.

Summoned at last by a master's mate who was himself glad to escape below, even for a few moments, he climbed up to a fearful scene. The rage of battle was the worst he had ever known. But first of all, as he peered out of the quarter-deck hatch, he was aghast at the brightness of his Admiral.

Nelson was clad in no more than the blue and white of his working rig. Yet the gold braid caught the sun, the hat proclaimed his rank, the blaze of stars on his breast attested to four orders of knightly splendour. A cockatoo among crows could not have made a better target for a hunter.

Beyond him, the decks of the *Victory* had already been brought close to ruin. Though poor Scott's body, with many others, had been thrown overboard, much grim evidence remained. The smell of death was not honest powder, but the foulness of scattered entrails half-way through their work.

Amid a shroud of smoke, it could be seen that their topmasts were gone, and their sails in scorched tatters. On the quarter-deck, the wheel had been shot away to splinters; the flag-ship was being steered from below, by ropes and tackles which needed the strong arms of forty men to make one turn of the tiller.

Nelson, observant of this shambles, remained cool to iciness, pale as the ghost of Hamlet, yet possessed by his own demon of courage: ready to die, aflame to live and conquer—and watchful of everything. He gave Matthew a keen glance as he drew near.

'Mr Lawe, what are you at?'

'I was helping the surgeon, my lord.'

'Well . . . Is Mr Beatty busy?'

'Unhappily yes, my lord.'

'He will be busier yet . . . This work is too warm to last. Scott is gone from us. Pasco is wounded and below. So, transform yourself into a signal-lieutenant, and a secretary.' Suddenly he smiled, in that fashion which made men, for ever after, remember him as the one to love. Matthew felt that his Admiral had divined the truth and, as with little midshipmen at their first blood-taste of battle, or old comrades who faltered under a heavy load, was ready to place an instant comforting arm round their shoulders. 'Transform yourself also to a friend.'

'Willingly, my lord.'

'It is all I need.'

At one o'clock, amid a continuing, frightful uproar, Captain Hardy approached. He pointed to the *Bucentaure*, which was silent but not yet surrendered, to another ship of the second French line which seemed ready to engage, and last to the *Redoutable*, now bearing down on them, breathing fire and smoke like a dragon which truly knew its business.

'My lord,' Hardy said, cool as his Admiral, but more intent on the plain timber of instant action, 'we must lay alongside one of them. Which shall it be?'

'It does not matter,' Nelson answered, and thus chose his death.

Hardy made straight for the *Redoutable*, thought to be the fiercest Frenchman of all, and within minutes they were grappled together, their rigging entangled as if with claws, their sturdy hulls prepared to bruise each other to pieces and their men to board like tigers.

Redoutable, of which Captain Lucas had made a hive of infamy —and of boundless zeal—fired langrel shot, poured out musket-balls, dropped flaming torches, and tossed down hand-grenades upon the *Victory*'s deck. Along the yards her marksmen took aim, and fired, and laughed, and sometimes tumbled down to their death. The *Victory* also had her ardent sportsmen.

At forty minutes past one o'clock—the time scale had now shrunk down to such little segments of eternity—Matthew, attending his Admiral and noting his orders, signals, and entries for a later battle-journal, saw a blue-coated French marine, perched in the mizzen-top of the *Redoutable* no more than fifteen yards away, take steady aim at himself—or at Nelson by his side.

In pure terror, he could do nothing honourable, nor useful. He did not shout a warning, or try to draw his commander aside. Instead, as the *Victory* rolled to a chance bumping from her enemy, he stumbled away, took half a dozen steps, and found himself in safety behind the main-mast, to which Nelson had, long ago in peaceful Portsmouth, nailed a horse-shoe.

From this shameful lair he peeped out to see his Admiral, the true target, fall in a poor heap upon the deck, and heard the gasp of his wounded words to Captain Hardy:

'My backbone is shot through.'

After many battles, many blows, many false alarms and glimpses of mortality, he was right.

Down and down to the cockpit, that great terminal of pain, was borne Lord Nelson, with a handkerchief spread over his face and across the starry orders of his breast: not in mourning but, at his own wish, in disguise. There were fifty other sufferers below, and bearers coming and leaving, sometimes carrying back those who had died on their way down.

There was enough grief in the ship, without the dread news of their Admiral's wound, which might be mortal.

Amid the muted noise of battle, the shock and thud of cannon-balls striking and tearing at their ship's side, Mr Beatty the surgeon, with his two assistants, were very busy. The arrival of another crumpled man was nothing; he must take his turn, in accord with Fleet orders—Nelson's own Fleet orders.

So they hacked and sawed, bandaged and strapped, staunched gross wounds and let others take their fatal course. They did their swift best, and prayed mercy for their forced transgressions. Surgeons, though angels, were only men.

Then their chief became aware of an extraordinary murmuring round about him. It grew stronger than the shrieks of men under his treatment, invading the cockpit like some unearthly music. The handkerchief had fallen from Nelson's face and chest as he was laid down, and the shocked words of the wounded, themselves sick with suffering, became clear:

'Mr Beatty, Lord Nelson is here!'

'Sir, he is wounded!'

'Look to him first!'

The music of pity could not, need not be denied . . . Quickly and mercifully a new patient was stripped and carried to a cot, and Beatty, with great tenderness, made his examination. Though he began: 'I will not hurt you more, my lord,' he went on to do so, as a necessary turning and probing pursued its course. In ten anguished minutes he had traced, as well as he could, the path of a musket-ball which had pierced the shoulder slightly from above, passed through the chest, struck the spine, and lodged in the muscles below the other arm.

But a surgeon must know more. 'My lord, can you hear me?'

'Yes, Beatty.'

'Pray tell me what you feel.'

Nelson, rallying from his lumpish journey downwards, spoke remarkably clear: 'A gush of blood within the breast, every few minutes . . . No feeling in my lower parts . . . Pain—' he gasped suddenly, 'much pain in the backbone . . . Send for Hardy.'

'My lord, you must rest.'

'I must die, and you know it. Send for Hardy.'

While they waited, and Beatty was called away, and men died and others cried out and yet others sank into peaceful sleep, and Dr Scott the chaplain—'the other Scott', as Nelson always called him—rubbed the tortured bosom and fanned when the Admiral panted for air, two watchers in the darkness of the cockpit kept a humble vigil.

They were Matthew Lawe, who had followed down, and a friend of long ago, Gaetano Spedilo, the faithful valet of Sir William Hamilton and now, by natural inheritance, the proud body-servant of Lord Nelson. But Gaetano was no longer proud. He was already reduced to bitter weeping. Matthew was not proud, nor ever would be so again.

Mostly they were silent together. They observed, with pain which almost matched that of the wounded man, the necessary probings of Dr Beatty. They followed the loving ministrations of the Reverend Dr Scott, no more than a diffident clergyman afloat, as he overcame his revulsion at the blood and wounds of the cockpit, and patiently fashioned a fan from torn strips of paper, and gave his lordship constant sips of lemonade, and rubbed the cold flesh of his chest, which must have been, for him, akin to that of Christ in the tomb.

When Matthew and Gaetano talked, it was in whispers. If their cheeks were wet, they only foretold the tears of a world of sailors.

'*Tenente*, my eyes are so poor, I cannot see.' The face of a tall, whiskered, merry-minded Neapolitan had surrendered to grief, and was in desperate, childish disorder. 'Is he gone?'

'No, no,' Matthew answered, and wished to believe it. 'He moves his head. Dr Scott is giving him every care. He would not comfort a dead man.'

'But he must die?'

'By his look, yes.'

'Did you see it happen?'

'Aye.'

'What then?'

'A shot, from a French musketeer.'

'God curse them all!'

'Well, it was an honest battle.'

'But to kill an Admiral!'

'They could not have chosen a better.'

'And to think that I put such a polish on his stars and orders, not five hours back. They have been the greatest burden of my life, all these years . . . And now, and now—sir, do you think that the medals were the mark?'

Matthew could only answer: 'Yes.'

'*Maledizione!*'

Then the two guilty servants of a great man began to witness his dissolution.

It seemed to start with the arrival of a new attendant, Flag-Captain Thomas Masterman Hardy, who was at last freed from his urgent duty on deck to comfort a loved commander. With the greatest irony of all, it was accompanied by the sound of hearty cheering from above.

As Hardy bent over Nelson, the Admiral opened his eyes and asked: 'Is it Hardy? Thank God! What is the cheering?'

'Fresh surrenders, my lord.'

'But none of ours, I hope.'

'Not a stick!'

'How many of the French?'

'Fourteen for certain, perhaps more. And Villeneuve has struck his flag, and is prisoner on board the *Mars*.'

'I had hoped for twenty.'

'Twenty will come . . . How is your lordship?'

'They have done for me, Hardy.'

Hardy, looking down on a face already beset with the ghostly pallor of death, could still answer: 'Not so. Never so! I hope and pray you will reach London, to report a glorious victory.' Then he became Hardy the work-horse sailor again. 'My lord, I must leave you now. It is not yet over. There are prizes to be had, before we are done. And the French may rally. There are brave men afloat today.'

'Hardy, you are the best fellow in the world.' But Nelson, exhausted by the exchange, now had to whisper: 'Bend near.'

'What is it, my lord?'

A distressed voice answered: 'Do not throw me overboard.'

'Never!'

'Let Lady Hamilton have my hair.'

The tall Flag-Captain pressed his hand and left him, bending his head as much in sorrow as from the menace of the deck-beams so close above; and ordinary care, the same for wounded seamen as for stricken Admirals, became the rule. An hour passed, as it passed for all, in fanning his brow, wiping cold sweat from his forehead, putting a cup to his lips, trying to ease the bursting pain in his back and breast. Then Mr Beatty, streaked with the blood of a hundred other men, came to his side once more.

He had just taken an arm from the socket of a gunner who immediately died; now he returned to what must certainly be another death: death by pale lantern-light, in the heart of a rolling war-torn ship, death inevitable, piteous, and brave.

Nelson's anxious mind was still upon the battle. 'Send for Hardy again,' he told Beatty. 'I have ordered him down four times. Why does he not come? He is killed—I know it!'

Beatty touched a fluttering heart, and then the lower half of a body which had lost all sensation. 'My lord, I beg you, do not concern yourself. Captain Hardy will come as soon as he can be spared. I know him to be alive. But he cannot leave his command.'

'I will die before he is here!'

'Indeed, my lord, I hope not.'

The form of words must have pierced a cloudy brain. Nelson murmured: 'Beatty, tell me the truth.'

The answer came in the moving poetry of grief as the surgeon answered: 'My lord, unhappily for our country, nothing can be done for you,' and then turned away to hide his tears.

'Christ, he is weeping!' the man in the shadows said.

'Who is weeping?' Gaetano asked.

'The surgeon.'

'Then the Admiral is gone?'

'No. But soon.'

'God have mercy.'

Though he was not yet gone, 'soon' was a true judgement. Now the scale of time was to be measured in mere seconds, by the beat of a faltering pulse. As Hardy returned to the cockpit, Lord Nelson began to die in all his three estates, as sailor, man, and lover.

To Captain Hardy he was an Admiral, with the care of a Fleet, and a beloved ship, still tormenting his brain. Under his poor back he had felt the *Victory* rising and falling to a westerly swell, the forerunner of a storm. He remembered the hungry shoals of Trafalgar, waiting to catch them if they drifted too far.

'Anchor, Hardy. Signal to anchor!'

'My lord, do you intend that I should give the order to the Fleet? Admiral Collingwood will——'

'While I live, I command! Do *you* anchor, Hardy!'

'Aye, sir.'

'Use your judgement . . . That, I trust for ever . . . I am going soon. Kiss me, Hardy.'

Hardy said: 'Farewell, my lord,' and kissed him twice, once as he had been ordered and once again, after a long watchful moment, as a strong man alive saluting one who could no longer be numbered in the same company. Then he climbed, as he must, back to his sea-service.

To the chaplain, Dr Scott, still ceaselessly rubbing a tortured chest which was growing icy cold, Lord Nelson made his last claim on heaven: 'Doctor, I have not been such a *great* sinner.'

To any who might still hear him: 'Take care of poor Lady Hamilton . . . Thank God I have done my duty.'

At half-past four a noble heart cracked and ceased to beat; and to the sound of distant gun-fire, after forty-seven years of faithful endeavour, time gave up its value.

With the coming of dusk on a dreadful day, Matthew crept out on deck to survey a scene sublime and tragic. It was a prospect too great for his own grief and shame, which must await another day, another age. All that he could bear to do, this night, was to gaze, and remember, and mourn.

Stillness had fallen on a mighty battlefield, though an eerie churchyard wind persisted, ruffling the water, making what remained of their shrouds and rigging sigh in protest. The westward Atlantic swell which presaged a storm was steadily building, and they must soon collect themselves, and go.

There would be much to collect, many to nurse, some to lose on the way . . . Sombrely he allowed his eyes to wander. Fifty wounded ships were riding, with little sign of life, on a sluggish, troubled, tainted sea. Ships still burning, ships with their sides beaten in. Ships which were mere hulks, with not a spar showing above their decks. Ships with four hundred killed.

But twenty of the French and Spanish were prizes; and though the *Victory* had fallen cold and silent, there was nearby the triumphant *Téméraire* with two captives, one of them the fatal *Redoutable*, lashed on either side.

Would ships be valued more precious than men?

Beyond all this there waited for them, growling, the shoals of Trafalgar, which Lord Nelson, with a sailor's instinct, the inner ear of command, had foreseen as he cautioned: 'Anchor, Hardy, anchor.'

When darkness came, and the far ships faded, the coldest words of all began to go round the Fleet, with a midshipman's whispered report:

'No Admiral's lights on board the *Victory*.'

Victory.